T0259158

GENESIS, PROPERTIES AND UTILIZATION

Further Titles in this Series

Developments in Soil Science 21

VOLCANIC ASH SOILS

GENESIS, PROPERTIES AND UTILIZATION

SADAO SHOJI

Faculty of Agriculture, Tohoku University, 1-1 Tsutsumidori Amamiyamachi, Aobaku, Sendai 981, Japan

MASAMI NANZYO

Faculty of Agriculture, Tohoku University, 1-1 Tsutsumidori Amamiyamachi, Aobaku, Sendai 981, Japan

and

RANDY DAHLGREN

Land, Air & Water Resources, 151 Hoagland Hall, University of California, Davis, CA 95616, U.S.A.

ELSEVIER   Amsterdam — London — New York — Tokyo   1993

ELSEVIER SCIENCE PUBLISHERS B.V.
Sara Burgerhartstraat 25
P.O. Box 211, 1000 AE Amsterdam, The Netherlands

Library of Congress Cataloging-in-Publication Data

Shoji, Sadao.
    Volcanic ash soils : genesis, properties, and utilization / Sadao
Shoji, Masami Nanzyo, and Randy Dahlgren.
        p.   cm. -- (Developments in soil science ; 21)
    Includes bibliographical references and index.
    ISBN 0-444-89799-2
    1. Andosols--Congresses.   I. Nanzyo, Masami.   II. Dahlgren,
Randy.   III. Title.   IV. Series.
    S592.17.V65S48   1993
    631.4--dc20   93-4551
                                                                    CIP

ISBN: 0-444-89799-2

Transferred to digital printing 2006

Printed and bound by CPI Antony Rowe, Eastbourne

PLATE 1. Bekkai soil: Thaptic Udivitrand (see page 13)

PLATE 2. Sitka soil: Andic Placocryod (see page 17), courtesy of C.L. Ping

PLATE 3. Lanco soil: Typic Durudand (see page 20)

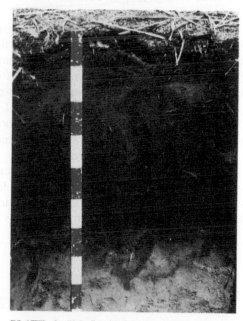

PLATE 4. Tohoku University Farm soil: Alic Pachic Melanudand (see page 24)

PLATE 5. Tsutanuma soil: Acrudoxic Fulvudand (see pages 27 and 46)

PLATE 6. Vegetation of Tsutanuma soil: *Fagus crenata* (see pages 27 and 46)

PLATE 7. Yunodai soil: Acrudoxic Melanludand (see pages 27, 46 and 260)

PLATE 8. Vegetation of Yunodai soil: *Miscanthus sinensis* (see pages 27, 46 and 260)

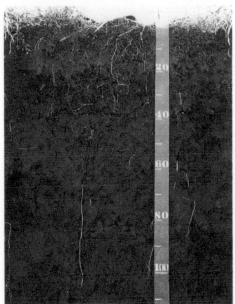

PLATE 9. Hilo soil: Acrudoxic Hydrudand (see page 31), courtesy of H. Ikawa

PLATE 10. Tsukuba soil: Hydric Hapludand (see page 253)

PLATE 11. Findley Lake soil: Andic Humicryod (see pages 64 and 256)

PLATE 12. Abashiri soil: Typic Hapludand (see pages 66 and 258)

PREFACE

Volcanic eruptions are awesome and destructive; however, this natural phenomenon is beneficial in the long term. One of the earliest known written records of a natural disaster concerned the eruption of Pompeii in A.D. 79, and was recorded by Pliny the Younger in his letters to the historian Tacitus (Jashemski, 1979). Volcanic eruptions, ranging in intensity and numbering between 17 and 27 per year during the past decade (Bullard, 1984), continue to remind us of their potential impacts on the environment. Violent volcanic eruptions range from complete obliteration of the landscape near the volcano, to a mere dusting of tephra at great distances. The most obvious effect is the destruction caused by catastrophic ejection of ash, lava, pyroclastic flows and/or mudflows. These agents of destruction may result in loss of wildlife, vegetation, and even human lives.

There is a tendency to view volcanoes primarily as agents of destruction and to overlook their beneficial contributions. The beneficial effects of volcanic eruptions are often more subtle; occurring on a geologic time-scale rather than during the lifetime of an individual. They include building of continents, oceanic islands, and the ocean floor; creation of magnificent scenery and recreation areas; and the development and rejuvenation of soils which provide an environment favorable to the eventual establishment of lush vegetation and the ecology of organisms, including human beings. The periodic additions of volcanic ash renew the long-term fertility status by providing a source of nutrients from the rapid weathering of ash.

Soils formed in volcanic ash have many distinctive properties that are rarely found in soils derived from other parent materials. It is estimated that soils derived from volcanic ejecta are distributed over approximately 0.8% of the earth's surface (Leamy, 1984). These soils have a high potential for agricultural production as illustrated by the fact that many of the most productive regions of the world are located near active or dormant volcanoes and the most densely populated areas in regions, such as Indonesia, are found near volcanoes. To maximize the productivity of volcanic ash soils and to minimize the deterioration of these soils, proper management, based on an understanding of the unique physical, chemical, and mineralogical properties of these soils, must be practiced.

Volcanic ash soils did not receive worldwide recognition among soil scientists until the middle of this century; however, a considerable understanding of the genesis, unique properties, and productivity of these soils was established in Japan and New Zealand several decades earlier. It was only in 1960 that volcanic ash soils were recognized as a distinct category of soils with unique properties in the international system of soil classification proposed by the Soil Survey Staff, Soil Conservation Service, United Sates Department of Agriculture. This recognition stemmed from an urgency to determine the potential agricultural productivity of the world's soils, necessitated by the increased demand for food production to feed the rapidly increasing world population.

It is interesting to note that the discovery, characterization, and significance of short-range-order minerals and noncrystalline colloidal materials, such as imogolite, allophane, laminar opaline silica, ferrihydrite, and Al/Fe humus complexes, owe their recognition to the intensive studies of volcanic ash soils associated with efforts to classify these soils. These achievements have greatly contributed to the establishment of the concept and science of variable charge soils.

A renewal and upsurge of interest in volcanic ash soils was initiated by the proposal for an "Andisol" soil order in Soil Taxonomy, first proposed by Guy Smith in 1978. Development, testing and refinement of criteria to define the Andisol soil order brought together, for the first time, many of the prominent soil scientists working on volcanic ash soils. After 12 years of study, the Andisol soil order was finally adopted as the 11th soil order in Soil Taxonomy (Soil Survey Staff, 1990). During these 12 years, a wealth of ideas and knowledge was exchanged and debated.

The purpose of this book is to synthesize and integrate our current understanding of volcanic ash soils including the topics of morphological and physical characteristics, chemical and mineralogical properties, soil genesis and classification, and productivity and utilization of these soils. The information and experience on volcanic ash soils shared by the authors are mostly limited to soils formed in the middle latitudes or temperate regions. Thus, this book is written largely from this perspective. The authors realize, however, that in order to present a more balanced and complete international understanding, the related literature on volcanic ash soils of the tropical regions should be addressed. More than half of the volcanic ash soils of the world are located in the tropics where climatic conditions, such as temperature and precipitation, are significantly different from those of the temperate regions. Since climate greatly affects soil forming processes, the soils formed in tropical regions are expected to differ considerably from those formed in the temperate regions. Unfortunately, the limited literature on tropical volcanic ash soils available to the authors makes it impractical to rigorously discuss this group of soils in the present monograph.

September, 1992   SADAO SHOJI,
                                                                   MASAMI NANZYO
                                                              and RANDY DAHLGREN

REFERENCES

Bullard, F.M., 1984. Volcanoes of the Earth, 2nd revised edition. University of Texas Press, Austin, Texas.

Jashemski, W.F., 1979. Pompei and Mount Vesuvius A.D. 79. In: P.D. Sheets and D.K. Grayson (Editors), Volcanic Activity and Human Ecology. Academic Press, New York, London, Toronto, Sydney and San Francisco, pp. 587–622.

Leamy, M.L., 1984. Andisols of the world. In: Congresco international de Suelos Volcanicos. Communicaciones. Universida de La Laguna Secretariado de Publicaciones, serie informes 13, pp. 368–387.

Soil Survey Staff, 1990. Keys to Soil Taxonomy, 4th edition. AID, USDA-SMSS Technical Monograph No. 19. Blacksburg, Virginia.

ACKNOWLEDGEMENTS

We would like to acknowledge all those who have contributed to the content of this book through their research in volcanic ash soils. We are also indebted to our many colleagues who have encouraged us to write this monograph and for their advice and editorial suggestions on this monograph.

Many people helped and inspired the authors in their studies of volcanic ash soils. Among them are I. Yamada (Chief, Kyushu National Agricultural Experiment Station), M. Saigusa (Associate Professor, Tohoku University), T. Ono (Chief, Iwate Prefectural Agricultural Experiment Station), T. Takahashi (Assistant Professor, Akita Prefectural Agricultural College.), T. Ito and K. Kimura (Instructors, Tohoku University), F.C. Ugolini (Former Professor, University of Washington, presently Professor, University of Florence, Italy), C.L. Ping (Associate Professor, University of Alaska), M. Otowa (Professor, Hirosaki University), T. Egawa (Emeritus Professor, Meiji University), K. Wada (President, Kyushu University), N. Yoshinaga (Professor, Ehime University), Y. Kato (Emeritus Professor, Shizuoka University) and Y. Oba (Emeritus Professor, Tsukuba University).

The senior author (Shoji) was provided with many opportunities to study volcanic ash soils in various countries including volcanic ash soils in North and South Americas by R. Arnold (Director, Soil Survey Division, USDA-SCS) and H. Eswaran (National Leader, World Soil Resources, USDA-SCS), in Hawaii by G. Uehara (Professor, University of Hawaii), and H. Ikawa (Associate Soil Scientist, University of Hawaii), in Alaska by C.L. Ping, in Washington by F.C. Ugolini, in Oregon by B.P. Warkentin (Professor, Oregon State University), in New Zealand by the late M.L. Leamy (Chairman, International Committee on the Classification of Andisols), in Indonesia by M. Sudjadi (Former Director, Center for Soil Research), and K. Igarashi (Former Leader, Japan-Indonesia Agricultural Research Strengthen Project), and in Korea by H. Kim (President, Cheju National University).

He was also offered opportunities to collect valuable information on volcanic ash soils and farming in Japan by many colleagues as follows: In Hokkaido, by T. Sasaki, E. Tomioka, and M. Katayama (Former Staff of Hokkaido National Agricultural Experiment Station), and A. Nishimune (Chief, Hokkaido National Agricultural Experiment Station), in Aomori by H. Nasu (Director, Aomori Upland Cropping and Horticulture Experiment Station), in Ibaraki, by K. Ogawa (Chief, Ibaraki Prefectural Agricultural Experiment Station), in Kanagawa, by

S. Fujiwara (Chief, Kanagawa Prefectural Agricultural Experiment Station), in Kumamoto, Kyushu, by T. Iizuka (Chief, Kyushu National Agricultural Experiment Station), and Y. Murakami (Head, Kumamoto Prefectural Agricultural Experiment Station), and in Miyazaki, by Y. Kobayashi (Chief, Kyushu National Agricultural Experiment Station) and S. Maki (Head, Miyazaki Prefectural Agricultural Experiment Station).

Valuable information and suggestions on the genesis and classification of volcanic ash soils were offered by many soil scientists: J.E. Witty (National Leader of Soil Classification, USDA-SCS), J.E. Kimble (Research Soil Scientist, USDA-SCS), F.C. Ugolini, C.L. Ping, the late M.L. Leamy, R.L. Parfitt and B. Clayden (Pedologists, Division of Land and Soil Science, DSIR, New Zealand), M. Otowa, H. Otsuka (Chief, Hokkaido National Agricultural Experiment Station), T. Wakatsuki (Associate Professor, Shimane University), and H. Ikawa. The new Keys to Soil Taxonomy was provided by J.E. Witty.

Critical reading of the manuscript of this book that helped to greatly improve it, was provided by the following people: A.T. Gandeza (visiting soil scientist of Tohoku University, presently National Tobacco Administration Sub-center, Philippines) (Chapters 1–8), J.E. Witty (Chapters 1 to 4), I. Yamada (Chapter 3), S. Iwata (Professor of Ibaraki University) (Chapter 7), and M. Saigusa (Chapter 8).

E. Makino (Former Staff of Tohoku University) did word-processing for the manuscript of the book and H. Kanno (Technical Staff of Tohoku University) made many of the figures that are included in this book. C. Heusner (California, USA) technically reviewed the manuscript. Drs. F. Wallien (Publishing Editor, Elsevier) and Elsevier Science Publishers offered the fine support and technical assistance to publish the book.

We wish to express hearty thanks to these people described above.

September, 1992   Sadao Shoji, Masami Nanzyo and Randy Dahlgren

THE AUTHORS

Sadao Shoji

Professor of Soil Science in the Faculty of Agriculture, Tohoku University, Sendai, Japan. He holds B. Agric. and Ph.D. degrees from Tohoku University and a M.S. degree from Michigan State University. He has served as a technical researcher at Hokkaido National Agricultural Experiment Station, 1954–1962, and on the staff of the Faculty of Agriculture, Tohoku University, 1962 to present. He is president of the Japanese Society of Soil Science and Plant Nutrition. He has published many articles on the genesis, classification, properties and agronomy of volcanic ash soils.

Masami Nanzyo

Assistant Professor of Soil Science in the Faculty of Agriculture, Tohoku University, Sendai, Japan. He holds B. Agric., M.S., and Ph.D. degrees from Tohoku University. He has served as a technical researcher at the National Institute of Agricultural Sciences, 1977–1983, National Institute of Agro-Environmental Sciences, 1983–1989, and on the staff of the Faculty of Agriculture, Tohoku University, 1989 to present.

Randy Dahlgren

Assistant Professor of Soil Science in the Soils and Biogeochemistry section of the Department of Land, Air and Water Resources, University of California, Davis, CA USA. He obtained his B.S. degree in Soil Science at North Dakota State University and his M.S. and Ph.D. degrees in Forest Soils at the University of Washington. He was a research associate in the Department of Environmental Engineering at Syracuse University (1987–1988) prior to joining the faculty at the University of California, Davis (1989 to present).

CONTENTS

Chapter 1

TERMINOLOGY, CONCEPTS AND GEOGRAPHIC DISTRIBUTION OF VOLCANIC ASH SOILS

S. SHOJI, R. DAHLGREN and M. NANZYO

1.1. TERMINOLOGY

The term "volcanic ash soils" is commonly used to designate soils formed from tephras or pyroclastic materials. Since most of these soils have unique properties inherited from or associated with the properties of tephra, the general term "volcanic ash soils" is often used to denote Kurobokudo (The Third Division of Soils, 1973), Andosols (FAO/Unesco, 1974), and Andisols (Soil Survey Staff, 1990, 1992). However, not all volcanic ash soils are Kurobokudo, Andosols or Andisols and vice versa. As described in Chapter 3, there are tephra-derived Spodosols, Inceptisols, Mollisols, Oxisols, etc., and there are some Kurobokudo, Andosols or Andisols formed from nontephritic materials such as volcanic rocks, sedimentary rocks, mixed materials of tephra and loess, etc.

A variety of soil names connotative of the volcanic ash soils have been proposed in Japan as follows: Volcanogenous loams (Seki, 1913), Prairie-like brown forest soils (Kamoshita, 1955), Volcanogenous soils (Yamada, 1951), Volcanogenous black soils (Uchiyama et al., 1954), Humic Allophane soils (Kanno, 1961), and Kurobokudo (The Third Division of Soils, 1973). The FAO/Unesco (1974) also lists several soil names to describe volcanic ash soils throughout the world: Yellow-brown loams and Yellow-brown pumice soils (New Zealand); Humic Allophane soils, Trumao soils (Chile); Acid Brown Forest soil, Acid Brown Wooded soil (Canada); Sols bruns tropicaux sur materiaux volcaniques (Zaire); Andosols (France); Andosols (Indonesia); Kurobokudo (Japan); Andepts (U.S.A.); Volcanic soils (former U.S.S.R.). It is interesting to note that soil names, such as "Andosols" and "Andepts" originated from "Ando soils", while there is no Japanese soil name coined from the term "Ando" (An; dark: do; soils), because "Ando" is an uncommon Japanese word literally translated as "dark soil".

"Ando soils" was first introduced in 1947 during the reconnaissance soil surveys in Japan by American soil scientists (Simonson, 1979). The principal characteristics of volcanic ash soils which were the basis for recognizing the new Ando great soil group in the intrazonal order (Simonson, 1979) are described below:

(1) The A1 horizons are usually thick and dark, and the soils lack the E horizons expected in Podzols. Amounts of organic matter in A1 horizons commonly range

up to 15% and reach 30% in the extreme cases. The C/N ratios of the organic matter range from 13 to 25.

(2) Where not cultivated, the soils are moderately to strongly acid with pH ranges of 4.5 to 5.1 in A horizons and 5.0 to 5.7 in deeper profiles. Levels of exchangeable bases are low, ranging from 2 to 9 $cmol_c$ kg^{-1} of soil material, whereas exchangeable Al (in M NH_4Cl) exceeds 3 $cmol_c$ kg^{-1} generally and ranges up to 8 $cmol_c$ kg^{-1}.

(3) The silica–sesquioxide ratios of colloid fractions range from 1.3 to 2.0 in A horizons and from 0.75 to 0.90 in deeper profiles of these soils on Honshu. In Kyushu, the corresponding ranges in ratios are 0.4 to 1.13 in A horizons and 0.67 to 1.03 in deep profiles.

The term "Ando soils" appeared for the first time in the journal Soil Science in 1949 and then the Ando group was identified in widely separated regions of the world (Simonson, 1979).

Though the term "Ando soils" is still widely used, it was replaced by "Andepts (currently the Andisol order)" as a suborder of Inceptisols in Soil Taxonomy (Soil Survey Staff, 1975, 1990, 1992). A part of the original name "Ando soils" was retained in coining the new names of Andosols and Andisols.

1.2. CENTRAL CONCEPTS

The Japanese name "Kurobokudo" means black (Kuro) fluffy (boku) soils (do) and it has been traditionally used in Japan. The central concept for Kurobokudo includes the following (The Third Division of Soils, 1973, 1982):

(1) The parent material is volcanic ejecta.

(2) The horizon sequence commonly shows dark-colored A horizons over brown to yellowish B horizons.

(3) The soil material is vitric and/or rich in allophane.

(4) The allophanic soils have a low bulk density, a high exchange capacity, and high phosphate adsorption.

Andepts were recognized as immature soils (Inceptisols) that have not developed features diagnostic for other soil orders (Soil Survey Staff, 1975). The central concept can be summarized as follows:

(1) The parent material is mostly pyroclastic material.

(2) The soil material contains an appreciable amount of allophane and shows low bulk density, or

(3) The soil material is mostly vitric.

By comparing the two central concepts described above, it appears that the central concept of Kurobokudo places emphasis on the existence of dark-colored A horizons while that of Andepts includes some nontephritic materials such as pyroclastic rocks, sedimentary rocks, or basic extrusive igneous rocks. Both of these concepts suggest that the unique chemical and physical properties are primarily attributable to allophane. Andosols described by Duchaufour (1982),

FitzPatrick (1971), etc., have a central concept similar to that of Andepts.

Before 1978, Andisols were regarded as soils dominated by allophanic mineralogy. The discovery of Andisols with nonallophanic mineralogy by Shoji and Ono (1978) initiated a reevaluation of the central concepts of Kurobokudo, Andosols, and Andepts, because the nonallophanic soils share many of the unique chemical and physical properties exhibited by allophanic soils. The name "Andepts" was changed to "Andisols" in the Andisol proposal (Smith, 1978), and the central concept of "Andisols" was based on the presence of active forms of Al and Fe, such as the noncrystalline materials, allophane, imogolite, ferrihydrite or aluminum–humus complexes (Leamy et al., 1988).

As described above, there are many soil names referring to volcanic ash soils. However, none of them are defined as precisely as the Andisols described in Chapter 4.

1.3. GEOGRAPHIC DISTRIBUTION OF VOLCANIC ASH SOILS

The distribution of soils derived from volcanic materials closely parallels the global distribution of active and recently active volcanoes. Because of this association, the distribution of volcanic ash soils is geographically predictable. Volcanic ash soils cover approximately 124 million hectares or 0.84% of the world's land surface (Leamy, 1984). Approximately 60% of volcanic ash soils occur in tropical countries. While volcanic ash soils comprise a relatively small extent of the world's surface, they represent a very important land resource due to the disproportionately high human populations living in these regions as described in Chapter 8. This section briefly describes the geographical distribution of volcanoes and Andisols throughout the world.

1.3.1. Geographic distribution of volcanoes

The great majority of active and recently extinct volcanoes are concentrated in the circum-Pacific Ring of Fire. This ring includes many volcanoes in New Zealand, Melanesia, the Philippines, Japan, Kamchatka, Aleutian, Alaska, and the western coast of North America to South America (Bullard, 1984).

The volcanoes in the Pacific Ring of Fire are associated with geologically young or still growing volcanic mountains and show high explosion indices of approximately 80 percent (Katsui, 1978). Their rocks and tephras are mostly andesite belonging to the Calc-alkali-rock series. Thus, the volcanoes of this belt disperse andesitic tephras resulting in a wide distribution of Andisols or volcanic ash soils in their proximity.

The Alpine–Himalayan volcanic belt extends from the Canary and Madeira Islands through the Mediterranean region including Italy and Sicily, the Turkish–Armenian border and Mount Demavend in Iran. The volcanic belt reappears in Burma and is traced to Sumatra, Java and the Lesser Sunda Islands in Indonesia (Bullard, 1984). There are many active andesitic volcanoes along this belt.

Many active volcanoes occur in the vicinity of the Red Sea to central Africa being associated with the great Rift Valleys. They show a mean explosion index of approximately 40 percent and their ejectas are various alkali rocks characterized by high K_2O content (Katsui, 1978).

About 17 percent of the known active volcanoes of the world occur within the true ocean basins (Bullard, 1984). The ejectas of these volcanoes are largely basalt and the explosion indices are very low (for example, 1–3 percent for basaltic volcanoes in Hawaii) (Katsui, 1978). Thus, the tephra dispersal of these volcanoes is not extensive and Andisols formed from these ejectas under perhumid, isohyperthermic conditions as in Hawaii are very rich in noncrystalline Al and Fe hydroxides, reflecting the high content of these elements in the parent material.

1.3.2. Andisols on a regional basis

Andisols are distributed preferentially in regions where active and recently extinct volcanoes are located as shown in the Map of "Major Soil Regions of the World" with a scale of 1:30,000,000 which has been prepared by Eswaran et al. (1992). This soil resources map is based on a recent map produced by FAO in 1990. The mapping units are defined by the suborders of Soil Taxonomy and the temperature regime. Thus, the distribution of five suborders of Andisols such as Cryands, Xerands, Vitrands, Ustands and Udands are shown on the map.

The principal regions of the world where Andisols are found were summarized by Leamy (1984) as follows:

– *In Europe*: Italy, Sicily, Sardinia, France (Massif–Central).

– *In Africa and the Indian Ocean*: Kenya, Rwanda, Tanzania, Ethiopia, Cameroon, Malagasy, Reunion, Canary Islands, Uganda, Sudan, Zaire.

– *In America*: Alaska, British Columbia, Washington, Oregon, California, Mexico, Costa Rica, Panama, Honduras, Guatemala, El Salvador, Nicaragua, West Indies, Ecuador, Colombia, Peru, Chile, Argentina, Bolivia.

– *In Asia and the Pacific*: Hawaii, Aleutian Islands, Kamchatka, Japan, Korea, Micronesia, Philippines, Indonesia, Papua New Guinea, Solomon Islands, Vanuatu, Fiji, Samoa, Tonga, New Zealand.

Dudal et al. (1983) reported an estimate of the extents of Andisols in the major regions of the developing world as follows: 5,424,000 ha for Africa, 59,000 ha for southwest Asia, 7,353,000 ha for southeast Asia, 30,421,000 ha for South America, and 13,526,000 ha for Central America.

REFERENCES

Bullard, F.M., 1984. Volcanoes of the Earth, 2nd revised edition. University of Texas Press, Austin.
Duchaufour, P., 1982. Andosols. In: T.R. Paton (translated), Pedology: Pedogenesis and Classification. George Allen and Unrwin, London, Boston, and Sydney, pp. 196–210.

Dudal, R., Haggins, G.M. and Pecrot, A., 1983. Utilization of soil resource inventories in agricultural development. Proc. 4th Int. Soil Classification Workshop, Rwanda, 2 to 12 June, 1981. Part 1: Papers, ABOS-AGCD, Brussels, Belgium. pp. 6–17.

Eswaran, H., Bliss, N., Lytle, D. and Lammers, D., 1992. The 1:30,000,000 Map of Major Soil Regions of the World (in preparation).

FAO/Unesco, 1974. Soil Map of the World, 1:5,000,000. Vol. 1, legend. Unesco-Paris.

FitzPatrick, E.A., 1971. Andosols. In: Pedology, A Systematic Approach to Soil Science. Oliver and Boyd, Edinburgh, pp. 165–167.

Kamoshita, Y., 1955. Principal soil types (great soil groups) in Japan. Soil and Plant Food, 1: 99–101.

Kanno, I., 1961. Genesis and classification of main genetic soil types in Japan, I. Introduction and Humic Allophane soils. Bull. Kyushu Agr. Exp. Sta, 7: 13–185 (in Japanese, with English abstract).

Katsui, Y., 1978. Distribution of volcanoes. In: Volcanoes. Urban Kubota, No. 15, Osaka, pp. 30–40 (in Japanese).

Leamy, M.L., 1984. Andisols of the world, In: Congresco International de Suelos Volcanicos, Communicaciones, Universida de La Laguna Secretariado de Publicaciones, serie informes 13, pp. 368–387.

Leamy, M.L., Clayden, B., Parfitt, R.L., Kinloch, D.I. and Childs, C.W., 1988. The Andisol proposal 1988, Final proposal of the International Committee on the Classification of Andisols (ICOMAND). New Zealand Soil Bureau, Lower Hutt, New Zealand.

Seki, T., 1913. Zwei vulkanogene Lehms aus Japan. Landw. Versuch. Sta. 79/80: 871–890.

Shoji, S. and Ono, T., 1978. Physical and chemical properties and clay mineralogy of Andosols from Kitakami, Japan. Soil Sci., 126: 297–312.

Simonson, R.W., 1979. Origin of the name "Ando soils". Geoderma, 22: 333–335.

Smith, G.D., 1978. A preliminary proposal for the reclassification of Andepts and some Andic subgroups (The Andisol proposal, 1978). New Zealand Soil Bureau, Lower Hutt, New Zealand.

Soil Survey Staff, 1975. Soil Taxonomy, A basic system of soil classification for making and interpreting soil surveys. USDA-SCS Agric. Handb. 436. U.S. Gov. Print. Office, Washington, D.C.

Soil Survey Staff, 1990. Keys to Soil Taxonomy, 4th edition. AID, USDA-SMSS Technical Monograph No. 19, Blacksburg, Virginia.

Soil Survey Staff, 1992. Keys to Soil Taxonomy, 5th edition. AID, USDA-SMSS Technical Monograph No. 19, Blacksburg, Virginia.

The Third Division of Soils, 1973. Criteria for making soil series and a list of soil series. The first approximation. Nat. Inst. Agr. Res., Japan (in Japanese).

The Third Division of Soils, 1982. Classification of cultivated soils in Japan. Nat. Inst. Agr. Res., Japan.

Uchiyama, N., Abe, K. and Tsuchiya, T., 1954. Research on soil types of arable land, Tochigi Prefecture. Bull. Nat. Inst. Agr. Sci., B3: 43–139 (in Japanese, with English abstract).

Yamada, S., 1951. A method to survey volcanogenous soils and volcanogenous soils in Hokkaido. Bull. Hokkaido Nat. Agr. Exp. Sta., 44: 1–93 (in Japanese).

Chapter 2

MORPHOLOGY OF VOLCANIC ASH SOILS

S. SHOJI, R. DAHLGREN and M. NANZYO

2.1. INTRODUCTION

There is no defined diagnostic horizon central to the concept of the Andisol order as described by Leamy et al. (1980). However, local names such as Kurobokudo (black-fluffy-soil) in Japan (Otowa, 1986) and Trumao (soil from volcanic ash with light stones breaking down in time to a black acid soil) in Chile (Wright, 1964; Leamy et al., 1980) suggest that Andisols have some general morphological characteristics. This chapter describes macromorphology of Andisols determined from the field observation of the soil profile. It relies much on the description and characterization data of Andisol pedons presented at meetings of the International Committee on the Classification of Andisols (ICOMAND), etc. (Pollok et al., 1981; Beinroth et al., 1985a, b; Kimble, 1986; Otowa and Shoji, 1987; Ping et al., 1989).

2.2. MORPHOLOGICAL CHARACTERISTICS

2.2.1. *Soil horizons*

Andisols may have various surface horizons such as histic, mollic, ochric, melanic, fulvic or umbric. Of these, the melanic and fulvic horizons have been introduced as distinctive surface horizons for Andisols with thick humus horizons (Otowa, 1985; Leamy et al., 1988; Otowa et al., 1988; Shoji, 1988; Soil Survey Staff, 1990, 1992). The melanic epipedon has requirements on the depth, thickness, color, melanic index and organic carbon content as presented in Table 4.6 in Chapter 4.

The melanic Andisols are among the blackest of all the soils in the world. The dark color (moist Munsell value and chroma of 2 or less) is attributable to organic matter with a predominance of A-type humic acid which shows the highest degree of humification (Shoji, 1988). The fulvic surface horizon must satisfy all the requirements for the melanic epipedon except the melanic color requirement and the melanic index. The organic matter of fulvic Andisols is dominated by fulvic acid, and the humic acid is mostly P-type, which shows an absorption spectrum distinctly different from that of A-type humic acid. Thus, the separation of melanic and fulvic Andisols is conducted based on the soil color and melanic index (Shoji, 1988).

Genesis of the two surface horizons has been elucidated by biosequential studies on Andisols in Japan (Kawamuro and Osumi, 1988; Shoji et al., 1988b). For example, in northeastern Japan, the melanic Andisols are formed under Japanese pampas grass (*Miscanthus sinensis*) and the fulvic Andisols under beech (*Fagus crenata*) forest vegetation (Shoji et al., 1988b) (Plates 5 to 8). It is common for Andisols to have very thick dark humus horizons (6 percent organic carbon; mollic color; thickness of 50 cm or more). These Andisols are classified into the pachic subgroups (Soil Survey Staff, 1990, 1992). In addition to the effect of grass vegetation, the intermittent thin ash deposition contributes significantly to the formation of the pachic epipedon due to burial of humus-rich horizons as described in Chapter 3.

Various subsurface horizons are also observed in Andisols. Andisols having placic and duric diagnostic subsurface horizons are classified into the placic and duric great groups, respectively (Soil Survey Staff, 1990, 1992).

Andisols may have AC, ABC, or multisequum of these horizon sequences. For example, the common profile types of Andisols in northeastern Japan are summarized in Fig. 2.1. Young Andisols formed from thick ash, pumice or scoria (cinder) show the AC profile (Fig. 2.1a). They are vitric and have low 1500 kPa water retention. Intermittent tephra deposition and subsequent soil formation result in the development of Andisols with a multisequum profile (Fig. 2.1b). These soils typically key out into the thaptic subgroups of Andisols (Soil Survey Staff, 1990, 1992). The effect of vegetation on the development of Andisols is clearly observed for the biosequence of Melanudands and Fulvudands (Fig. 2.1c

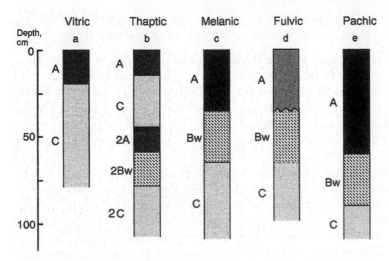

Fig. 2.1. Schematic representation of selected types of Andisol profiles.

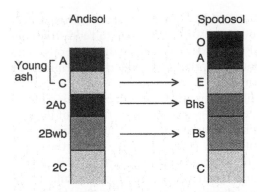

Fig. 2.2. Morphology of a Spodosol and an Andisol with a thin layer or C horizon.

and Fig. 2.1d). In addition to the effect of grass vegetation, intermittent thin ash depositions during the last several thousand years have significantly contributed to the formation of pachic Melanudands (Fig. 2.1e).

Some multisequum Andisols have C horizons or light-colored ash layers in the upper profile as shown in Fig. 2.2. This C horizon or light-colored ash layer can be easily confused with an albic horizon (Ito et al., 1991b). If the C horizon of an Andisol showing a horizon sequence of A-C-2Ab-2Bwb is misidentified as an albic horizon, the 2Ab horizon (buried humus horizon) could be regarded as a spodic horizon.

Careful field observation of soil profiles is necessary to differentiate a spodic horizon from a buried A horizon. To confirm field observations, a chemical method was proposed by Ito et al. (1991b). These workers studied the differences in the properties of organic matter between buried A horizons and spodic Bhs horizons that reflect the different soil-forming processes. They showed that the combination of (pyrophosphate-extractable C)/(organic C) ratio $\geq$ 0.5 and (fulvic acid C)/(pyrophosphate-extractable C) ratio $\geq$ 0.5 is a reliable criterion for differentiating spodic horizons from a buried A horizon.

2.2.2. Soil color

The color of Andisols is determined by the type of tephra, the type and amount of soil organic matter and composition of weathering products. Fresh tephras show various colors ranging from white to black according to their chemical composition and vesicularity. Vesicular tephras of rhyolite, dacite, and andesite composition are commonly white to gray in color because they are dominated by noncolored glass and low concentrations of mafic minerals. When they are subjected to weathering in nonhumus horizons with a udic moisture regime, they become yellow to red depending on the type of iron minerals formed. Nonhumus horizons of well-drained Andisols in the temperate and cold regions are yellow to reddish brown, reflecting the formation of goethite and ferrihydrite. Some tropical

Andisols have bright reddish color (5 YR or redder) due to presence of hematite.

Tephras of basalt and basaltic andesite composition are black to reddish brown because they contain a large amount of colored glass and mafic minerals. The color of these weathered tephras in nonhumus horizons is similarly determined by the type and amount of iron minerals formed.

The color of humus horizons in Andisols is primarily determined by the content and properties of the soil organic matter. In general, the darkness of humus horizons tends to increase with increasing organic carbon content, humic acid to fulvic acid ratio and degree of humification of humic acid. In northeastern Japan, Andisols containing organic $C \geq 6.0$ percent are very dark if their humus is dominated by A-type humic acid characteristic for Kurobokudo, while those containing the same amount of organic C are dark brown if the humus is dominated by fulvic acid and the humic acid is P-type. Thus, the former will meet the melanic color requirements and the latter the fulvic color requirements. The exact separation of melanic and fulvic Andisols is determined using not only soil color but also the melanic index as described earlier. If surface humus horizons fail to meet melanic or fulvic criteria, they typically classify as humic Andisols if they are dark colored, or as ochric epipedons if the colors are light (Soil Survey Staff, 1992).

2.2.3. Soil texture

Andisols show a wide variation in soil texture depending on the type and particle size of tephras, mode and degree of weathering, etc. It is, however, common that there is a significant difference between soil texture determined in the field and in the laboratory by mechanical analysis because noncrystalline materials often inhibit dispersion of mineral particles as discussed in detail in Chapter 7. Thus, modifiers that replace names of particle-size classes are employed (Soil Survey Staff, 1990, 1992). They reflect a combination of particle size and mineralogy.

The substitutes for the fragmental classes include "pumiceous" and "cindery" and those for the nonfragmental classes, "ashy", "ashy-pumiceous", "ashy-skeletal", "medial", "medial-pumiceous", "medial-skeletal", "hydrous", "hydrous-pumiceous", and "hydrous-skeletal". Of these, "ashy", "medial", and "hydrous" require measurement of 1500 kPa water retention which approximately reflects the quantities of mineral and organic colloids in the soil. These nonfragmental classes are defined as presented in Table 4.8 in Chapter 4.

Studies of allophanic Andisols from eastern Hokkaido by Ito et al., (1991a) obtained the following relationship between water retention and the concentrations of noncrystalline materials and organic matter:

1500 kPa water retention of undried samples (%) =

$\quad$ 1.93 (1.7 × Org. C %) + 1.47 (8 × Si_o% + 2 × Fe_o%) − 0.92

$$r = 0.944^{***} \ (n = 37)$$

The strong correlation shows that both allophanic clays and humus contribute to the 1500 kPa water retention.

2.2.4. Consistence

Andisols often contain a large amount of noncrystalline materials which influence consistence and significantly contribute to the development of excellent physical properties for both cultivation and plant root growth.

Consistence in Andisols is markedly affected by water content. Wet consistence, determined at or slightly greater than field capacity, is characterized by stickiness and plasticity. As described in the selected examples of profile morphology in this chapter, Andisols which are rich in humus and/or allophanic clays are typically slightly sticky to nonsticky and slightly plastic to nonplastic. In contrast, nonallophanic Andisols with low organic C content are sticky and plastic, reflecting the physical properties of layer silicates (see the Alic Melanudand at Tohoku University in the selected examples of profile morphology on page 24).

Moist consistence, measured at moisture contents between dryness and field capacity is notable especially for the upper horizons of Andisols. Moist consistences are typically very friable to friable, reflecting development of very porous aggregates of granular or subangular blocky structure. The massive lower horizons are generally observed to be firm.

Smeariness reflects the thixotropic property of noncrystalline materials. It is a characteristic for hydric Andisols as described for the Hilo soil in Hawaii, U.S.A. (see the selected examples of profile morphology on page 31). Only weak pressure is required for the soil material of hydric Andisols to become liquid.

2.2.5. Structure

Andisols have soil structure which reflects the abundance of noncrystalline materials and organic materials and contribute to low bulk density. The surface soils of Andisols primarily have granular structure and occasionally subangular blocky structure. The size and grade of these structures show a wide variation reflecting the kind of soil material, cultivation, and climate (drying and wetting, and thawing and freezing). Cultivation of Andisols tends to promote formation of subangular blocky structure.

The subsurface horizons (Bw horizons) of Andisols typically have subangular blocky structure which is larger in size and weaker in strength compared to the same structure found in the surface horizons. Some Andisol B horizons show prismatic structure. The subsurface horizons of young Andisols formed from coarse tephras often lack structure and are single grained.

Development of soil structure in Andisols is closely related to the unique physical properties of Andisols such as high water retention, large total porosity, and good drainage which are favorable for plant root growth.

2.3. EXAMPLES OF PROFILE MORPHOLOGY

Andisol suborders are classified mainly according to soil moisture and temperature regimes while distinct morphological properties are employed as differentiae
to classify the great groups. The wide range of morphological properties requires
a variety of great groups (Soil Survey Staff, 1990, 1992). Thus, vitric, placic, duric,
melanic, fulvic, and hydric Andisols are selected as examples to show the profile
morphology with description and characterization data.

2.3.1. Vitric Andisols

Vitric Andisols, called Vitrandepts in the 1975 Soil Taxonomy (Soil Survey
Staff, 1975), are weakly weathered and have the lowest amounts of weathering
products among Andisols. Their genesis is primarily determined by age, chemical,
mineralogical, and textural properties of the tephra and climatic conditions. They
accommodate recent and coarse textured tephras such as pumice and scoria. They
are found predominantly in dry climates as indicated by the fact that all Torrands
are regarded as vitric great groups (Soil Survey Staff, 1990, 1992). According
to the chemical kinetics of weathering in young Andisols in northeastern Japan,
Udivitrands can develop in volcanic ash dominated by noncolored glass within 200
to 300 years following deposition at a mean annual temperature of 10°C (Shoji et
al., 1993). They can form much faster in colored glass-rich volcanic ash than in
noncolored glass-rich ash due to the greater weathering rates of colored volcanic
glass.

Vitric Andisols are further classified at the suborder, great group, and sub-
group levels. The Vitrand suborder keys out after Xerands and before Ustands.
Therefore, vitric great groups are provided in the suborders of Aquands, Cryands,
Torrands, and Xerands.

The Bekkai soil from eastern Hokkaido, Japan (Plate 1) is an example of a
vitric Andisol and is described below. It has formed from intermittently deposited
rhyolitic ash layers ranging in age from a few hundred to less than 1000 years
old. The soil moisture and temperature regimes were udic and frigid, respectively.
The pedon has a multisequum profile consisting of AC horizons overlying four
ABwC horizon sequences and shows a low degree of weathering, especially in the
upper part of the profile. Thus, it is locally classified as Stratic Regosolic Andosol
according to the soil classification of agricultural land in Hokkaido (Hokkaido Soil
Classification Committee, 1979).

The A horizons are relatively thin (7–13 cm), failing the requirements for
melanic, fulvic, and humic surface soils and the Bw horizons are weakly developed.
The low degree of development is reflected in the ochric and cambic diagnostic
horizons. The unique consistency properties of high friability, moderately smeary
and low stickiness are observed for almost all the soil horizons, reflecting the
presence of noncrystalline materials.

DESCRIPTION OF BEKKAI SOIL (PLATE 1)

Source	Otowa and Shoji, 1987
Location	Bekkai, Nokke country, Hokkaido, Japan
	Latitude = 43°27'50"N; Longitude = 144°44'10"E
Classification	Ashy over medial, frigid, Thaptic Udivitrand
Physiography	Terrace
Slope characteristics	4% convex
Elevation	138 m M.S.L.
Air temperature (°C)	Ann: 5, Sum: 14, Win: −6
Soil temperature (°C)	Ann: 7, Sum: 14, Win: 1, Frigid temperature regime
Precipitation	1090 mm, Udic moisture regime
Water table	Not observed
Drainage	Well drained
Permeability	Moderate
Land use	Forestland not grazed
Vegetation	*Quercus mongolica, Quercus dentata, Sasa nipponica*
Erosion or deposition	Slight erosion and no deposition
Parent material	Several strata of rhyolitic volcanic ash
Diagnostic horizons	14–39 cm Ochric, 24–35 cm Cambic
Described by	T.D. Thorson and G. Holmgren
Date	September, 1985

O	2–0 cm:	Partially decomposed organic matter. Decaying mat of sasa and oak leaves.
A	0–8 cm:	Black (10 YR 2/1) mucky silt loam; dark gray (10 YR 4/1) dry; moderate very fine granular structure; soft, very friable, moderately smeary, slightly sticky, slightly plastic; many very fine and fine roots throughout and many medium roots throughout; many very fine and fine interstitial pores; abrupt smooth boundary.
C	8–14 cm:	70 percent pale brown (10 YR 6/3) and 30 percent strong brown (7.5 YR 5/6) very fine sandy loam; 70 percent very pale brown (10 YR 7/4), 20 percent very pale brown (10 YR 8/3), and 10 percent reddish yellow (7.5 YR 7/6) dry; massive; soft, very friable, moderately smeary, slightly sticky, slightly plastic; many very fine and fine roots throughout and many medium roots throughout; few very fine interstitial pores; abrupt smooth boundary.
2Ab	14–24 cm:	Black (10 YR 2/1) mucky silt loam; very dark grayish brown (10 YR 3/2) dry; moderate very fine and fine subangular blocky structure; soft, very friable, moderately smeary, slightly sticky, slightly plastic; many very fine roots throughout and common fine roots throughout; common very fine and fine interstitial pores; abrupt smooth boundary.
2Bwb	24–35 cm:	50 percent brown to dark brown (10 YR 4/3) and 50 percent strong brown (7.5YR 5/6) very fine sandy loam; 50 percent light yellowish brown (10YR 6/4) and 50 percent reddish yellow (7.5YR 7/6) dry; massive; soft, very friable, moderately smeary, slightly sticky, slightly plastic; common fine roots throughout and few very fine roots throughout; few very fine interstitial and few very fine continuous tubular pores; clear wavy boundary.
2C	35–48 cm:	80 percent grayish brown (10YR 5/2) and 20 percent yellowish brown (10YR 5/6) very fine sandy loam; 80 percent light gray (10 YR 7/2) and 20 percent yellow (10 YR 7/6) dry; massive; soft, friable moderately smeary, slightly

sticky, slightly plastic; common fine roots throughout and few very fine roots throughout; few very fine interstitial pores; abrupt smooth boundary.

3Ab   48–61 cm: Black (10 YR 2/1) mucky silt loam; very dark grayish brown (10 YR 3/2) dry; moderate very fine and fine subangular blocky structure; soft, very friable, moderately smeary, slightly sticky, slightly plastic; common fine roots throughout and few medium roots throughout; common very fine interstitial pores; clear wavy boundary.

3Bwb   61–78 cm: Very dark brown (10 YR 2/2) mucky silt loam; grayish brown (10 YR 5/2) dry; moderate fine and medium subangular blocky structure; slightly hard, very friable, moderately smeary, slightly sticky, slightly plastic; common very fine and fine roots throughout; common very fine interstitial pores; clear wavy boundary.

4Ab   78–85 cm: Black (10 YR 2/1) mucky silt loam; very dark grayish brown (10 YR 3/2) dry; weak very fine and fine subangular blocky structure; soft, very friable, moderately smeary, slightly sticky, slightly plastic; common fine roots throughout; common very fine interstitial pores; clear wavy boundary.

4Bwb   85–90 cm: Dark brown (10 YR 3/3) silt loam; pale brown (10 YR 6/3) dry; weak fine and medium subangular blocky structure; slightly hard, friable, moderately smeary, slightly sticky, slightly plastic; few fine roots throughout; few very fine discontinuous tubular pores; 5 percent pumice from ejecta; clear wavy boundary.

5Ab   90–98 cm: Very dark grayish brown (10 YR 3/2) silt loam; dark grayish brown (10 YR 4/2) dry; massive; soft, very friable, moderately smeary, slightly sticky, slightly plastic; common fine roots throughout; few very fine continuous tubular pores; clear wavy boundary. Structure grades to very weak fine and medium subangular blocky.

5Bwb   98–108 cm: Dark brown (10 YR 3/3) loam; pale brown (10 YR 6/3) dry; massive; soft, friable, weakly smeary, slightly sticky, slightly plastic; common fine roots throughout; common very fine continuous tubular pores; abrupt wavy boundary. Structure grades to very weak fine and medium subangular blocky.

Characterization data for selected horizons of the Bekkai soil are presented in Table 2.1. The upper horizons (0–48 cm) show a high content of volcanic glass (95–97%) and low values for clay content (<6%), acid oxalate-extractable components, and P retention. The acid oxalate-extractable components and P retention tend to increase with depth. The large difference between the clay and allophane contents in the lower horizons indicates incomplete dispersion of clay particles by the particle-size determination method. Bulk densities are somewhat high and water retention values at 1500 kPa are mostly low. From the foregoing, it is obvious that the pedon is vitric.

Though the depth of A horizons for the Bekkai pedon is relatively thin, as noted earlier, these horizons show high organic carbon concentrations (>6.0%). The range of CEC values is wide (3–30 $cmol_c$ kg^{-1}) and soil organic matter contributes greatly to the higher CEC values. The upper two A horizons (A and 2Ab) are very strongly acid (pH = 4.5–4.7) and contains high KCl-extractable Al (4 $cmol_c$ kg^{-1}) while all the underlying horizons are strongly acid and show a

TABLE 2.1

Characterization data of the Bekkai soil

Depth (cm)	Horizon	Total (%)			Bulk density (g cm^{-3})	Water retention at 1500 kPa (%)		Organic carbon (%)	KCl-extr. Al (cmol$_c$ kg^{-1})	CEC (cmol$_c$ kg^{-1})
		Clay <2 μm	Silt 2–50 μm	Sand 50 μm–2 mm		Air-dry	Field moist			
0–8	A	5.5	59.2	35.3		23.7	28.2	10.3	4.3	30.2
14–24	2Ab	3.2	54.8	42.0	0.80	13.6	21.2	6.6	3.8	20.9
35–48	2C	3.7	49.1	47.2	0.90	3.2	5.4	0.7	tr	2.6
48–61	3Ab	5.1	49.6	45.3		14.0	26.1	8.3	tr	30.3
61–78	3Bwb	3.9	47.8	48.3	0.61	11.3	24.0	4.3	tr	17.7
78–85	4Ab	4.5	39.2	56.3		18.2	35.5	7.6	tr	32.7
90–98	5Ab	3.9	40.6	55.5		15.0	30.7	4.5	tr	21.3
98–108	5Bwb	3.4	39.7	56.9	0.83	12.3	24.0	2.2	tr	11.7

Depth (cm)	Horizon	Base sat. (%)	pH H$_2$O	Acid oxalate extr. (%)				Allophane (%)	P ret. (%)	Glass content in 0.02–2 mm fraction (%)
				Al$_o$	Fe$_o$	Al$_o$ + 1/2 Fe$_o$	Si$_o$			
0–8	A	28	4.5	0.6	0.5	0.8	0.1	0.6	39	96
14–24	2Ab	6	4.7	1.2	0.8	1.6	0.2	1.4	66	95
35–48	2C	8	5.7	0.6	0.9	1.1	0.2	1.9	33	97
48–61	3Ab	21	5.6	1.9	1.0	2.4	0.4	3.2	83	
61–78	3Bwb	17	5.7	2.2	0.9	2.7	0.7	5.8	78	
78–85	4Ab	16	5.7	3.3	1.1	3.9	1.1	8.5	92	
90–98	5Ab	14	5.9	3.4	1.0	3.9	1.3	10.2	90	
98–108	5Bwb	11	5.9	3.2	0.8	3.6	1.4	11.0	85	

Source: Otowa and Shoji, 1987.

virtual absence of KCl-extractable Al. These differences between the two groups of soil horizons reflect nonallophanic clay mineralogy of the surface horizons and allophanic clay mineralogy of the lower soil horizons, respectively.

According to the morphological properties and characterization data described above, the Bekkai soil is classified as ashy over medial, frigid Thaptic Udivitrand according to Keys to Soil Taxonomy (Soil Survey Staff, 1990, 1992).

2.3.2. Placic Andisols

Placic Andisols, called Placandepts in the 1975 Soil Taxonomy, have a thin, black to dark reddish pan commonly cemented with iron and occur primarily in very humid climates. Their genesis is attributed to redox mechanisms and acid leaching of soluble organic matter. Most placic horizons are composed of ferrihydrite, poorly crystalline goethite and organic matter (Clayden et al., 1990). The placic horizon strongly impedes water movement and seriously inhibits root development. The placic property is employed as differentia at the great group level of Aquands and Udands, and at the subgroup level of Hydrocryands (Soil Survey Staff, 1990, 1992).

The Sitka soil from Alaska, U.S.A. (see Plate 2) is presented as an example of a volcanic ash soil containing a placic horizon. As described below, the pedon has developed on intermittently deposited volcanic ash and pumice layers under coniferous vegetation. It shows a multisequum profile having three E horizons (0–2 cm; 80–83 cm; 118–120 cm), a spodic horizon (2–24 cm) and a placic horizon (34–37 cm). These morphological features indicate that podzolization is prevailing in tephras under the very humid cold climate. Thus, the Sitka soil used to be classified as a Cryic Placohumod.

As noted later in the characterization data of the Sitka soil, tephra-derived Spodosols and Andisols share many important chemical and mineralogical properties (Shoji et al., 1988a; Ping et al., 1989; Shoji and Ito, 1990; Shoji and Yamada, 1991). Since these properties make it difficult to separate the two orders, various proposals on the separation problem have been presented (Shoji and Ito, 1990). Thus, the International Committee on the Classification of Spodosols (ICOMOD) has proposed the criteria to separate the two orders (Rourke, 1991). According to this proposal, the E horizon of the Sitka soil fails the thickness requirement for the albic horizon (greater than 2.5 cm). Mottles are common in the Bs subhorizons overlying the placic horizon suggesting the aquic moisture regime. Thus, the pedon is classified into Aquands.

Based on the intense discussion of the Spodosol–Andisol separation problem, Soil Survey Staff (1992) has proposed new Spodosol criteria. The Sitka soil is reclassified as a Spodosol (Andic Placocryod) according to this proposal.

DESCRIPTION OF SITKA SOIL (PLATE 2)

Source	C.L. Ping and J.P. Moore, 1986 (unpublished)
Location	Sitka, Alaska, U.S.A.
	Latitude = 57°3'N, Longitude = 135°20'W
Classification	Medial shallow, Typic Cryaquand (Andic Placocryod according to 1992 Keys to Soil Taxonomy)
Slope characteristics	Slightly undulating glacial moraine in lowland
Air temperature (°C)	Ann: 5.7, Sum: 11.0, Win: −0.5, Cryic temperature regime
Precipitation	2400 mm
Drainage	Poorly drained
Land use	Forest
Vegetation	Coniferous forest (*Tsuga heterophylla*, *Vaccinum* spp., *Sphagnum* spp. *Pinus contorta*)
Parent material	Several strata of volcanic ash and pumice
Diagnostic horizons	0–2 cm Albic, 2–24 cm Spodic, 34–37 cm Placic
Described by	C.L. Ping and J.P. Moore
Date	May 23, 1986

Oi	14–7 cm:	Partially decomposed organic matter (fabric peat).
Oa	7–0 cm:	Dusky red (2.5 YR 2.5/2), decomposed organic matter (sapric muck).
E	0–2 cm:	Gray (5 YR 6/2) fine sandy loam; massive to weak medium granular structure; very friable, weak smeary, nonsticky, nonplastic; few medium and common fine roots; abrupt irregular boundary.
Bhs	2–7 cm:	Black (2.5 YR 2.5/0) silt loam; moderate medium subangular blocky structure; very friable, moderately smeary, slightly sticky, nonplastic; few medium and common fine roots; abrupt wavy boundary.
Bs1	7–15 cm:	Dark reddish brown (2.5 YR 3/4) silt loam; common medium reddish brown (5 YR 3/4) mottles; moderate medium subangular blocky structure; friable, moderately smeary, slightly sticky, slightly plastic; common fine roots; clear smooth boundary.
Bs2	15–18 cm:	Dark reddish brown (2.5 YR 3/4) silt loam; common medium yellowish red (5 YR 5/6) mottles; moderate medium subangular blocky structure; friable, moderately smeary, slightly sticky, slightly plastic; few fine roots; abrupt smooth boundary.
Bs3	18–24 cm:	Reddish yellow (7.5 YR 6/6) fine sandy loam; common medium dark reddish brown (5 YR 3/3) mottles; massive; friable, moderately smeary, slightly sticky, nonplastic; few fine roots; abrupt smooth boundary.
BC	24–34 cm:	60 percent dark brown (7.5 YR 4/4), 30 percent dark reddish brown (5 YR 3/3), 10 percent very dark gray (5 YR 3/1); sandy loam; massive; firm, weakly smeary nonsticky, nonplastic; abrupt smooth boundary.
Csm	34–37 cm:	85 percent dark yellowish brown (10 YR 4/8), and 15 percent black (N 2/0) very fine sandy loam; moderate to strong coarse platy structure; extremely firm, weakly smeary, nonsticky, nonplastic; abrupt smooth boundary.
C1	37–40 cm:	60 percent strong brown (7.5 YR 5/6), 30 percent yellowish red (5 YR 5/8), and 10 percent dusky red (2.5 YR 3/2) coarse sandy loam; common dusky red (2.5 YR 3/2) mottles; massive; friable, nonsticky, nonplastic; clear smooth boundary.

2C2	40–80 cm:	Variegated color; very gravelly coarse sand (40 percent pumice in 2 mm–76 mm fragments); single grain; loose, nonsticky, nonplastic; abrupt smooth boundary.
3Eb	80–83 cm:	80 percent dark grayish brown (10 YR 4/2) and 20 percent light gray (10 YR 7/2) silt loam; massive; firm, moderately smeary, slightly sticky, nonplastic; abrupt smooth boundary.
3Bsb	83–86 cm:	Variegated color (red and black) coarse sand; single grain; loose, nonsticky, nonplastic; abrupt smooth boundary.
Ob	86–89 cm:	Black (7.5 YR 2/0); massive; friable, nonsticky, nonplastic; abrupt smooth boundary.
4Bsb	89–118 cm:	Variegated color, sand; single grain; loose, nonsticky, nonplastic; abrupt smooth boundary.
5Eb	118–120 cm:	Very dark brown (7.5 YR 2/2) fine sand; massive; friable, nonsticky, nonplastic; abrupt smooth boundary.
5Bsb	120–123 cm:	Variegated color, sand; massive to single grain; loose, nonsticky, nonplastic; abrupt smooth boundary.
5Cb	123–127 cm:	Dark grayish brown (10 YR 4/2) very fine sandy loam; massive; firm, nonsticky, nonplastic; abrupt irregular boundary.

Characterization data of the Sitka soil are presented in Table 2.2. The clay percentage is highest in the upper horizons while it is very low in the lower horizons compared with the allophane content, indicating incomplete dispersion of clay particles. Though bulk density data is only partially available for this pedon, it is believed to be low in all the horizons. Water retention at 1500 kPa is moderate, meeting the medial criteria within the nonfragmental class. The upper part of the pedon (0–34 cm) contains organic carbon greater that 6.0 percent, but shows soil color lighter than moist Munsell value and chroma of 2, thus it fails to meet the melanic color requirement (Soil Survey Staff, 1990, 1992).

Soil acidity is extremely acid in the E horizon (pH = 4.0) and tends to become weaker with depth (pH = 4.5–6.0). Accordingly, KCl-extractable Al tends to decrease with depth. It is noted that the horizons from the Bs1 to BC show extreme to very strong acidity, even though they contain considerable amounts of allophane.

The analyses of acid oxalate extractable components provide evidence as to the mechanism of iron accumulation in the Csm horizon or placic horizon. Concentrations of acid oxalate extractable Fe are small in all the horizons overlying the Csm horizon and show a dramatic increase in the Csm horizon. In contrast, acid oxalate extractable Al shows high concentrations in all the horizons except the E horizon. The high ratio of acid oxalate extractable Fe to acid oxalate extractable Al in the Csm horizon as compared with all other horizons indicates that Fe has been selectively accumulated within the Csm horizon. According to these data, it is concluded that the placic horizon has been formed by redoxy processes.

It is interesting to observe that the 3Eb horizon contains a considerable amount of noncrystalline material (allophane = 20.6 percent). This fact indicates that

TABLE 2.2

Characterization data of the Sitka soil

Depth (cm)	Horizon	Total [a] (%) Clay <2 µm	Silt 2–200 µm	Sand 200 µm–2 mm	Bulk density (g cm^{-3})	Water retention at 1500 kPa (%) Air-dry	Field moist	Organic carbon (%)	pH H$_2$O
0–2	E	13	49	48	0.43			17.5	4.0
2–7	Bhs	–	–	–				–	–
7–15	Bs1	27	35	38	0.41	52.7	94.5	20.6	4.3
15–18	Bs2	25	37	38	0.53	54.7	82.6	15.0	4.6
18–24	Bs3	13	25	62	0.55	34.7	82.3	7.9	4.8
24–34	BC	9	25	66	0.73	27.3	48.0	6.3	4.9
34–37	Csm	7	18	75		19.3	39.0	2.9	5.1
37–40	C1	–	–	–					
40–80	2C2	4	8	88				0.4	6.1
80–83	3Eb	7	36	57				1.1	5.7
83–86	3Bsb	4	5	91				0.3	5.9

Depth (cm)	Horizon	KCl-extr. Al (cmol$_c$ kg^{-1})	Acid oxalate extr. (%) Al$_o$	Fe$_o$	Al$_o$ + 1/2 Fe$_o$	Si$_o$	Fe$_o$/Al$_o$	Allophane (%)	P ret. (%)
0–2	E	16.8	1.3	0.9	1.7	tr.	0.74	tr.	82
2–7	Bhs	–	–	–	–	–	–	–	–
7–5	Bs1	18.8	5.9	0.5	6.2	1.2	0.09	9.8	97
15–18	Bs2	7.6	7.6	0.1	7.7	2.2	0.01	17.4	97
18–24	Bs3	3.0	6.4	0.1	6.5	2.3	0.01	18.1	98
24–34	BC	1.9	6.5	0.2	6.6	2.4	0.03	19.4	98
34–37	Csm	0.2	4.8	13.0	11.3	2.6	2.71	20.6	98
37–40	C1	–	–	–	–	–	–	–	
40–80	2C2	0.2	2.0	1.2	2.6	1.3	0.61	10.3	68
80–83	3Eb	0.1	4.2	2.1	5.2	2.6	0.51	20.6	92
83–86	3Bsb	0.2	2.3	2.3	3.5	1.7	0.97	13.8	74

Source: Ito, Shoji, and Ping, 1991 (unpublished).
[a] Dispersed by sonication and pH adjustment using dilute HCl or NaOH.

strong *in situ* weathering, primarily by carbonic acid, is taking place in the lower part of pedon. Thus, it appears that the properties of the pedon are primarily influenced by three processes; podzolization, redoxy processes and *in situ* weathering.

According to the foregoing, the Sitka soil is classified as a medial shallow, Typic Cryaquand according to the 1990 Keys to Soil Taxonomy. Since both placic and fulvic subgroups are not provided for Cryaquands (Soil Survey Staff, 1990), the pedon cannot be classified rationally. The "shallow" family is used to indicate the shallow rooting due to the root-limiting placic horizon. As described earlier, however, the Sitka soil is reclassified as Andic Placocryod according to the 1992 Keys to Soil Taxonomy (Soil Survey Staff, 1992).

2.3.3. Duric Andisols

Duric Andisols, called Durandepts in the 1975 Soil Taxonomy, have a duripan within 1 m of the soil surface. These soils occur in Central and South America where climatic conditions display a pronounced warm dry season (Soil Survey Staff, 1975). Duripans, amorphous silica- or opaline silica-cemented subsurface horizons are formed in volcanic materials by the following mechanism: release of silica to the soil solution by rapid weathering of volcanic glasses, percolation of the silica-rich soil solution downward and precipitation of the silica near the wetting front or at a lithologic discontinuity to form the duripan (Chadwick et al., 1987). Duripans impede water movement and severely reduce root development. Duric great groups are provided in the Aquands, Ustands, and Udands, and duric subgroups in some great groups of Aquands, Torrands and Udands (Soil Survey Staff, 1990, 1992).

The Lanco soil described below from Chile (see Plate 3) is an example of a duric Andisol. It is a shallow Andisol and has a strongly cemented duripan (4Bqlm) at a depth of 45–51 cm. The parent material consists of volcanic ash (0–35 cm), mixed lithology of volcanic ash and glacial outwash (35–45 cm) and glacial outwash (>45 cm). Pedogenesis of the pedon is considered to have begun under mesic temperature and udic soil moisture regimes. However, the silica cementation in the lower horizons has impeded water movement, resulting in the development of an aquic moisture regime in the pedon.

The horizons from A1 to 3Bw2 show morphological properties common to most allophanic Andisols, such as friable to very friable, slightly sticky and slightly plastic consistence. On the other hand, the Bq1m and 5Bq2 horizons are cemented with silica released from the weathering volcanic ash in the overlying horizons.

DESCRIPTION OF LANCO SOIL (PLATE 3)

Source	Beinroth et al., 1985a
Location	Latitude = 39°42′S, Longitude = 72°50′W
Classification	medial, mesic, Typic Durudand
Geomorphic position	Terrace
Slope characteristics	2% planar

Elevation	50 m M.S.L
Air temperature (°C)	Ann: 13.8, Sum: 17, Win: 8
Precipitation	1950 mm, Aquic moisture regime
Water table	>10 m
Drainage	Poorly drained, impeded below 45 cm
Permeability	Moderate
Land use	Pasture
Erosion or deposition	None
Parent material	Volcanic ash over unrelated cemented gravelly glacial outwash at 45 cm
Diagnostic horizons	45–51 cm Duripan
Described by	T. Cook, W. Luzio, R. Honorato, G. Galindo, W. Vera, R. Grez
Date	January 7, 1983

A1   0–19 cm: Dark reddish brown (5 YR 3/3) and dark reddish brown (5 YR 3/4) sandy loam, strong brown (7.5 YR 5/4) dry; weak fine and medium subangular blocky structure; slightly hard, friable, slightly sticky and slightly plastic; weakly smeary, very moist or wet; common very fine and fine roots, upper 1–3 cm is a dense mat of roots in the mineral soil; many very fine tubular, pores; 2 percent 2–5 mm gravel; pH = 5.3, strongly acid; abrupt smooth boundary.

2Bw1   19–35 cm: Dark reddish brown (5 YR 3/4) loam, strong brown (7.5 YR 5/6) dry; weak medium prismatic parting to moderate fine and medium subangular blocky structure; soft, very friable, slightly sticky and slightly plastic; weakly smeary, very moist or wet; common very fine roots; many very fine tubular and interstitial pores; 1 percent cobble, 2 percent gravel; pH = 5.7, medium acid; clear smooth boundary.

3Bw2   35–45 cm: Dark brown (7.5 YR 3/4 crushed) very gravelly sandy loam; structureless; friable, slightly sticky and slightly plastic; many very fine roots; many very fine interstitial and tubular pores; coarse fragments, 55 percent gravel, 2 percent cobble mixed lithology; pH = 5.7, medium acid; abrupt wavy boundary.

4Bq1m   45–51 cm: Dark red (2.5 YR 3/6) and yellowish red (5 YR 4/6) indurated duripan; very strongly cemented; coarse fragments, 60 percent gravel mixed lithology; pH = 5.7, medium acid; abrupt wavy boundary. Laminar cap on surface of pan.

5Bq2   51–65 cm: Black (10 YR 2/1) and very dark gray (10 YR 3/1); weakly cemented; coarse fragments, 60 percent gravel, 3 percent cobble from mixed lithology; pH = 5.5, strongly acid.

The characterization data of the Lanco soil are presented in Table 2.3. The clay content determined by mechanical analysis was very low and was obviously underestimated as compared with the allophane content obtained by chemical analysis because of incomplete dispersion of clay particles. Bulk densities of the A1 and 2Bw1 horizons are lower than 0.9 g cm^{-3}. Water retention at 1500 kPa ranged from 35 to 41 percent, meeting the medial criteria within the nonfragmental class.

The organic carbon content is very high only in the surface horizon. The CEC values are fairly high and base saturation is very low in all the horizons except the 5Bq2 horizon. However, these horizons show weak acidity and a virtual absence of KCl-extractable Al reflecting the allophanic clay mineralogy.

TABLE 2.3

Characterization data of the Lanco soil

Depth (cm)	Horizon	Total (%) Clay <2 μm	Silt 2–50 μm	Sand 50 μm–2 mm	Bulk density (g cm⁻³)	Water retention at 1500 kPa (%) Air-dry	Organic carbon (%)	KCl-extr. Al (cmol_c kg⁻¹)	CEC (cmol_c kg⁻¹)
0–19	A1	0.3	64.6	35.1	0.64	34.5	8.33	0.1	26.2
19–35	2Bw1	–	43.4	56.6	0.53	40.8	3.41	0.1	16.8
35–45	3Bw2	–	45.7	54.3		35.0	4.80	tr	18.3
45–51	4Bq1m	8.0	20.4	71.6		13.4	0.61	tr	12.7
51–65	5Bq2	2.2	3.7	94.1		4.8	0.08	tr	4.0

Depth (cm)	Horizon	Base sat. (%)	pH H₂O	Acid oxalate extr. (%) Al_o	Fe_o	Al_o + 1/2Fe_o	Si_o	Allophane (%)	P ret. (%)
0–19	A1	9	5.8	5.0	1.2	5.6	1.9	15.2	99
19–35	2Bw1	2	5.8	6.7	0.6	7.0	3.2	25.6	98
35–45	3Bw2	1	5.8	5.5	0.3	5.7	2.5	20.0	98
45–51	4Bq1m	4	5.9	2.7	0.8	3.1	1.7	13.6	94
51–65	5Bq2	22	6.2	0.8	0.7	1.2	0.7	5.6	38

Source: F.H. Beinroth et al., 1985a.

All the horizons derived from volcanic ash and mixed volcanic ash and glacial outwash (A1 to 4Bqlm horizons) show high acid oxalate extractable Al and a half Fe and a large P retention. According to the comparison of acid oxalate extractable Al and Si in the duripan with those of the overlying horizons, it appears that the silica of the duripan has a very low solubility in acid oxalate reagent.

Based on the morphological and characterization data described above, the Lanco soil is classified as medial mesic, Typic Durudand by Keys to Soil Taxonomy (Soil Survey Staff, 1990, 1992).

2.3.4. Melanic Andisols

Many Andisols in Japan have a thick dark-colored humus horizon on which the central concept of Kurobokudo has been traditionally based. Such an epipedon is significantly different in color and humus characteristics from histic, umbric, and mollic epipedons. Thus, ICOMAND (Leamy et al., 1988) has proposed the melanic epipedon as a diagnostic horizon which was introduced in Keys to Soil Taxonomy (Soil Survey Staff, 1990).

Studies on a biosequential relationship between Melanudands and Fulvudands in Japan have shown that the Japanese pampas grass ecosystem (*Miscanthus sinensis*) greatly contributes to the genesis of Melanudands and that soil organic matter of Melanudands is dominated by A-type humic acid showing the highest degree of humification (Kawamuro and Osumi, 1988; Shoji et al., 1988b). Intermittent thin ash deposition also significantly contributes to the formation of a thick humus horizon by burial of surface humus-rich horizons(Shoji and Otowa, 1988).

Melanic great groups were first provided for Borands and Udands by Smith (1978). They are identified in the Aquand, Cryand, Xerand, and Udand suborders in Keys to Soil Taxonomy (Soil Survey Staff, 1990, 1992).

The soil of Tohoku University Farm in northeastern Japan (see Plate 4) is an example of Melanudands which are classified as Thick High-humic Kurobokudo in Japan. As described below, it has a distinct morphological feature of a very dark thick humus horizon (0–57 cm) formed from cumulative ash of 10,000 to 1000 yr B.P. Soil organic matter is largely extracted with dilute NaOH solution and the extract shows a melanic index of less than 1.70. From the foregoing, it is obvious that the melanic color of Andisols is determined by high soil organic matter content and humus with a high degree of humification.

There are significant differences in consistence between the humus horizons and nonhumus horizons. All the humus horizons show friable, slightly sticky and slightly plastic consistence common for allophanic Andisols. Since these horizons are nonallophanic and have a clay fraction dominated by chloritized 2:1 minerals (Shoji et al., 1985), it appears that their consistence is substantially determined by soil organic matter which occupies more than 20 percent of the soil. In contrast, the nonhumus horizons exhibit firm, sticky and plastic consistence which is largely attributable to an abundance of chloritized 2:1 minerals.

DESCRIPTION OF TOHOKU UNIVERSITY FARM SOIL (PLATE 4)

Source	Otowa and Shoji, 1987
Location	Mukaiyama, Tohoku University Farm
	Latitude = 38°42′18″N, Longitude = 140°33′24″E
Classification	Medial mesic, Alic Pachic Melanudand
Physiography	Upland slope in mountains or deeply dissected plateaus
Geomorphic position	On the crest of mountain
Slope characteristics	5% plane, southeast facing
Elevation	500 m M.S.L
Air temperature (°C)	Ann: 10, Sum: 20, Win: 0
Soil temperature (°C)	Ann: 11, Sum: 22, Win: 1, Mesic temperature regime
Precipitation	1500 mm, Udic moisture regime
Water table	Not observed
Drainage	Well drained
Permeability	Moderate
Land use	Forest land not grazed
Vegetation	*Miscanthus sinensis*
Erosion or deposition	Slight
Parent material	Volcanic ash from acidic-ash material
Diagnostic horizons	0 to 57 cm Melanic, 57 to 160 cm Cambic
Described by	T.D. Thorson and George Holmgren
Date	September, 1985

Oe	5–3 cm:	Partially decomposed organic matter. Decaying mat of Japanese pampas grass leaves.
Oa	3–0 cm:	Partially decomposed organic matter. Decomposed leaves containing many very fine roots.
A1	0–14 cm:	Black (10 YR 2/1) mucky silt loam; very dark gray (10 YR 3/1) dry; moderate fine subangular blocky structure; slightly hard, friable, moderately smeary, slightly sticky, slightly plastic; many very fine and fine roots throughout; many very fine and fine interstitial pores; clear wavy boundary.
2A2	14–30 cm:	Very dark brown (10 YR 2/2) mucky silt loam; very dark grayish brown (10 YR 3/2) dry; weak fine and medium subangular blocky structure; slightly hard, friable, moderately smeary, slightly sticky, slightly plastic; many fine roots throughout and common very fine roots throughout; many very fine and fine interstitial pores; clear wavy boundary.
3A3	30–57 cm:	Black (10 YR 2/1) mucky silt loam; very dark brown (10 YR 2/2) dry; moderate fine and medium subangular blocky structure; slightly hard, very friable, moderately smeary, slightly sticky, slightly plastic; common fine roots throughout and few very fine roots throughout; many very fine and fine interstitial and few very fine discontinuous tubular pores; clear wavy boundary.
4Bwb1	57–80 cm:	50 percent brown to dark brown (7.5 YR 4/4), 30 percent dark brown (10 YR 3/3), and 20 percent black (10 YR 2/1) clay loam; 80 percent yellowish brown (10 YR 5/4) and 20 percent very dark grayish brown (10 YR 3/2) dry; moderate medium and coarse subangular blocky structure; hard, firm, sticky, plastic; common fine roots throughout and few very fine roots throughout; common very fine discontinuous tubular and few fine discontinuous tubular pores; clear irregular boundary.

4Bwb2	80–126 cm:	Strong brown (7.5 YR 5/6) silty clay; reddish yellow (7.5 YR 7/6) dry; weak medium and coarse subangular blocky structure; very hard, firm, sticky, plastic; few very fine and fine roots throughout; common very fine discontinuous tubular and few fine discontinuous tubular pores; gradual wavy boundary.
4Bwb3	126–160 cm:	Strong brown (7.5 YR 5/6) clay loam; reddish yellow (7.5 YR 7/6) dry; weak coarse subangular blocky structure; hard, firm, sticky, plastic; few very fine roots throughout; many very fine continuous tubular and common fine continuous tubular pores; gradual wavy boundary.
4C	160–200 cm:	Strong brown (7.5 YR 5/6) clay loam; 70 percent pink (7.5 YR 7/4) and 30 percent reddish yellow (7.5 YR 7/6) dry; massive; slightly hard, friable, sticky, plastic; few very fine roots throughout; many very fine continuous tubular and common fine continuous tubular pores.

The characterization data of the Tohoku University Farm soil are presented in Table 2.4. Comparison of the clay contents determined by two methods of mechanical analysis indicates that clay particles are insufficiently dispersed with sodium hexametaphosphate dispersing agent even though the clay fraction is dominated by chloritized $2:1$ minerals. Large amounts of active Al and Fe components extracted by acid oxalate reagent inhibit complete dispersion of clay particles. The pedon exhibits low bulk densities of less than 0.9 g cm^{-3} and water retention values at 1500 kPa which are regarded as medial in particle-size classes.

The analytical data of soil organic matter indicate that the soil of Tohoku University Farm is rich in humus characteristic of melanic Andisols. The pedon shows organic carbon concentrations greater than 6.0 percent, a predominance of A-type humic acid and a melanic index less than 1.70 in the humus horizons (0–57 cm). Thus, it meets not only the melanic epipedon requirements, but also the pachic subgroup requirement for Andisols (Soil Survey Staff, 1990, 1992).

The CEC values range from 52 to 14 cmol$_c$ kg^{-1}. Comparison of the clay and organic carbon contents shows that soil organic matter greatly contributes to the high CEC values. Because the soil is intensely leached, the base saturation is extremely low (<5%). There are notable differences in the pH(H$_2$O) and KCl-extractable Al values between the humus horizons (A1 to 3A3) and nonhumus horizons (4Bwb1 and 4Bwb2). The former shows extreme to very strong acidity and high KCl-extractable Al values which are severely toxic to certain plant roots. On the other hand, the nonhumus horizons are strongly acid and have KCl-extractable Al less than 2.0 cmol$_c$ kg^{-1}. These differences reflect nonallophanic mineralogy dominated by chloritized $2:1$ minerals in the humus horizons and mixed mineralogy of chloritized $2:1$ minerals and allophane in the nonhumus horizons.

All the soil horizons contain acid oxalate extractable Al and a half Fe values greater than 2.0 percent and mostly show P retention larger than 90 percent. The high ratios of sodium pyrophosphate-extractable Al to acid oxalate extractable Al in the humus horizons indicate that the active Al in these horizons is largely complexed with soil organic matter.

TABLE 2.4

Characterization data of the Tohoku University Farm soil

Depth (cm)	Horizon	Total (%) Clay <2 μm		Silt 2-50 μm	Sand 50 μm-2 mm	Bulk density (g cm^{-3})	Water retention at 1500 kPa (%) Air-dry	Field moist	Organic carbon (%)
		I[b]	II[b]	I	I				
0-14	A1	17.6	40	48.7	33.7	0.48	30.5	66.1	17.8
14-30	2A2	5.3	39	49.7	45.0	0.54	22.8	60.1	13.5
30-57	3A3	10.4	38	60.9	28.7	0.54	23.8	57.1	14.3
57-80	4Bwb1	7.9	41	27.9	64.2		17.8	44.2	2.8
80-126	4Bwb2	15.7	32	29.3	55.0	0.82	19.4	41.8	0.9

Depth (cm)	Horizon	Melanic index[a]	Ratio of humic acid to fulvic acid[a]	Type of humic acid	KCl-extr. Al (cmol$_c$ kg^{-1})	CEC (cmol$_c$ kg^{-1})	Base sat. (%)
0-14	A1	1.65	1.54	A	11.0	52.2	4
14-30	2A2	1.65	1.48	A	8.0	49.1	1
30-57	3A3	1.61	2.11	A	8.0	56.1	1
57-80	4Bwb1	1.59			1.9	18.4	1
80-126	4Bwb2				0.7	13.7	1

Depth (cm)	Horizon	Acid oxalate extr. (%) Al$_o$	Fe$_o$	Si$_o$	Al$_o$ + Fe$_o$/2	Pyrophos.-extr. Al$_p$ (%)	Al$_p$/Al$_o$ ratio	Allophane (%)	P ret. (%)
0-14	A1	1.6	1.2	0.1	2.2	1.5	0.94	0.5	83
14-30	2A2	2.8	1.7	0.2	3.7	2.5	0.88	1.5	93
30-57	3A3	3.1	1.4	0.2	3.8	2.7	0.87	1.7	93
57-80	4Bwb1	2.6	1.4	0.6	3.3	0.9	0.34	4.9	93
80-126	4Bwb2	2.5	1.4	0.7	3.2	0.6	0.23	5.9	94

Source: Otowa and Shoji, 1987.

[a] Provided by Soil Science Laboratory, Tohoku University.

[b] I: Dispersed using Na-hexametaphosphate.

[b] II: Dispersed by sonication and pH adjustment using dilute HCl or NaOH.

Based on the morphological and characterization data described above, the Tohoku University Farm soil is classified as a medial, mesic Alic Pachic Melanudand according to Keys to Soil Taxonomy (Soil Survey Staff, 1990, 1992).

2.3.5. Fulvic Andisols

Fulvic great groups were first proposed by Otowa (1985) to classify Andisols having a thick humus horizon similar to the melanic epipedon, but having lighter colors than for melanic epipedons. As already described, fulvic Andisols have a biosequential relationship with melanic Andisols in northeastern Japan; Melanudands form under Japanese pampas grass (*Miscanthus sinensis*) and Fulvudands under beech forest (*Fagus crenata*) (Shoji et al., 1988b). Characteristics of humus accumulated under the beech forest substantially determine the color of the surface horizon. It is dominated by fulvic acid and the humic acid is P-type with a relatively low degree of humification. Intermittent ash deposition also contributes to the formation of the thick epipedon of Fulvudands as observed for Melanudands. Fulvic great groups are provided in the Aquands, Cryands and Udands. Melanic great groups are proposed in the Xerand suborder but fulvic great groups are not.

The Tsutanuma soil from northeastern Japan (see Plates 5 and 6) is described below as an example of fulvic great groups. It has some notable morphological features. It shows composite morphology formed from cumulative tephra deposits of different ages and has a thick cumulative humus horizon (0–48 cm). However, it fails the melanic color requirement of moist Munsell value and chroma of 2 or less throughout (Soil Survey Staff, 1990, 1992). It also exhibits horizon boundaries significantly different from those of Melanudands (Shoji et al., 1988a). It has irregular boundaries between the 2Ab2 and 2Bwb horizons and a broken boundary between the 2Bwb and 3C horizons. These boundaries strongly reflect the process of tree-throw which results in mixing of the profile. The Tsutanuma soil shows soil structure, consistence, pore distribution etc. which are common to allophanic Melanudands.

DESCRIPTION OF TSUTANUMA SOIL (PLATES 5 AND 6)

Source	Shoji et al., 1987, 1988a
Location	Tsutanuma, Towada, Aomori
	Latitude = 40°35′20″N, Longitude = 140°57′50″E
Classification	Medial over cindery, mesic, Acrudoxic Fulvudand
Physiography	Mountainside
Geomorphic position	Backslope
Slope characteristics	12% undulating, east facing
Elevation	430 m M.S.L
Air temperature (°C)	Ann: 8, Sum: 19, Win: −3
Soil temperature (°C)	Ann: 9, Sum: 20, Win: −2, Mesic temperature regime
Precipitation	1860 mm, udic moisture regime
Water table	Not observed

Drainage   Well drained
Permeability   Moderate
Land use   Forest land not grazed
Vegetation   *Fagus crenata, Sasa kulirensis*
Erosion or deposition   Slight
Parent material   Several strata of tephras (Towada-a ash of 1000 yr B.P. in the depth of 0–20
                                cm; Chuseri ash of 5000 yr B.P. in the depth of 20–88 cm; Nanbu ash of 8600
                                yr B.P. in the depth below 88 cm)
Diagnostic horizons   48–70 cm Cambic
Described by   S. Shoji et al.

Oi   4–0 cm: Partially decomposed organic matter; decaying mat of beech leaves.

A1   0–11 cm: Very dark brown (7.5 YR 2.5/2) silt loam; moderate fine granular structure;
                            very friable, slightly sticky, slightly plastic; many very fine and fine roots
                            throughout and few medium roots throughout; common very fine interstitial
                            pores; gradual wavy boundary.

A2   11–20 cm: Very dark brown (7.5 YR 2.5/3) silt loam; moderate medium subangular
                            blocky structure; very friable, slightly sticky, slightly plastic; common very
                            fine and fine roots throughout and few medium and coarse roots throughout;
                            common very fine interstitial pores; gradual wavy boundary.

2Ab1   20–33 cm: Dark brown (7.5 YR 3/3) loam; weak medium subangular blocky structure;
                            very friable, slightly sticky, slightly plastic; common fine and medium roots
                            throughout and few coarse roots throughout; many very fine and fine inter-
                            stitial pores; clear wavy boundary.

2Ab2   33–48 cm: Dark brown (7.5 YR 3/4) loam; weak medium subangular blocky structure;
                            very friable, slightly sticky, slightly plastic; common fine and medium roots
                            throughout; common very fine and fine interstitial pores; clear irregular
                            boundary.

2Bwb   48–70 cm: Brown (7.5 YR 4/4) sandy loam; weak medium subangular blocky structure;
                            friable, slightly sticky, slightly plastic; common medium roots throughout
                            and few coarse roots throughout; many very fine and fine interstitial pores;
                            10 percent pumiceous gravels; abrupt broken boundary.

3C   70–88 cm: Brownish yellow (10 YR 6/6) gravelly coarse sand (pumice); single grain;
                            loose, nonsticky, nonplastic; few very fine and fine roots; many very fine
                            and fine interstitial and many medium interstitial pores; 25 percent gravels
                            (pumice); broken boundary.

4Ab   88–100+ cm: Dark yellowish brown (10 YR 3/4) silty clay loam.

Table 2.5 shows the characterization data for the Tsutanuma soil. Clay particles
were effectively dispersed by sonication and pH adjustment as indicated by the
clay content of all the soil horizons, except 3C, being greater than the allophane
content calculated using the acid oxalate extractable silica. Bulk densities were less
than 0.6 g cm^{-3}, meeting one of the andic soil properties. Water retention values
at 1500 kPa are mostly in the range of 29–12 percent for the air-dry samples and
in the range of 72–37 percent for the field moist samples, indicating that the soil
is medial in the particle-size/mineralogy family class.

TABLE 2.5

Characterization data of the Tsutanuma soil

Depth (cm)	Horizon	Total[a] (%) Clay <2 μm	Silt 2–50 μm	Sand 50 μm–2 mm	Bulk density (g cm⁻³)	Water retention[b] at 1500 kPa (%) Air-dry	Field moist	Organic carbon (%)
0–11	A1	17.0	31.5	51.5	0.51	28.8	71.6	12.4
11–20	A2	14.3	30.4	55.3	0.51	21.4	49.3	8.3
20–33	2Ab1	21.6	22.4	56.0	0.55	16.9	46.7	4.9
33–48	2Ab2	19.6	20.8	59.6	0.55	16.9	46.7	4.5
48–70	2Bwb	12.5	12.2	75.3		11.8	36.6	2.6
70–88	3C	4.9	2.5	92.6		4.7	13.5	0.5

Depth (cm)	Horizon	Melanic index	Ratio of humic acid to fulvic acid	Type of humic acid	KCl-extr. Al (cmol_c kg⁻¹)	CEC (cmol_c kg⁻¹)	Base sat. (%)
0–11	A1	2.16	0.71	P	5.2	27.0	4
11–20	A2	2.16	0.61	P	2.4	19.8	3
20–33	2Ab1	2.20	0.33	P	0.4	13.0	10
33–48	2Ab2	2.22	0.24	P	tr	11.9	10
48–70	2Bwb		0.23	P	tr	7.7	10
70–88	3C		0.13	P	tr	2.6	8

Depth (cm)	Horizon	pH H₂O	Acid oxalate extr. (%) Al_o	Fe_o	Si_o	Al_o + Fe_o/2	Allophane (%)	P ret. (%)
0–11	A1	4.6	1.2	0.6	0.2	1.5	1.6	85
11–20	A2	4.7	1.9	0.6	0.4	2.2	3.5	92
20–33	2Ab1	5.6	2.9	1.0	1.1	3.4	8.9	96
33–48	2Ab2	5.7	3.3	1.1	1.4	3.8	11.0	97
48–70	2Bwb	5.9	2.5	0.7	1.2	2.9	9.4	93
70–88	3C	6.0	1.2	0.4	0.6	1.6	5.0	61

Source: Shoji et al., 1987; 1988b.
[a] Dispersed by sonication and pH adjustment.
[b] Cited from Otowa and Shoji, 1987.

Soil organic carbon is markedly accumulated in the upper part of the soil profile with more than 6 percent as a weighted average in the depth of 0–30 cm. It decreases significantly with depth, contrasting with the gradual decrease with depth characteristic of organic carbon distribution in Melanudands (Shoji et al., 1988b). This difference in the distribution of organic carbon between the two soils is mainly due to differences in how organic matter is added to the soil by the vegetations; addition of organic matter to the soil surface under the forest ecosystem and to both the soil surface and the solum under the grass ecosystem.

The dilute NaOH solution extracts a large proportion of soil organic matter. The extracts show melanic indices greater than 1.70 and are dominated by fulvic acid as shown by the low ratios of fulvic acid to humic acid. The humic acid belongs to the P-type, indicating a relatively low degree of humification. These facts indicate that the soil organic matter accumulated under the beech forest has the characteristic features of fulvic Andisols.

The CEC values differ widely (3–27 $cmol_c$ kg^{-1}), reflecting the accumulation of soil organic matter and clay formation. The amounts of exchangeable bases are very low leading to base saturation values less than 10 percent in all horizons. The upper humus horizons (A1 and A2) show $pH(H_2O)$ of 4.6–4.7 and contain a considerable amount of KCl-extractable Al, reflecting the abundance of chloritized 2:1 minerals in the clay fraction. In contrast, all the horizons underlying the A1 and A2 horizons show $pH(H_2O) > 5.5$ and a virtual absence of KCl-extractable Al. These properties are attributed to the predominance of allophane in the clay fraction. All the horizons except the A1 and 3C show acid oxalate extractable Al and a half Fe > 2.0 percent and P retention > 85 percent, satisfying those criteria for andic soil properties.

According to the profile description and characterization data shown above, the Tsutanuma soil is classified as medial over cindery, mesic Acrudoxic Fulvudand (Soil Survey Staff, 1992).

2.3.6. Hydric Andisols

Hydric Andisols, considered as Hydrandepts in the 1975 Soil Taxonomy mostly occur in regions of very high, well-distributed rainfall with a perudic moisture regime (Soil Survey Staff, 1975). They show 1500 kPa water retention more than 100 percent and 33 kPa water retention of several hundred percent. They have unique consistence such as strong thixotropy and become liquid with only weak pressure. The soil material contains large amounts of a gelatinous weathering product that dehydrates irreversibly into aggregates of various sizes. These distinct properties are attributable to very large amounts of noncrystalline weathering products and organic matter.

Hydric great groups are identified in the Cryand and Udand suborders. They must have 1500 kPa water retention for undried samples of 100 percent or more as a weighted average throughout a thickness of 35 cm or more within 100 cm of the mineral soil surface or from the upper boundary of an organic layer with andic soil properties, whichever is shallower (Soil Survey Staff, 1990, 1992).

The Hilo soil from Hawaii, U.S.A. is an example of a hydric Andisol (Hydrudand) with a perudic moisture regime and is described below. It was classified as thixotropic, isohyperthermic, Typic Hydrandept according to the 1975 Soil Taxonomy. It has very thick B horizons with dark reddish to reddish brown color indicating large quantities of iron oxides. These horizons have subangular blocky structure with various sizes and grades.

The Hilo soil is characterized by unique consistence and the presence of large amounts of noncrystalline weathering products. It shows moderate to strong thixotropy that is indicated by the smeariness observed in the field. The ped surfaces of the Bw horizons have translucent gelatinous coatings, indicating intense formation of noncrystalline materials under the isohyperthermic, perudic conditions.

DESCRIPTION OF HILO SOIL (PLATE 9)

Source	USDA Soil Conservation Service, Pedon Narrative Description (July 13, 1990)
Location	Island of Hawaii, Hawaii, U.S.A. Above Hilo on Mauna Kea Sugar Latitude = 19°43'51"N, Longitude = 155°06'33"W
Classification	Hydrous, isohyperthermic, Acrudoxic Hydrudand
Physiography	Upper backslope (planar)
Elevation	121 m M.S.L.
Soil temperature (°C)	Ann: 22, Isohyperthermic temperature regime
Precipitation	4010 mm, Perudic moisture regime
Drainage	Well drained
Land use	Sugarcane field
Vegetation	Freshly planted sugarcane
Erosion or deposition	Slight
Parent material	Volcanic ash material (basaltic)
Described by	S. Nakamura, C. Smith, and N. Simmons
Date	June, 1989
Hydraulic Conductivity	Moderate

Ap1	0–15 cm:	Dark brown (7.5 YR 3/4) silty clay loam; moderate very fine granular structure; friable, slightly sticky, slightly plastic, moderately smeary; common very fine and fine roots; many very fine interstitial pores; few 4–10 cm clods; gritty due to irreversible drying; gradual wavy boundary.
Ap2	15–38 cm:	Dark brown (7.5 YR 3/4), and dark reddish brown (5 YR 3/4) silty clay loam; moderate very fine and fine subangular blocky structure; friable, slightly sticky, slightly plastic, moderately smeary; common very fine and fine roots; many very fine pores; few 4 to 10 cm clods; gritty due to irreversible drying; clear wavy boundary.
Bw1	38–63 cm:	Dark brown (7.5 YR 3/4), dark brown (7.5 YR 3/4), and dark reddish brown (5 YR 3/4) silty clay loam; strong medium prismatic structure parting to moderate fine and medium subangular blocky; friable, slightly sticky, slightly plastic, moderately smeary; common very fine and fine roots; many very fine pores; gradual wavy boundary.

Bw2   63–93 cm: Dark brown (7.5 YR 3/4), and dark reddish brown (5 YR 3/4) silty clay loam; moderate medium prismatic structure parting to moderate fine and medium subangular blocky; friable, slightly sticky, slightly plastic, moderately smeary; few very fine and fine roots; many very fine pores; gradual wavy boundary.

Bw3   93–124 cm: Dark brown (7.5 YR 3/4), dark reddish brown (5 YR 3/4), and dark reddish brown (2.5 YR 3/4) silty clay loam; moderate medium prismatic structure parting to moderate fine and medium subangular blocky; friable, slightly sticky, slightly plastic, smeary; few very fine and fine roots; many very fine pores; 2.5 YR colors appear as weathered bands of ash about 2 cm wide; gradual wavy boundary.

Bw4   124–152 cm: Dark brown (7.5 YR 3/4), dark reddish brown (5 YR 3/4), and dark reddish brown (2.5 YR 3/4) silty clay loam; moderate medium prismatic structure parting to moderate fine and medium subangular blocky; firm, moderately sticky, slightly plastic, smeary; few very fine and fine roots; many very fine pores; 2.5 YR colors appear as weathered bands of ash about 2 cm wide; gradual wavy boundary.

Bw5   152–200 cm: Dark brown (7.5 YR 3/4), dark reddish brown (5 YR 3/4), and dark reddish brown (2.5 YR 3/4) sandy clay loam; strong medium prismatic structure parting to moderate fine and medium subangular blocky; firm, moderately sticky, slightly plastic, smeary; few very fine and fine roots; many very fine, and few fine pores; 2.5 YR colors appear as weathered bands of ash about 2 cm wide.

The characterization data of the Hilo soil are presented in Table 2.6. The pedon shows very low bulk densities (<0.5 g cm^{-3}), 1500 kPa water retention greater than 100 percent, and 33 kPa water retention of several hundred percent. Thus, it is obvious that very high water retention significantly contributes to the low bulk density due to the high porosity of the soil material.

Organic carbon does not exceed 6 percent as for a melanic epipedon, but it is present in relatively large amounts ($>1.7\%$) to a depth of over 2 m. The sum of exchangeable bases is small, reflecting the advanced stage of weathering of the soil material and intense base leaching. In spite of the low base saturation, all the horizons show only slight acidity. The abundance of noncrystalline materials resulted in high P retention values ($>95\%$). Finally, the Hilo soil is classified as hydrous, isohyperthermic, Acrudoxic Hydrudand (Soil Survey Staff, 1992).

TABLE 2.6

Characterization data of the Hilo soil

Depth (cm)	Horizon	Total (%)			Bulk density (g cm^{-3})	Water retention at 1500 kPa (%)		Organic carbon (%)	CEC (cmol$_c$ kg^{-1})	Base sat. (%)	pH H$_2$O	P ret. (%)
		Clay <2 μm	Silt 2–50 μm	Sand 50 μm–2 mm		Air-dry	Field moist					
0–16	Ap1	3.7	31.4	64.9	–	53.4	56.6	5.67	29.3	28	6.2	100
16–39	Ap2	4.1	24.6	71.3	0.54	75.7	79.8	4.81	29.4	11	5.6	95
39–63	Bw1	–	9.3	90.7	0.29	129.6	150.3	2.96	29.7	3	5.6	98
63–94	Bw2	22.7	43.9	33.4	0.30	126.0	158.1	2.33	26.3	5	6.0	–
94–125	Bw3	–	10.0	90.0	0.29	128.7	153.3	1.99	23.7	4	5.8	–
125–152	Bw4	0.4	15.6	84.0	0.28	136.4	177.9	1.86	24.5	4	5.7	–
152–200	Bw5	–	26.4	73.6	–	120.2	142.9	1.70	21.0	4	–	–

Source: USDA Soil Conservation Service, National Soil Survey Laboratory, Lincoln, Nebraska, 1990.

REFERENCES

Beinroth, F.H., Luzio L.W., Maldonado P.F. and Eswaran, H. (Editors), 1985a. Proc. 6th International Soil Classification Workshop, Chile and Ecuador, 2. Tour guide for Chile. Sociedad Chilena de la Ciencia del Suelo, Santiago, Chile.

Beinroth, F.H., Luzio L.W., Maldonado P.F. and Eswaran, H. (Editors), 1985b. Proc. 6th International Soil Classification Workshop, Chile and Ecuador, 3. Tour guide for Ecuador. Sociedad Chilena de la Ciencia del Suelo, Santiago, Chile.

Chadwick, O.A., Hendricks, D.M. and Nettleton, W.D., 1987. Silica in duric soils, 1. A depositional model. Soil Sci. Soc. Am. J., 51: 975–982.

Clayden, B., Daly, B.K., Lee, R. and Mew, G., 1990. The nature, occurrence, and genesis of placic horizons. In: J.M. Kimble and R.D. Yeck (Editors), Proc. 5th Int. Correlation Meeting (ISCOM). USDA-SCS, Lincoln, NE, pp. 88–104.

Hokkaido Soil Classification Committee, 1979. Soil classification of agricultural land in Hokkaido, 2nd approximation. Misc. Publ. Hokkaido Nat. Agr. Exp. Sta., 17: 1–89 (in Japanese, with English abstract).

Ito, T., Shoji, S. and Saigusa, M., 1991a. Classification of volcanic ash soils from Konsen district, Hokkaido according to the last Keys to Soil Taxonomy (1990). Jap. J. Soil Sci. Plant Nutr., 62: 237–247 (in Japanese, with English abstract).

Ito, T., Shoji, S., Shirato, Y. and Ono, E., 1991b. Differentiation of a spodic horizon from a buried A horizon. Soil Sci. Soc. Am. J., 55: 438–442.

Kawamuro, K. and Osumi, Y., 1988. Melanic and fulvic Andisols in central Japan, soil pollen and plant opal analyses. In: D.I. Kinloch, S. Shoji, F.H. Beinroth and H. Eswaran (Editors), Proc. 9th Int. Soil Classification Workshop, Japan, 20 July to 1 August, 1987. Publ. by Jap. Committee for 9th Int. Soil Classification Workshop, for the Soil Management Support Services, Washington D.C., U.S.A., pp. 389–401.

Kimble, J.M. (Editor), 1986. First Int. Soil Correlation Meeting (ISCOM), Tour guide for Idaho, Washington, and Oregon, U.S.A. SMSS, NSSL, and USDA-SCS, U.S.A.

Leamy, M.L., Smith, G.D., Colmet-Daage, F. and Otowa, M., 1980. The morphological characteristics of Andisols. In: B.K.G. Theng (Editor), Soils with Variable Charge. DSIR, Lower Hutt, pp. 17–34.

Leamy, M.L., Clayden, B., Parfitt, R.L., Kinloch, D.I. and Child, C.W., 1988. Final proposal of the International Committee on the Classification of Andisols (ICOMAND). New Zealand Soil Bureau, DSIR, Lower Hutt.

Otowa, M., 1985. Criteria of great groups of Andisols. In: F.H. Beinroth, W. Luzio L., F. Maldonado P. and H. Eswaran (Editors), Proc. 6th Int. Soil Classification Workshop, Chile and Ecuador, I. Papers. Sociedad Chilena de la Ciencia del Suelo, Santiago, Chile, pp. 191–208.

Otowa, M., 1986. Morphology and classification. In: K. Wada (Editor), Ando Soils in Japan. Kyushu University Press, Fukuoka, Japan, pp. 3–20.

Otowa, M. and Shoji, S. (Editors), 1987. Ninth Int. Soil Classification Workshop, Tour guide for Kanto, Tohoku, and Hokkaido, Japan. Sendai, Japan.

Otowa, M., Shoji, S. and Saigusa, M., 1988. Allic, melanic, and fulvic attributes of Andisols. In: D.I. Kinloch, S. Shoji, F.H. Beinroth and H. Eswaran (Editors), Proc. 9th Int. Soil Classification Workshop, Japan, 20 July to 1 August, 1987. Publ. by Jap. Committee for 9th Int. Soil Classification Workshop, for the Soil Management Support Services, Washington D.C., U.S.A., pp. 192–202.

Ping, C.L., Shoji, S., Ito, T., Takahashi, T. and Moore, J.P., 1989. Characteristics and classification of volcanic ash-derived-soils in Alaska. Soil Sci., 148: 8–28.

Pollok, J.A., Parfitt, R.L. and Furkurt, R.J. (Editors), 1981. Soils with Variable Charge Conference, Guide book for Tour 4. Post-conference North Island, New Zealand. Wellington, New Zealand.

Rourke, R., 1991. International Committee on the Classification of Spodosols. Circular letter No. 10, Maine.

Shoji, S., 1988. Separation of melanic and fulvic Andisols. Soil Sci. Plant Nutr., 34: 303–306.

Shoji, S. and Ito, T., 1990. Classification of tephra-derived Spodosols. Soil Sci., 150: 799–815.

Shoji, S. and Otowa, M., 1988. Distribution and significance of Andisols in Japan. In: D.I. Kinloch, S. Shoji, F.H. Beinroth and H. Eswaran (Editors), Proc. 9th Int. Soil Classification Workshop, Japan, 20 July to 1 August, 1987. Publ. by Jap. Committee for 9th Int. Soil Classification Workshop, for the Soil Management Support Services, Washington D.C., U.S.A., pp. 13–24.

Shoji, S. and Yamada, H., 1991. Comparisons of mineralogical properties between tephra-derived Spodosols from Alaska and nontephra-derived Spodosols from New England. Soil Sci., 152: 162–183.

Shoji, S., Ito, T., Saigusa, M. and Yamada, I., 1985. Properties of nonallophanic Andosols from Japan. Soil Sci., 140: 264–277.

Shoji, S., Takahashi, T., Ito, T. and Ping, C.L., 1988a. Properties and classification of selected volcanic ash soils from Kenai Peninsula, Alaska. Soil Sci., 145: 395–413.

Shoji, S., Takahashi, T., Saigusa, M., Yamada, I. and Ugolini, F.C., 1988b. Properties of Spodosols and Andisols showing climosequential and biosequential relation in southern Hakkoda, northeastern Japan. Soil Sci., 145: 135–150.

Shoji, S., Takahashi, T., Saigusa, M. and Yamada, I., 1987. Morphological properties and classification of ash-derived soils in South Hakkoda, Aomori Prefecture, Japan. Jpn. J. Soil Sci. Plant Nutr., 58: 638–646. (in Japanese)

Shoji, S., Nanzyo, M., Shirato, Y. and Ito, T., 1993. Chemical kinetics of weathering in young Andisols from northeastern Japan using soil age normalized to 10°C. Soil Sci., 155: 53–60.

Smith, G.D., 1978. A preliminary proposal for the reclassification of Andepts and some andic subgroups (The Andisol proposal, 1978). New Zealand Soil Bureau, DSIR, Lower Hutt, New Zealand.

Soil Survey Staff, 1975. Soil Taxonomy. A basic system of soil classification for making and interpreting soil survey. Agr. Hdbk. No.436. USDA, Washington, D.C., USGPO.

Soil Survey Staff, 1990. Keys to Soil Taxonomy, 4th edition. AID, USDA-SMSS Technical Monograph No. 19, Blacksburg, Virginia.

Soil Survey Staff, 1992. Keys to Soil Taxonomy, 5th edition. AID, USDA-SMSS Technical Monograph No. 19, Blacksburg, Virginia.

Wright, C.A., 1964. The "Andosols" or "Humic Allophane" Soils of South America. In: Meeting of the Classification and Correlation of Soils from Volcanic Ash, Japan, 1964. FAO-Unesco, pp. 9–22.

Chapter 3

GENESIS OF VOLCANIC ASH SOILS

S. SHOJI, R. DAHLGREN, and M. NANZYO

3.1. INTRODUCTION

Volcanic ash or tephra is commonly unconsolidated, comminuted materials containing a large quantity of volcanic glass which shows the least resistance to chemical weathering. Therefore, tephras weather rapidly resulting in formation of large amounts of noncrystalline materials. This process occurring preferentially in tephras was first called "andosolization" by Duchaufour (1984).

Development of A horizons by andosolization is characterized by accumulation of organic matter, organic matter stabilization by active Al and Fe, carbonic acid weathering (allophanic) versus organic acid weathering (nonallophanic), and formation of laminar opaline silica. Formation of B horizons proceeds primarily by carbonic acid weathering with no significant translocation of Al, Fe, and dissolved organic carbon. Therefore, preferential formation of noncrystalline materials such as allophane, imogolite, laminar opaline silica, ferrihydrite, and Al/Fe humus complexes is a characteristic feature of the process of andosolization.

Such formation of noncrystalline materials is not specific to Andisols. It is also widely observed for tephra-derived Spodosols making it difficult to separate Andisols and tephra-derived Spodosols according to solid-phase chemical criteria (Shoji and Ito, 1990). However, the major difference between andosolization and podzolization occurring in tephras was successfully deciphered by soil solution studies of Ugolini et al. (1988) and Ugolini and Dahlgren (1991). These workers showed that andosolization is characterized by an accumulation of Fe, Al, and dissolved organic carbon in the A horizon with little subsequent leaching of these components to the B horizon, and that formation of the B horizon is dominated by *in situ* weathering. On the other hand, podzolization is a process driven in the upper profile (O, E and Bhs) by organic acids originating in the canopy and humus layer. These organic acids play significant roles such as lowering of pH, preventing dissociation of carbonic acid, formation of mobile complexes with Fe, Al and other metals and migration as soluble metal–humus complexes to the B horizon where they are arrested (Ugolini et al., 1977; Ugolini and Dahlgren, 1991).

Andisols can form in tephras by andosolization in a relatively short time under most climates throughout the world. However, not all Andisols are derived from tephra and not all tephra-derived soils are Andisols. There are some nontephra-derived Andisols (Garcia-Rodeja et al., 1987; Hunter et al., 1987) and transition of Andisols to most soil orders can occur, reflecting especially time and climate factors affecting soil formation.

Many reviews describing the formation of volcanic ash soils have been published (e.g. Kanno, 1961; FAO/Unesco, 1964; Takehara, 1964; Gibbs et al., 1968; Mohr et al., 1972; Flach et al., 1980; Kato, 1983; Shoji, 1983; Tan, 1984; Wada, 1985; Lowe, 1986; Kato et al., 1988; Matsui, 1988; Oba and Nagatsuka, 1988; Van Wambeke, 1992).

3.2. FORMATION OF ANDISOLS

3.2.1. State factors of Andisols

A soil is a dynamic system showing a variety of energy and mass fluxes and transformations, and various conceptual models have been proposed to describe these processes. Although there are no totally acceptable models as reviewed by Smeck et al. (1983), the state factor theory by Jenny (1941) has commanded general popularity. This theory relates soil properties to state factors such as climate, organisms, topography, parent material, and time that control genesis and behavior of ecosystems and their soils. It has been further extended to ecosystems containing humans and their soils by Amundson and Jenny (1991).

Since Jenny's conceptual model provides a useful framework to relate properties of Andisols to the state factors, the following section discusses the soil forming process of andosolization in terms of the state factor model.

3.2.2. Parent material

Genesis of Andisols is so strongly influenced by the properties of the parent material that the general term volcanic ash soils, denoting the kind of parent material, is often used instead of Andisols, Ando soils, Kurobokudo, etc. Of the various properties of tephras as a parent material, the chemical and mineralogical properties, texture, and depositional features are especially important. A comprehensive review on this subject matter was published by Lowe (1986).

(1) Rates of chemical weathering and humus accumulation

Rates of chemical weathering in tephras are often determined by measuring the quantity of clay formed. This estimate is subject to errors from measurement of clay content because mechanical analysis underestimates clay content due to incomplete clay dispersion. The concentration of acid oxalate extractable Al is also employed to estimate weathering rates in allophanic Andisols instead of clay content as described below.

Aluminum released from tephras is incorporated mainly into allophane, imogolite and Al-humus complexes in young Andisols, and it can be preferentially extracted by the acid oxalate solution. Since Al is one of the major elements present in tephras, Al contained in such noncrystalline materials is used to estimate the rate of chemical weathering in Andisols when soil age is known (Shoji et al., 1993). In addition, the humus content of Andisols shows a close linear relationship with Al which is complexed with humus. Thus, the rate of humus accumulation can also be estimated by Al release rates from tephras or the relationship between soil age and acid oxalate extractable Al.

Shoji et al. (1993) applied a zero-order reaction to describe chemical weathering in the surface horizons of young Andisols having mesic to frigid temperature and udic moisture regimes in northeastern Japan. These soils formed in noncolored glass-rich ashes on well-drained sites in grasslands and forests lacking podzolizers.

Since these soils developed under various temperature regimes, their ages were transformed to "10°C-normalized soil ages" that is defined as the number of years a soil has been subjected to pedogenesis under a mean monthly soil temperature of 10°C. Acid oxalate extractable Al was selected as the most reliable element to evaluate chemical weathering of the bulk soils. The calculation was made using a temperature dependency coefficient, Q_{10} (increase in the rate of a chemical reaction when the temperature increases by 10°C) of 1.5 for Al release which was obtained by laboratory dissolution experiments using the separated glass fraction.

The relationship between 10°C-normalized soil ages and acid oxalate extractable Al (Al_o) is expressed by a zero-order equation as follows:

$$Y = 0.000705 \, X + 0.57 \qquad r = 0.94^{***} \ (n = 18)$$

where X stands for 10°C-normalized soil ages (number of years) and Y, for acid oxalate extractable Al ($Al_o\%$).

The sum of acid oxalate extractable Al plus a half acid oxalate extractable Fe ($Al_o + 1/2\,Fe_o$) has been employed to develop the criteria of andic soil properties such as ($Al_o + 1/2\,Fe_o$) $\geq$ 2.0 percent (Soil Survey Staff, 1990, 1992). According to the equation for soil age and Al_o, ($Al_o + 1/2\,Fe_o$) of volcanic ash dominated by noncolored glass attains 2.0 percent within 1280 years. This calculation is based on the following relationship between Al_o and ($Al_o + 1/2\,Fe_o$) (Shoji et al., 1992):

$$(Al_o + 1/2\,Fe_o) = 1.36 \times Al_o$$

On the other hand, the age of humus horizons having ($Al_o + 1/2\,Fe_o$) of 0.4 percent, that separates Entisols and Andisols, is estimated to be a 10°C-normalized soil age of about 200 years. However, the weathering rate of tephras at the very early stage is so great that the real soil age is considerably less than 200 years.

The rate of chemical weathering is much greater in Andisols derived from colored glass-rich tephras (basaltic tephras) than from noncolored glass-rich tephras. The rate of release of Al and Si from colored glass is observed to be 1.5 and 2 times greater, respectively, than that from noncolored glass (Shoji et al., 1993). Thus, Andisols can rapidly form in basaltic tephras after their deposition.

. Humus also rapidly accumulates by stabilization through complexation with Al in Andisols from northeastern Japan having mesic to frigid temperature and udic moisture regimes as expressed by the following equation:

$$Y = 5.96\,X + 0.95 \qquad r = 0.84^{***}\ (n = 92)$$

where X stands for the content of pyrophosphate-extractable Al ($Al_p\%$) and Y, for soil organic carbon content (%) (Andisol TU Database, 1992). This equation was obtained using humus horizons below the uppermost horizons because considerable amounts of fresh organic carbon are contained in the surface horizon. This equation shows that soil organic carbon reaches 6 percent at a pyrophosphate-extractable Al concentration of 0.82 percent. The soil age required for the accumulation of this organic carbon content was determined to be a 10°C-normalized age of 360 years. This age was obtained from the soil age–acid oxalate extractable Al equation described above with the assumption that all the Al released from the parent material forms Al-humus complexes.

(2) Glass composition of tephras and clay formation

Glass is the most abundant component in volcanic ash and shows the least resistance to chemical weathering among all primary minerals (Shoji, 1986). Thus, glass weathering primarily determines the genesis of volcanic ash soils. Kirkman (1980) and Kirkman and McHardy (1980) showed that weathering of andesitic and colored glasses is very rapid compared to that of rhyolitic or noncolored glass and that the fine, soft structure and extensive substitution of Al for Si in andesitic glass favor the rapid weathering.

Shoji and Fujiwara (1984), and Shoji (1986) proposed to divide tephras into two groups according to their influence on clay formation in the surface horizons of Andisols from northeastern Japan. One tephra type is dominated by colored glass having refractive indices greater than 1.52 (tephras of basaltic andesite and basalt). This tephra group favors formation of allophanic soils because more Ca and Mg are released which maintain higher soil $pH(H_2O)$ values as compared with the other group. The other tephra group is rich in noncolored glass having refractive indices between 1.49 and 1.52 (tephras of rhyolite, dacite and andesite). This type of tephra often weathers to form nonallophanic soils (Shoji and Fujiwara, 1984; Saigusa and Shoji, 1986).

For example, Saigusa and Shoji (1986) observed distinct differences in the clay mineralogy of surface soils derived from Zao-a ash of basaltic andesite composition (<1000 years, colored glass-rich) and Towada-a ash of rhyolite composition (1000 years, noncolored glass-rich). The former has allophanic clay mineralogy in the *Fagus crenata* (1200–800 m in elevation) and *Castanea crenata–Quercus serrata* zones (800–100 m in elevation) and has nonallophanic clay mineralogy only in the subalpine zone (1800–1200 m in elevation). In contrast, the latter has nonallophanic mineralogy in all elevation zones except a part of the *Castanea crenata–Quercus serrata* zone where mean annual precipitation was lowest (1000–900 mm).

The difference in the clay mineralogy described above reflects the properties of the parent materials which, in part, determine soil acidity through the release of bases as indicated below:

$$Y = 0.366\,X + 3.44 \qquad (r = 0.877^{***},\ n = 10)\ \text{for Zao-a soil}$$

$$Y = 0.706\,X + 1.16 \qquad (r = 0.921^{***},\ n = 10)\ \text{for Towada-a soil}$$

where X stands for pH(H_2O) values of Oa horizons and Y, for pH(H_2O) values of the underlying A1 horizons. These equations show that the pH(H_2O) values of A1 horizons are significantly higher in the Zao-a soil than in the Towada-a soil when both soils have the same pH(H_2O) values in the Oa horizons. The pH(H_2O) range of these horizons is 4–6 (For example, the Zao-a soil shows pH(H_2O) of 5.3, and the Towada-a soil, pH(H_2O) of 4.7 when both soils have pH($H_2$0) of 5.0 in the Oa horizons). Thus, the formation of allophanic humus horizons is favored in the Zao-a soil while the Towada-a soil tends to form nonallophanic humus horizons whenever both soils show similar weathering conditions.

The significance of pH(H_2O) on clay formation in A1 horizons is clearly indicated by the relationships between pH(H_2O) values and the allophane content of both Zao-a and Towada-a soils as presented in Fig. 3.1. It appears that a pH(H_2O) of 4.9 is the critical value with regard to allophane formation. Allophane formation proceeds with increasing pH(H_2O) above 4.9 while it is inhibited by formation of Al-humus complexes (anti-allophanic reaction) below 4.9 as already described. Similar pH(H_2O) values separating soils with and without allophane were shown by Parfitt and Kimble (1989) using the data from the New Zealand Soil Bureau National Soils Database and the Andisol Database.

Nonallophanic humus horizons are also common in Andisols under cold and humid climates. Under such environmental conditions in the Aleutian Islands and Alaska Peninsula, U.S.A., allophanic humus horizons have been observed in some

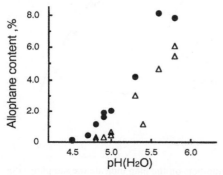

Fig. 3.1. Relationship between pH(H_2O) values and allophane content in the A1 horizons of Zao-a volcanic ash soils (triangle) and Towada-a volcanic ash soils (circle) (figure prepared by the authors using data of T. Takahashi and S. Shoji (unpublished, 1992) and Saigusa and Shoji, 1986).

Andisols containing large amounts of colored glass. This example depicts the influence of chemical and mineralogical properties of the parent material on clay formation (Ping et al., 1988).

(3) Depositional features of tephras and pedogenesis

Intermittent tephra deposition, dispersal patterns, and depth, size and mineralogy of tephras substantially determine the distribution, genesis, and properties of soils derived from tephra (Wada and Aomine, 1973; Shoji and Saigusa, 1977; Saigusa et al., 1978; Katayama et al., 1983; Lowe, 1986; Kato et al., 1988).

Dispersal patterns of tephra depend largely on the wind directions, violence of the volcanic eruption, and type of volcanic ejecta. Since the high-altitude winds in the middle and high latitudes are commonly strong westerly, materials blown into the air by violent volcanic eruptions fall predominantly east of the volcanoes forming very elongated ellipses. On the contrary, the high-altitude winds in the low latitudes are easterly, influencing the tephra distribution in these areas. When the eruptions are relatively quiet, the tephras tend to fall in the vicinity of the source volcano. However, the direction of dispersal of ashy materials is more or less irregular.

Dispersal patterns of tephras in relation to the types of explosions and kinds of materials are illustrated for the Towada tephras in Fig. 3.2. The Towada caldera

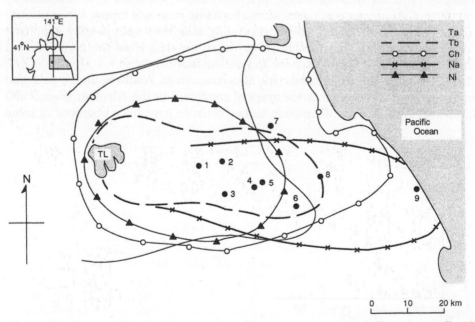

Fig. 3.2. Distribution of Towada Holocene tephras. The numbers on the map indicate the sampling sites. *TL* = Towada Lake Caldera, *Ta* = Towada a (1000 yr B.P.), *Tb* = Towada b (2000 yr B.P.), *Ch* = Chuseri (5000 yr B.P.), *Na* = Nanbu (8600 yr B.P.), and *Ni* = Ninokura tephra (10,000 yr B.P.) (adapted from Oike, 1972).

northeastern Japan had five strong eruptions during the Holocene period. The first deposit (Ninokura tephra, Ni; 10,000 yr B.P.) was blown out by a strombolian type of eruption and consists mainly of scoria having the composition of basaltic andesite. It is distributed in the vicinity of the source, although the principal axis of the deposit is east of the Towada caldera. The second deposit (Nanbu tephra, Na; 8600 yr B.P.) is andesite pumice from a plinian eruption and is distributed to distances far from the source, showing a very elongated ellipsis. The third deposit (Chuseri tephra, Ch; 5000 yr B.P.) consists of dacite pumice and ash formed by a plinian outburst. The earlier pumice deposit shows an elongated ellipsis while the subsequent ashy deposit is irregular and widely distributed. The fourth (Towada-b tephra, Tb; 2000 yr B.P.) and fifth (Towada-a tephra, Ta; 1000 yr B.P.) deposits consist largely of rhyolitic ash. The distribution of the former shows a relatively small ellipsis, while that of the latter is wide and irregular.

Tephra deposits show lateral and vertical variations in their particle size and mineralogy. The largest and heaviest particles tend to fall nearest the source volcanos, while progressively smaller and lighter particles fall at increasing distances. The size and depth of deposits commonly can be expressed by a negative exponential equation of distance from the vents. Furthermore, volcanic explosions are repeated intermittently, normally resulting in the cross section of tephra deposits being composite and highly variable in regard to the depth of each material.

Such characteristics of tephra deposits described above are closely related to the morphology of Andisols, and their classification as observed for Towada Andisols. Figure 3.3 shows nine profile sections, along a transect away from the caldera, indicating the variations in thickness and particle size of tephras, and morphology (Saigusa et al., 1978) and classification of these Andisols.

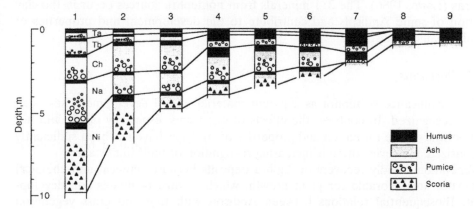

Fig. 3.3. Profile sections of Towada Andisols. The numbers on the sections indicate the sampling sites as described in Fig. 3.2. (figure prepared by the authors using data of Saigusa et al., 1978; copyright Elsevier, with permission).

Sections 1–3, occur nearest the caldera and show deep pentasequum profiles formed from five tephra deposits. Sections 4–6 in the middle area show trisequum profiles because Towada-a and Towada-b ashes are thin and are totally enriched with humus. Sections 7–9, being farthest from the caldera, show a total tephra thickness of 1 to 2 m. These sections show a monosequum or bisequum profile and have a pachic melanic epipedon consisting of Towada-a, Chuseri and Nanbu tephras or Towada-a, Towada-b and Chuseri tephras. These Andisols classify as Udivitrands in the area nearest to the caldera, Hapludands in the middle area, and Pachic Melanudands in the area farthest from the caldera.

Lahar deposits are also an important parent material in tropical regions where intensive rainfall is common. For example, lahars are occurring in Central Luzon, the Philippines, especially in areas surrounding Mt. Pinatubo where 1991 eruptions are believed to be one of the largest in this century. These lahars transport a tremendous amount of dacitic pyroclastic materials to the lowland areas. It is interesting to note that various soils derived from the old lahar deposits from previous eruptions are observed in the same areas (M. Nanzyo and S. Shoji, 1992, unpublished).

(4) Additions of minerals from nontephra sources

In addition to primary minerals formed by solidification of the magma, accessory or accidental minerals and minerals from eolian dusts contribute significantly to the mineralogical and chemical properties of Andisols. For example, some tephras in Japan contain clay minerals hydrothermally formed in the volcanic craters prior to the eruption (Ossaka, 1982) and eolian dusts transported from Asian interiors to Japan. These eolian dusts consist mainly of mica, kaolinite, quartz, and feldspar (Inoue and Naruse, 1990). In New Zealand, 2:1 minerals are rarely reported in "pure" tephras but they are relatively common in soils formed on loess derived from sedimentary rocks and alluvium that contains admixed tephras (Lowe, 1986). The 2:1 minerals from nontephra sources occur in the clay fraction of some Andisols and contribute to the development and properties of nonallophanic Andisols.

3.2.3. Organisms

The significance of tephra as a parent material in soil development has been widely recognized. In contrast, the effects of organisms, especially plants and human beings, on the formation and properties of Andisols have not been sufficiently emphasized. However, there is increasing recognition of their importance.

Vegetation rapidly recovers on tephra deposits because physical and chemical properties are favorable for plant growth, which in turn facilitates soil development. Biosequential relations between Andisols with tree and grass vegetation have been noted. For instance, Japanese pampas grass (*Miscanthus sinensis*) is a strong andisolizer and has been observed to contribute greatly to the formation of very dark, humus-rich epipedons or melanic Andisols (Plates 5 to 8). Human beings also conspicuously alter Andisols in various ways and often create new soils when intensive farming is practiced. Thus, processes of soil change by human beings will be described in a following section (3.2.5).

(1) Rapid recovery of plants on tephra deposits

Tephras are commonly fine-grained, vesicular and unconsolidated which contribute to a high plant available water holding capacity and high surface area which in turn lead to rapid release of nutrients by weathering. These conditions favor the rapid re-establishment of higher plants as pioneers. As described in Chapter 8, even phosphorus in young tephras is considerably soluble in dilute acid solutions due to release of P by weathering and the lack of active Al and Fe components which render P unavailable by sorption. Accordingly, the important limiting nutrient for revegetation is not phosphorus, but nitrogen, so N-fixing plants are the pioneer species.

When violent volcanic explosions take place, vegetation near the volcano is completely destroyed by strong blasts, direct hits and thick coverage by tephras, evolution of heat and poisonous gases, etc. However, under humid temperate and tropical climates, revegetation will soon start with invasion of higher plants whose seeds are transported by wind and animals. Vegetation damage by volcanic eruptions tends to decrease with increasing distance from the volcano. Recovery of vegetation in this area will proceed rapidly with revegetative growth of both the canopy and understory.

For example, rapid revegetation on a thick (>50 cm) dacitic tephra deposited by the 1977 eruption of Mt. Usu, northern Japan, started with invasion of higher plants (Ito and Haruki, 1984; Haruki, 1988). Vegetation consisting of immigrant species such as giant knotweed (*Polygonum sachalinense* Fr. Schm.), Japanese butterbur (*Petasites japonicus* Miq.), sheep sorrel (*Rumex acetosella* L.), pearly everlasting (*Anaphalis margaritacea* var. *angustior*), etc., has established in ten years since the eruption. In comparison, vegetative recovery of higher plants on a pumiceous deposit less than 50 cm in depth took place in the year following the eruption (Tomioka et al., 1990). Rapid recovery of vegetation consisting of higher plants was also observed shortly after the eruptions of Mt. Komagatake, northern Japan (Yoshioka, 1966), Mt. Sakurajima, southern Japan (Shinagawa, 1962), and Mt. St. Helens Washington, U.S.A. (Antos and Zobel, 1985, 1986).

From the foregoing, it is noted that pioneer plants on tephra deposits are distinctly different from those on consolidated rocks (commonly mosses), indicating that tephra deposits can rapidly create an excellent media for plant growth soon after deposition.

(2) Biosequence of Andisols in Japan

Many Andisols in Japan have a very dark A horizon (melanic epipedon) from which the term "Kuroboku" or "Ando" originated. The genesis and properties of such Andisols have attracted the interest of Japanese soil scientists for a long time. It is believed that grass vegetation, such as Japanese pampas grass (*Miscanthus sinensis*) is associated with the formation of the dark humus-rich horizon or melanic epipedon as reviewed by Shoji et al. (1990b). Forest vegetation exists naturally and is very common in Japan where a humid temperate climate prevails. In contrast, grass vegetation is produced by human activities, such as fires, grazing, cutting, etc.

The contribution of the grass vegetation to the formation of melanic Andisols is clearly indicated by biosequential studies on Andisols (Kawamuro and Osumi, 1988; Kawamuro and Torii, 1986; Shoji et al., 1982, 1988b). It is also confirmed by studies of the relationship between the content of plant opals of *M. sinensis* and humus accumulation in Andisols (Sase and Kato, 1976a, b; Kondo and Sase, 1986; Sase 1986b).

As shown in the pictures on the cover pages (Plates 5 to 8), Fulvudands and Melanudands have a biosequential relation in the montane zone of the southern Hakkoda Mountains, northeastern Japan. This zone has a mean annual temperature of 6–9°C and a mean annual precipitation of 1200–1800 mm. Shoji et al. (1982, 1988b) showed that the parent material common to all the pedons is Towada-a ash (1000 yr B.P.) and Chuseri ash (5000 yr B.P.) which are dominated by noncolored glass. Fulvudands and Melanudands are formed in *Fagus crenata* (Siebold's beech) forests and in *M. sinensis* grasslands, respectively. All the pedons show bisequa as a result of the two ash deposits.

Fulvudands and Melanudands have striking differences mainly in morphological and chemical properties relating to soil humus (Shoji et al., 1988b). Of the morphological properties, the difference between the two Andisols is most pronounced in the color of the humus horizons. Although the humus horizons of Fulvudands are thick and contain a high concentration (>6%) of organic carbon, the color of these horizons is dark brown (color values > 2, chromas ≥ 2). In contrast, Melanudands have darker colored humus horizons (color values ≤ 2, chromas < 2) as described in detail in Chapter 2.

There are remarkable differences in the distribution and chemical characteristics of humus between Fulvudands and Melanudands (Shoji et al., 1988b). As shown in Fig. 3.4, the Melanudands show high organic carbon concentrations that gradually decrease with increasing depth as compared to a lower organic carbon concentration and rapid decrease with depth in the Fulvudands. These differences are considered to reflect the source of organic matter in the two ecosystems: the input comes from both litterfall and underground components in *M. sinensis* ecosystems and mainly from litterfall in *F. crenata* ecosystems. The ratio of humic acid to fulvic acid ranges from 0.7 to 0.2 in the Fulvudands while they are greater than one in the Melanudands. This fact indicates that the humification process has been much greater in the Melanudands.

Classification of humic acids separated from the two Andisols is given in Fig. 3.5. The $\Delta \log K$ values are approximations of the slope of the absorption spectrum for humic acid versus wavelength (400 and 600 nm) and show an inverse relationship with RF values. The RF values are a measure of the relative color density of humic acids (Kumada et al., 1967). According to this diagram, humic

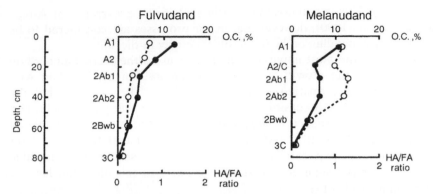

Fig. 3.4. Organic carbon content (solid line) and ratio of humic acid to fulvic acid (broken line) for a Fulvudand and Melanudand, southern Hakkoda, northeastern Japan (figure prepared by the authors using data of Shoji et al., 1988b; copyright Williams & Wilkins, with permission).

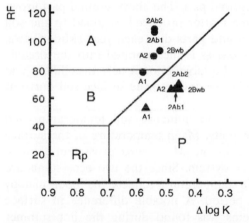

Fig 3.5. Types of humic acids separated from a Fulvudand (triangles) and Melanudand (circles), southern Hakkoda, northeastern Japan (figure prepared by the authors using data of Shoji et al., 1988b; copyright Williams & Wilkins, with permission).

acids from the Fulvudand are determined to be P-type which is common for less humified organic matter. In contrast, humus horizons in the Melanudand, except the A1 horizon, contain A-type humic acid which shows the greatest degree of humification. The large annual addition of fresh organic matter from the vegetation lowers the degree of humic acid humification in the A1 horizon of the Melanudand. It is these differences in the degree of humification that result in Fulvudands failing the melanic index criterion while Melanudands satisfy this index as described in Chapter 4.

The burning of *M. sinensis* has been proposed as a hypothesis to explain the formation of A-type humic acid in melanic Andisols (Kumada et al., 1985;

Shindo et al., 1986). This hypothesis may partly explain the occurrence of A-type humic acid (Tate et al., 1990); however, the major processes are considered to be biochemical, largely depending on high microbial activity in the soil.

Since *M. sinensis* is the most important andisolizer in northeastern Japan, Shoji et al. (1990b) studied the growth and chemical composition of the plant with special reference to the formation of melanic Andisols. They showed that the growth of *M. sinensis* is highly sensitive to temperature and that plant height, leaf number, and dry matter production are largely determined by the accumulated effective temperature above 10°C (AET). They also found that the upper limit of the *M. sinensis* ecosystem corresponds to the upper distribution of melanic Andisols in northeastern Japan and *M. sinensis* growing at this elevation has the minimum leaf number (12–13).

M. sinensis ecosystems produce large amounts of organic matter in both the above-ground and below-ground parts. The dry weight (g/m^2) at full heading time in the Kawatabi Farm, Tohoku University, northeastern Japan, was estimated as follows: mean, 568 and maximum, 3200 for the above-ground part, and mean, 1782 and maximum, 4448 for the below-ground part. The above-ground part grows and dies each year so that all the organic matter produced is added to the soil surface. Since one fourth of the below-ground parts dies each year (Midorikawa, 1978), one fourth of the below-ground biomass is incorporated into the organic matter pool in the rooting zone. Addition of a large amount of hemicellulose and cellulose-rich organic matter to the soil surface and into the surface soil horizons can increase activity of soil microorganisms.

The *M. sinensis* ecosystem was observed to influence soil temperature and moisture at Kawatabi Farm, Tohoku University. Mean temperature of the surface soils during April to June is 2°C higher in the *M. sinensis* ecosystem than in an adjacent *Quercus serrata* (oak)–*Sasa* ecosystem. Since the two ecosystems are at the same elevation, the difference in temperature reflects the poor canopy development of *M. sinensis* during this period. A notable difference in surface soil moisture between the two ecosystems was found during the hot summer months when precipitation is low: the soil moisture was greater in the *M. sinensis* ecosystem than in the *Quercus serrata–Sasa* ecosystem. This indicates that *M. sinensis* can decrease transpiration when soil moisture decreases. Therefore, microbial activity is enhanced in *M. sinensis* ecosystems by the addition of a large amount of hemicellulose and cellulose-rich organic matter to the soil surface and into the surface soil, and higher soil temperature and moisture. These conditions promote accumulation and humification of soil organic matter, resulting in the formation of a melanic epipedon under this ecosystem.

A biosequence of Andisols under grass vegetation and Spodosols under tree vegetation occurs on tephra deposits in south central Alaska, U.S.A. (Shoji et al., 1988a; Simonson and Rieger, 1967). Bluejoint grass (*Calamagrostis canadensis*) is a fire species that grows densely after forest fires. This plant like *M. sinensis* has a deep-rooting system and produces large amounts of dry matter in both above-ground and below-ground parts.

Bluejoint grass contributes to conversion of tephra-derived Spodosols to Andisols through additions of organic matter which mask the morphological and chemical properties of the albic horizon. It also increases the ratio of humic acid to fulvic acid and the degree of humification (Shoji et al., 1988a). Thus, bluejoint grass can also be regarded as an andisolizer.

(3) Paleovegetation and Andisols

The North Island of New Zealand is situated at almost the same latitude and has almost the same mean annual temperature and mean annual precipitation as the Tohoku region of Japan. However, the two locations have significantly different kinds of Andisols. The North Island has few Andisols with thick dark-colored humus horizons which are attributable to the abundance of A-type humic acid. On the other hand, this type of Andisol is common in Tohoku.

Andisols on the North Island rarely display a thickness of dark-colored humus horizons greater than 20 cm or contain organic carbon contents of 8–12 percent (Leamy et al., 1980). In comparison, Andisols in Tohoku have thick dark-colored humus horizons ranging from 22 to 50 cm thick and have organic carbon contents greater than 6 percent with a mode of 9 percent (Andisol TU Database, 1992). Since different vegetation types produce varying types of humic acids, the predominant difference between Andisols in these locations appears to be due to differences in the paleovegetation.

This observation was confirmed by the study of Shoji et al. (1987) on the characterization of humus from many Andisol pedons from the North Island. These workers found that most pedons have the P-type humic acid with the lowest degree of humification and that only two pedons have the A-type humic acid with the highest degree of humification. This observation was confirmed by Yamamoto et al. (1989), strongly suggesting that Andisols on the North Island have developed under forest ecosystems.

Opal phytolith analysis of Andisols has been successfully used to interpret the relationships between paleovegetation and formation of Andisols in Japan (Kondo et al. 1988). Using this method of analysis, it has been shown that melanic Andisols (Kurobokudo) in Tohoku have developed under the *M. sinensis* grass ecosystem. Sase (1986a) conducted opal phytolith analysis of selected Andisols from the North Island of New Zealand (Tirau silt loam, Taupo sand, Egmont black loam and Patua loam). The analysis showed that the plant opal particles in Tirau, Taupo, and Egmont soils originated primarily from broad-leaved trees while those from the Patua soil were derived mainly from ferns, although the present vegetation on all these soils is improved pasture grasses.

Sase et al. (1988) further investigated the paleovegetation for the last 20,000 years using the Te Ngae road tephra section, Rotorua Basin, North Island. The results of this study are summarized in Table 3.1. They found that the main source of plant opals was from grass vegetation during the period between 20,000 and 11,250 yr B.P. However, the low opal content of the soils formed before the Rerewhakaaitu ash deposition (14,700 yr B.P.) indicates that the soils developed

S. SHOJI, R. DAHLGREN, and M. NANZYO

TABLE 3.1

Soil-vegetation relationships at Te Ngae Road tephra section, Rotorua Basin, North Island, New Zealand (prepared from Sase et al., 1988)

Age(YBP) x1000	Volcanic ash soils	Main origins of plant opal	Notes
0	Black-colored A horizon	Grasses, trees and ferns	
1			← Polynesian settlement (deforestation)
3			
5		Trees	
7	Brown-colored A horizon		
10			Holocene
12		Grasses	Pleistocene
15			

under scattered grasslands or forests that produce few opals. Plant opals present in the soils which formed between 11,250 and 7330 yr B.P., originated from both trees and grasses. In contrast, plant opals in soils developed between 7300 and 930 yr B.P. are dominated by those originating from trees, indicating establishment of forest vegetation during this period of soil formation. On the other hand, since human activity was high in Tohoku during this period, melanic Andisols in this region were formed under grass vegetation.

An increase in the content of plant opals from grass species and ferns since 930 yr B.P. was considered to be caused by the Polynesian settlements reducing the forests and increasing grass and fern vegetation. Leamy et al. (1980) described the change from native podcarb forest to scrub dominated by bracken fern (*Pteridium aquilinum* var. *esculentum*) as marked by an abrupt increase in blackness of the topsoil. Thus, it appears that some dark-colored Andisols on the North Island have formed under the bracken fern vegetation which has annually produced large amounts of dry matter in both the above-ground and below-ground parts for the last 1000 years. Prior to this, there was no grass or fern vegetation to contribute to the development of thick dark-colored humus horizons on the North Island.

3.2.4. Climate

Formation of Andisols is strongly controlled by climatic factors and its importance is recognized in all suborders except Aquands and Vitrands. A close relationship between Andisols and the prevailing climate is also recognized at the great group and subgroup levels. For example, all Torrands formed under an aridic moisture regime are regarded to be vitric and only one great group and six subgroups are established. In contrast, Udands have six great groups and many subgroups (Soil Survey Staff, 1990, 1992).

Climate has many different components or aspects such as temperature, precipitation, evaporation, evapotranspiration, humidity, duration, seasons, etc. Its effect on soil formation is generally the integration of all these components; however, temperature and precipitation have a controlling influence on the formation and properties of Andisols.

(1) Temperature

The rate of chemical weathering in tephras increases remarkably with increasing temperature in Andisols in the udic moisture regime. For example, laboratory dissolution experiments showed that temperature exerts a marked influence on the release of chemical elements from tephras as indicated by temperature dependency coefficients, Q_{10}, of 1.5 for Al, 1.6–1.8 for Si, and 2 for Ca and Mg. The release of these elements shows a zero-order reaction (Shoji et al., 1993). When the mean annual temperature is 10°C and the parent material consists mainly of noncolored glass, it was determined that weathering processes are rapid enough to provide andic soil properties ($Al_o + 1/2 Fe_o \geq 2.0$ percent) within 1280 years as already described.

(2) Precipitation and soil moisture

The rate of chemical weathering in the soil notably decreases with decreasing soil moisture and leaching. The significance of soil moisture in soil formation is clearly recognized for Andisols by comparing great groups and subgroups between Torrands, Ustands, Xerands and Udands (Soil Survey Staff, 1990, 1992).

Torrands formed in arid climates are least weathered and all are regarded to be vitric so that only a vitric great group is established for this suborder. The number of subgroups is also the smallest among all the suborders. A calcic subgroup showing accumulation of calcium and magnesium carbonates is characteristic of Torrands.

Ustands, occurring in semiarid climates, have two great groups, namely duric and haplic. Since this suborder keys out after the Vitrand suborder, vitric Andisols in semiarid climates are excluded from Ustands and are included in Vitrands. Formation of Durustands is favored by release of large amounts of silica from tephras into soil water, incomplete leaching of the profile and evapotranspiration of soil water resulting in silica cementation.

Xerands in Mediterranean climates have vitric, melanic and haplic great groups. Pachic, thaptic, and humic subgroups as well as the melanic great group (Soil Survey Staff, 1992) indicate accumulation of large amounts of soil organic matter with a high degree of humification that may be associated with the climatic conditions of moist winters and dry summers.

Udands formed under humid or very humid climates have a variety of great groups and subgroups (Soil Survey Staff, 1990, 1992). They have six great groups consisting of placic, duric, melanic, fulvic, hydric and haplic. Placudands develop as a result of oxidation and reduction processes in humid climates. Fulvudands are formed under forest vegetation (Shoji et al., 1988b). Hydrudands, having high

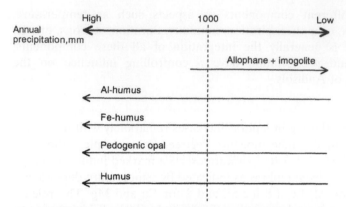

Fig. 3.6. Influence of annual precipitation on clay formation and humus accumulation in the surface horizons of young Andisols developed from noncolored glass-rich tephras in northeastern Japan (figure prepared by the authors using data of Shoji and Fujiwara, 1984; copyright Williams & Wilkins, with permission).

1500 kPa water contents contain a large amount of colloidal materials as a result of intense weathering in very humid climates, typically at high temperatures. Many subgroups are established within Udands; subgroups such as alic, hydric, acrudoxic, and pachic are commonly found for Udands.

In the climatic region of northeastern Japan (udic, mesic) where mean annual temperatures are similar throughout the entire region, precipitation shows a close relationship with chemical weathering and mineralogical properties in the surface horizons of Andisols as presented in Fig. 3.6 (Shoji and Fujiwara, 1984). When the parent materials are tephras dominated by noncolored glass and mean annual precipitation is greater than 1000 mm, the surface soils have strong acidity ($pH(H_2O)$ < 4.9), reflecting the strong leaching of bases from the soil. These surface soils are characterized by nonallophanic clay mineralogy that shows an abundance of 2:1 layer silicates, intense formation of Al and Fe-humus complexes, significant formation of laminar opaline silica, and a marked accumulation of humus. In contrast, soils receiving a lower annual precipitation (<1000 mm) and having pH (H_2O) > 4.9 are characterized by allophanic clay mineralogy and less abundant formation of Al/Fe-humus complexes (Shoji et al., 1982, Shoji and Fujiwara, 1984).

Shoji et al. (1982), and Shoji and Fujiwara (1984) explained the contrasting weathering sequences in the surface horizons according to differences in the reaction of Al with humus and Si in relation to soil acidity. Below $pH(H_2O)$ 4.9, soluble Al occurs primarily as Al ions with high charge and does not readily react with Si, instead favoring Al-humus complex formation (anti-allophanic reaction) when a plentiful supply of organic matter is present. In contrast, as pH increases above 4.9, Al forms hydroxy-Al species that may undergo polymerization and coprecipitation with Si. The plentiful formation of allophane and imogolite takes place under these conditions. On the other hand, Fe released from the parent

TABLE 3.2

Effect of mean annual rainfall and leaching on weathering of tephra of similar age and composition in New Zealand (Parfitt, 1990)

	Inceptisol Udic (near Ustic)	Andisol Udic	Andisol Udic (near Perudic)
Classification	Andic Dystric Eutrochrept	Typic Hapludand	Acric Hapludand
Mean annual rainfall (mm)	1200	1400	2600
Mean annual leaching (mm)	400	550	1600
Si in soil solution (g m^{-3})	15	7	2
Clay mineral (% of soil)	Halloysite (30%)	Allophane (15%) Halloysite (10%)	Allophane (26%)

material has a low affinity for Si and humus and thus primarily forms ferrihydrite.

Parfitt (1990) summarized the effect of mean annual precipitation and leaching intensity on a weathering sequence in similar aged tephras with an Andic Dystric Eutrochrept-Typic Hapludand-Acric Hapludand association in New Zealand (Table 3.2). Inceptisols occur in rhyolitic tephras where the annual precipitation is less than 1200 mm, while Andisols were found in similar tephras in regions receiving higher precipitation. He inferred that the low rainfall leads to the preferential formation of halloysite under the relatively weak leaching conditions that contribute to high Si concentrations. On the other hand, leaching of Si and bases becomes more intense with increasing precipitation, leading to the formation of allophane when soluble Si concentrations are lower.

(3) Climate change and formation of Andisols

Most Andisols in northeastern Japan have developed in Holocene tephras and are characteristically dark-colored Andisols or melanic Andisols. In contrast, all Andisols derived from late-Pleistocene tephras are not as dark, even though they occur in the same district (Yamada, 1986). Melanic Andisols formed in the late-Pleistocene period have been found in Kanto and Kyushu areas where the climate was warm enough for growth of M. sinensis and tephra deposition was intermittently repeated.

The andisolizer, M. sinensis requires an accumulated effective temperature above 10°C (yearly summation of mean monthly temperature minus 10°C) greater than 27°C for maintaining the ecosystem in Tohoku (Shoji et al., 1990b). Such a thermal condition prevails below 600 m elevation in Aomori Prefecture, northern Tohoku at the present time, but did not occur in the late-Pleistocene period when temperatures were much colder (Yasuda, 1987). The oldest Andisol in Aomori that has a dark-colored humus horizon is formed from Nanbu tephra deposited 8600 yr B.P. (Shoji and Saigusa, 1977). Sase and Kato (1976a; 1976b) observed an abundance of plant opals from *M. sinensis* in the humus horizons of soils formed in Nanbu tephra and the overlying tephras. Sase et al. (1990) also confirmed a predominance of plant opals from *M. sinensis* among the opal particles in the dark-colored humus horizons of Andisols formed 8000 to 2000 yr B.P. at the foot of Mt. Iwate, in Tohoku. Thus, formation of dark-colored Andisols or melanic Andisols in Tohoku started with the change from cold to warm climate and the subsequent appearance of *M. sinensis* vegetation induced by strong human or Jomon people's impact in the early Holocene period.

3.2.5. Human beings

The first strong influence of human beings on Andisols was due to the use of fire. They cut trees for firewood and set fires to maintain grasslands for hunting and grazing. Accidental fires destroyed a vast area of forests in Japan. The role of fire has contributed to the formation of the very dark-colored Andisols (melanic Andisols) in Japan by conversion of forest vegetation to *M. sinensis* (Japanese pampas grass) vegetation.

Dark-colored Andisols have not been found in Pleistocene tephra-derived soils, but have developed only in Holocene tephras in northeastern Japan. The age of the most intense formation of these soils is coincident not only with the history of Jomon people showing their high activity (Yamada, 1986), but also with the warmer climate during Jomon age (8000–2500 yr B.P.) under which *M. sinensis* could luxuriously grow (Yasuda, 1987; Ohmori and Yanagimachi, 1988). If human beings had not lived in northeastern Japan, Melanudands could not have developed and Fulvudands would be more extensively distributed.

The second strong effect of human beings on Andisols is land reclamation for farming. Land reclamation removes natural vegetation, drastically changes ecosystems, and may destroy natural Andisols. Repeated cultivation has a strong effect on soil physical properties and may accelerate soil erosion.

Ono et al. (1981) observed changes in the humus horizon depths of alic Melanudands in Rokuhara Farm, Iwate, northeastern Japan following 90 years of farming. Two fields, one used for mulberry-pasture (mean slope; 2.7 percent) and the other for rotation of common crops such as cereals (mean slope; 1.0 percent) were selected to illustrate the changes (Fig. 3.7). Both cultivated fields have the same mean depth of humus horizons (29 cm); however, the mulberry-pasture field shows a wide variation in depth ranging from 5 to 100 cm. Since the neighboring uncultivated Melanudands have a mean depth of about 40 cm, the depth of the humus horizons in cultivated Andisols has decreased 11 cm, possibly because of soil erosion, degradation of soil structure, and compaction due to heavy agricultural equipment. Cultivation of Melanudands has created andic Inceptisols, alic Hapludands and alic pachic Melanudands.

Farming also conspicuously changes chemical and biological properties of Andisols, such as decreasing organic carbon content, improvement of base status and soil acidity by liming, enrichment of phosphorous by fertilization, accumulation

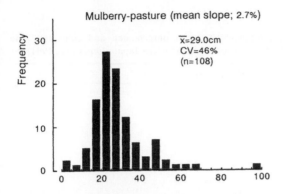

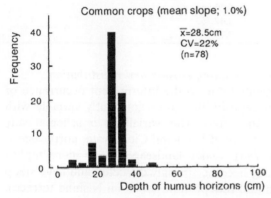

Fig. 3.7. Variations in the depth of humus horizons of an Alic Melanudand following 90 years of cultivation in Rokuhara Farm, Iwate Prefecture, Japan (adapted from Ono et al., 1981).

of heavy metals by spraying of agricultural chemicals, etc. For example, Table 3.3 shows changes in the humus and total phosphorus content in the surface horizons of an alic Melanudand at Rokuhara Farm 90 years after conversion for agriculture. The mean humus content has decreased about 20 percent while total phosphorus has increased two-fold during 90 years of farming. It was noted that copper, which forms stable complexes with humus, also accumulated, reaching three times the mean copper content of uncultivated soils (Ono et al., 1981).

3.2.6. Topography

Topography has a number of effects on soil-forming processes in Andisols. However, the relationship of topography to tephra deposition, erosion and redistribution of materials according to slope, and landscape distribution of moisture are especially important factors affecting the genesis and properties of Andisols.

TABLE 3.3

Comparison of humus content and total phosphorous between unreclaimed and cultivated Alic
Melanudands in Rokuhara Farm, Kitakami, Iwate Prefecture, northeastern Japan (Ono et al., 1981)

	Unreclaimed	Cultivated
Humus content (%)		
Number of samples	11	93
Maximum	28.7	20.0
Minimum	11.8	8.0
Mean	17.7	14.0
SD	5.8	2.9
Total phosphorus (% as P_2O_5)		
Number of samples	10	93
Maximum	0.19	0.94
Minimum	0.09	0.20
Mean	0.17	0.42
SD	0.08	0.14

(1) Landscape relationships to tephra deposition, erosion and redistribution

Tephra deposition is commonly repeated due to the intermittent occurrence of
volcanic eruptions resulting in tephra deposits that are significantly variable with
respect to age and features of the landscapes. This variability is schematically
shown for landscapes and soils in the vicinity of Kitakami City, Iwate, northeastern
Japan (Fig. 3.8) (Shoji and Ono, 1978b). Older landscapes have more tephra
deposits; however, these landscapes may become strongly dissected and have steep
slopes, leading to accelerated erosion and redeposition (soils on Nishine terrace).
Therefore, older landscapes may have both young and old soils intermixed on
the landscape. Younger landscapes receive less tephra resulting in formation of

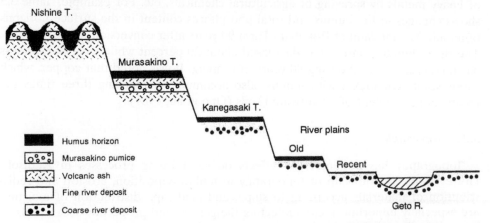

Fig. 3.8. Schematic representation of landscapes and soils in the vicinity of Kitakami City, Iwate, northeastern Japan (prepared by the authors using the data of Shoji and Ono, 1978b)

Andisols and andic subgroups of other soil orders (soils on Kanegasaki terrace and river plains). Thus, a variety of Andisols and/or andic intergrade soils can occur on different landscapes in a given region.

Shoji and Ono (1978b) and Ono and Shoji (1978) investigated in detail the relationship between topography and Andisol formation in the vicinity of Kitakami City, where nonallophanic Andisols were first observed (Shoji and Ono, 1978a). The landscape of this area consists of two recent river plains and three Pleistocene-age terraces (Fig. 3.8). The lower river plain lacks tephra deposits while the higher one is covered with a thin ash layer, forming locally Andic Fluvents. Kanegasaki terrace is the youngest among the three terraces and consists of fluvial deposits with a gentle slope as described in Fig. 3.9. Two recent tephras deposited 1000–5000 yr B.P. (M-1 and M-2) cover this terrace and

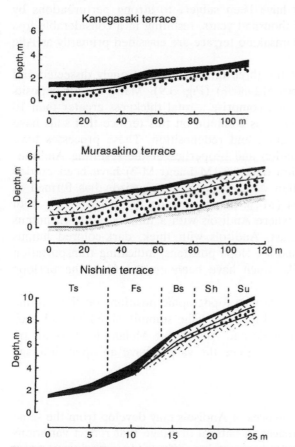

Fig. 3.9. Profile sections of Andisols on three Pleistocene terraces in the vicinity of Kitakami City, Iwate Prefecture, northeastern Japan. The legends are the same as shown in Fig. 3.8. Abbreviations: Su = summit, Sh = shoulder, Bs = back slope, Fs = foot slope, Ts = toe slope (adapted from Shoji and Ono, 1978b).

weather to typically form Alic Melanudands with a thick humus horizon underlain by fluvial deposits. Identification of individual tephra layers was conducted by determining the content of V and Zn in ferromagnetic minerals. This is one of the most useful methods for identifying the source of tephra layers (Shoji et al., 1974).

Murasakino terrace is the second youngest of the three. It consists of lacustrine deposits with a gentle slope and is slightly dissected (Fig. 3.9). Tephras occurring on this terrace have a total thickness of 2–3 m and show at least five volcanic ash layers (M1–M5) and one pumice layer often called the Murasakino pumice (M-6). The three upper tephras were mixed mainly by human activities (M-1, M-2, and M-3) and they constitute the parent material of the humus horizons. The subsurface horizon is highly weathered and clayey, contrasting with that of Andisols on the Kanegasaki terrace. Although Andisols are uniformly distributed on the Murasakino terrace, they have been subject to strong perturbations by human beings for the last several thousand years, resulting in a considerable loss of organic carbon. Soils on the Murasakino terrace are classified primarily as Alic Hapludands (Ono and Shoji, 1978).

Nishine terrace is the oldest of the three terraces. It is strongly dissected and has many valleys with steep slopes (12–30%) (Fig. 3.9). Many tephra deposits have accumulated on this terrace and comprise a total thickness greater than 10 m in the stable landscape positions. It is common on this terrace that soils have been subject to erosion, transportation and redeposition. These processes have substantially determined the formation and properties of the resulting Andisols. For example, the two recent tephra deposits (M-1 and M-2) have been eroded from the small summits and a thin dark-brown humus horizon has formed in the underlying subsurface horizon (M-3). Strong erosion has occurred on both shoulder and backslope positions where Andisols with a thin, dark-brown humus horizon have developed. In contrast, Andisols with thick, dark-colored humus horizons occur in the foot slope and toe slope positions, indicating transportation and redeposition of soil materials which have been eroded from the upslope positions (Fig. 3.9).

Thus, Andisols differ widely with varying topographic positions on the Nishine terrace. Alic Hapludands are observed mainly on the summit, shoulder and back slope positions, while Alic, Alic Thaptic and Alic Pachic Melanudands occur on the foot slope and toe slope positions. Since the valleys have an open drainage system, no Aquand occurs.

(2) Hydrosequence

Under humid climates, hydrosequences of Andisols may develop from the same parent tephras according to differences in natural drainage that reflect variations in topography. Even young Andisols show notable differences in their properties between the well-drained and poorly-drained members. Intermittent thin ash deposits and marked accumulation of humus in the poorly drained sites favor the formation of melanic Andisols.

Hydrosequences of Andisols are commonly observed on the late-Pleistocene terraces in Tokachi, Hokkaido (Kikuchi, 1981). The well-drained and poorly-drained members in this region are locally called "dry soil" and "wet soil", respectively. Saigusa et al. (1991) studied the properties of the two representative pedons which are classified as Typic Hapludand and Hydric Melanaquand. There are many notable differences between the two pedons as presented in Table 3.4. For example, the wet soil accumulates large amounts of highly humified organic matter and forms a melanic epipedon. The accumulation of organic matter is related to reducing conditions due to poor drainage. It is also closely related to the formation of Al-humus complexes as expressed by the following equation:

$$\text{Organic C } (\%) = 6.82 \times \text{pyrophosphate Al} (\%) - 0.20 \qquad r = 0.89^{***} \ (n = 13)$$

Thus, it appears that microbial decomposition of organic matter is retarded by seasonal reducing conditions, formation of Al-humus complexes, and humification of organic matter that effectively proceeds under the poorly drained conditions.

Clay mineralogical properties are distinctly different between the two pedons, reflecting the different soil environments. The major colloidal components of the wet soil are Al-humus complexes and smectite for the upper most horizons, Al-humus or Al-humus and allophane-imogolite for the middle part of the profile, and halloysite at the bottom of the profile. In contrast, the colloidal fraction of the dry soil is dominated by Al-humus complexes and chloritized 2:1 minerals for the uppermost horizons, and allophane-imogolite in the subsurface horizons. It was also noted that there is an inverse relationship between formation of allophane-imogolite and Al-humus complexes. Soil acidity and the supply of organic matter appear to be the key factors regulating formation of allophane-imogolite versus Al-humus complexes, while silica enrichment appears to provide a favorable environment for halloysite formation in the poorly drained subsurface horizons (Chapter 5).

3.2.7. Soil age

Time of tephra deposition is regarded as the time zero of Andisol formation. It is determined with reasonable accuracy by various procedures, such as historical records, C-14 dating, fission tracks, thermo-luminescence dating, archaeological remains, etc. An accurate soil age is extremely useful for understanding the relationships between time, soil genesis and soil properties of Andisols.

(1) Time zero of Andisols
Tephras are commonly dominated by volcanic glass particles with high vesicularity which can retain water and rapidly release mineral nutrients for plant uptake. Therefore, as described earlier, recovery of vegetation on barren landscapes destroyed by volcanic eruptions starts soon after tephra deposition. It is noted that the pioneer plants are commonly higher plants as described in the section on organisms.

TABLE 3.4

Selected properties of dry and wet Andisols showing a hydrosequential relation, Tokachi, Hokkaido (Saigusa et al., 1991)

Horizon	Depth (cm)	Color moist	Mottling (iron)	Parent material[a]	Water ret. at 1500 kPa (%)	Organic C (%)	pH H₂O	pH KCl	P ret. (%)
Fushiko soil (Typic Hapludand)									
A	0–13	10YR 2/2	–	Me-a, Ta-b	17	2.3	5.5	4.3	67
2A	13–22	10YR 2/3	–	Ta-c	27	3.6	5.7	4.6	94
2B	22–26	10YR 4.5/4	–	Ta-c	34	2.8	5.7	4.8	95
3B1	26–40	10YR 4/6	–	To-c2	34	2.0	5.8	4.8	96
3B2	40–51	10YR 5/6	–	To-c2	47	1.2	5.9	4.7	98
4B1	51–59	1.25Y 5/6	–	unknown	ND	0.5	6.0	4.4	87
Kitamotoimatsu soil (Hydric Melanaquand)									
A	0– 9	10YR 1.7/1	–	Me-a	23	10.8	4.2	3.7	69
2C–3C	9–18	mixed	–	Me-a2, Ta-b	9	2.2	5.0	4.2	49
4A	18–32	10YR 1.7/1	+	Ta-c	52	15.1	5.0	4.1	98
5A	32–49	10YR 1.7/1	+	To-c2	93	12.9	5.2	4.3	98
6B	49–61	1.25Y 5/3	++	To-c2	85	6.9	5.2	4.5	98
7B	61–76	2.5Y 6/3	+++	unknown	73	4.8	5.6	4.4	97
8B	76+	5Y 7/2	+++	unknown	ND	0.3	5.7	3.7	31

TABLE 3.4 (continued)

Horizon	Depth (cm)	Acid oxalate (%)			Al_y (%)	Al_p/Al_o [b]	Allo + Imogo (%)	Major colloidal components
		Fe	Al	Si				
Fushiko soil (Typic Hapludand)								
A	0–13	1.24	0.96	0.23	0.50	0.52	1.6	Al-humus, chloritized 2 : 1 min.
2A	13–22	0.96	2.92	1.18	0.59	0.20	8.4	Allo + imogo, Al-humus
2B	22–26	0.47	4.08	2.06	0.49	0.12	14.7	Allo + imogo
3B1	26–40	0.85	5.91	2.94	0.47	0.08	21.0	Allo + imogo
3B2	40–51	1.05	6.34	3.34	0.41	0.06	23.8	Allo + imogo
4B1	51–59	0.66	2.79	1.41	0.30	0.11	10.1	Allo + imogo
Kitamotoimatsu soil (Hydric Melanaquand)								
A	0– 9	0.48	0.72	0.05	0.57	0.79	0.4	Al-humus, smectite
2C–3C	9–18	0.42	0.48	0.10	0.36	0.75	0.7	Al-humus
4A	18–32	1.14	3.02	0.59	2.07	0.69	4.2	Al-humus
5A	32–49	2.26	6.43	2.04	2.10	0.33	14.6	Al-humus, Allo + imogo
6B	49–61	2.58	7.90	3.45	1.16	0.15	24.6	Allo + imogo, Al-humus
7B	61–76	2.20	5.92	2.59	0.86	0.14	18.5	Allo + imogo, Al-humus
8B	76+	0.23	0.42	0.10	0.09	0.21	0.7	Halloysite

[a] Me-a = Meakan-a ash (220–250 yr B.P.), Ta-b = Tarumae-b ash (320 yr B.P.), Ta-c = Tarumae-c ash (2200 yr B.P.), To-c2 = Tokachi-c2 ash (3800–4800 yr B.P.).

[b] Al_p = pyrophosphate-extractable Al; Al_o = oxalate-extractable Al.

(2) Development of horizon sequences of Andisols

Andisols show rapid change or development of horizon sequences with time. The rate of development in Andisol profiles is determined by various factors, such as soil moisture and temperature, rock types and texture of tephras, etc. It is greater in humid warmer climates than in dry cooler climates and in mafic tephras compared to felsic tephras.

The relationship between age of Andisols and their horizon sequences in northeastern Japan where soil temperature is mesic and soil moisture is udic can be summarized as follows:

Profile 1. A (humic)–C: several 100 years
Profile 2. A (humic)–Bw (colored)–C: several 100–1000 years
Profile 3. A (humic)–Bw (cambic)–C: 1000–several 1000 years
Profile 4. A (melanic or fulvic)–Bw (cambic)–C: several 1000 years or older

Profile 1 is the earliest stage of pedogenesis and lacks a B horizon. Since it is difficult to morphologically differentiate vitric Andisols from andic Entisols, the chemical criteria of acid oxalate Al + 1/2 Fe $\geq$ 0.4% and P retention > 25% are employed for this differentiation (Soil Survey Staff, 1990, 1992).

Profile 2 has a humic surface horizon showing an increase in the thickness and carbon content of the A horizon. Deep rooting plants such as *M. sinensis* contribute greatly to the development of these humus-rich horizons. The color of the Bw horizon is attributable to the formation of iron oxyhydroxides such as ferrihydrite, indicating the advance of andosolization. Andisols older than 1000 years have a humic and a cambic horizon as represented by profile 3. Usually considerable amounts of allophane and imogolite occur in the Bw horizon. Such Andisols typically meet the andic soil properties of acid oxalate Al + 1/2 Fe $\geq$ 2.0%, P retention $\geq$ 85% and bulk density $\leq$ 0.9 g cm^{-3} (Soil Survey Staff, 1990, 1992).

Many Andisols which are several thousand or more years old show a horizon sequence similar to profile 4. They have a melanic or fulvic surface horizon, the formation of which is favored by intermittent thin ash deposition. They also have the most weathered cambic horizon containing a large amount of allophane, imogolite and ferrihydrite. According to C-14 dating of the bottom portion of humus horizons of Andisols in Japan, most such Andisols are 4000–7000 years old (Yamada, 1986). Andisols having a melanic epipedon are regarded as typical Kurobokudo in Japan.

(3) Burial of Andisols

Andisols occurring in the vicinity of volcanoes often develop a multisequum as soil age increases since they are formed by intermittent tephra deposits. An example is shown schematically in Fig. 3.10 detailing the studies of Towada Andisols by Saigusa et al. (1978), Shoji and Saigusa (1977), and Shoji et al. (1982). In northeastern Japan (mesic, udic soil climate regimes), shallow soils (stages I and II) are subject to strong leaching. The intense leaching environment favors

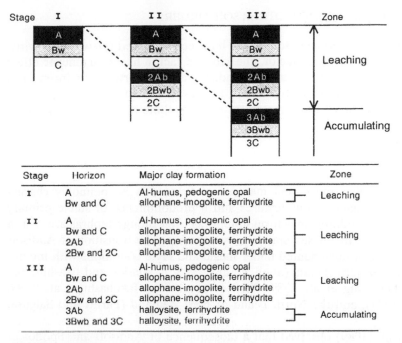

Fig. 3.10. Schematic representation of the relationship between overburden of tephras and development of Andisols in northeastern Japan.

the formation of allophane, imogolite and ferrihydrite in all the horizons except the uppermost humus horizon which shows preferential formation of Al/Fe-humus complexes and pedogenic opal (laminar opaline silica). After the humus horizon (2Ab) is buried by a new tephra deposit, clay formation in the buried horizon proceeds similar to that of nonhumus horizons. The tephra overburden prevents continued additions of organic matter from vegetation to the buried humus horizon and the pH(H_2O) of this horizon increases. Thus, the addition of tephra halts formation of Al/Fe-humus complexes and pedogenic opal and in turn favors the formation of allophane and imogolite.

Further additions of thick tephra deposits change the weathering conditions of the buried soils (stage III). Deeply buried soil horizons are subject to accumulation of bases and silica which migrate from the overlying soil horizons and favor the formation of halloysite in the buried soil. Thus, the profile section is divided into the leaching and accumulating zones as shown in Fig. 3.10. The boundary between the two zones is dependent on various factors, such as precipitation, evapotranspiration, water retention, drainage, vegetation, etc. The accumulating zone appears at a depth of 2.5 m or more in Towada Andisols and considerable halloysite formation is observed in the buried soil horizons irrespective of the kinds of soil horizons (Saigusa et al., 1978). If a dry climate prevails, the boundary appears nearer the surface and halloysite formation takes place at a shallower depth.

3.3. TRANSITION OF ANDISOLS TO OTHER SOIL ORDERS

Transition of Andisols to most soil orders has been observed (Hewitt and Witty, 1988). In addition to the time factor, the climate factor has a significant effect on the Andisol transition. For example, Andisols are converted into Inceptisols, Alfisols, Ultisols, and Oxisols in the humid tropics where strong weathering and soil formation prevail and transformation of noncrystalline materials to crystalline clays takes place.

3.3.1. Spodosols

It is well known that the parent material of Spodosols is commonly characterized by sandy to coarse loamy textures and a predominance of stable primary minerals. However, Spodosols may form in tephras even though tephras are rich in volcanic glass which is highly susceptible to weathering. The transition of Andisols to Spodosols has been documented under mesic and cryic conditions with intense leaching in northeastern Japan (Shoji et al., 1988b), Alaska and Washington, U.S.A. (Plate 11) (Ping et al., 1989; Shoji et al., 1988a; Takahashi et al., 1989; Ugolini et al., 1977), and the North Island of New Zealand (Parfitt and Saigusa, 1985).

Takahashi et al. (1989) observed that a biosequence of Andisols and Spodosols in Shimokita Peninsula, northeastern Japan is a typical transition. Since Andisols in this region are formed under broad-leaved tree vegetation, they do not have such thick, dark A horizons as those formed under *M. sinensis* vegetation and they are classified as Alic and Typic Hapludands. The Hiba tree (*Thujopsis dolabrata* var. *Hondai* Makino), a strong podzolizer, transforms Hapludands to Haplorthods or Placorthods as demonstrated in Fig. 3.11. For example, Andisol 1 shows a uniform distribution of pyrophosphate and acid oxalate iron. In contrast, Spodosol 1 exhibits distinct eluviation and illuviation as evidenced by the iron

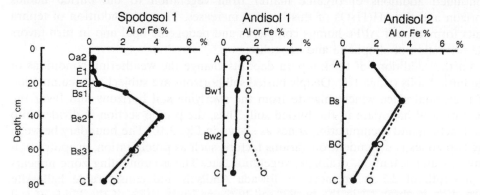

Fig. 3.11. Distribution of pyrophosphate-extractable iron (Fe_p, solid line) and acid oxalate extractable iron (Fe_o, broken line) in a Spodosol and two Andisols from Shimokita Peninsula, northeastern Japan (adapted from Takahashi et al., 1989; copyright Williams & Wilkins, with permission).

distribution (the maximum accumulation of iron amounts to about 4 percent in the Bs horizon).

The transition of Spodosols to Andisols is caused by the conversion of Hiba vegetation to broad-leaved tree vegetation. Changes in the morphological, chemical and mineralogical properties take place concurrently. For example, Andisol 3 shows iron accumulation in the Bs horizon, but it also contains a significant amount of iron in the A horizon compared with that in the corresponding horizon of the Spodosol.

Andisols (Cryands) and tephra-derived Spodosols (Haplocryods and Humicryods) also show a biosequential relation in south central Alaska (Simonson and Rieger, 1967; Shoji et al., 1988a). The former developed under bluejoint grass (*Calamagrostis canadensis*) while the latter formed under Sitka alder (*Alnus sinuata*) and Sitka spruce (*Picea sitchensis*) forests. Andisols seem to be transformed to Spodosols in a relatively short time after the change from grass vegetation to forest vegetation (Simonson and Rieger, 1967). In contrast, Spodosols are transformed into Andisols as a result of invasion of bluejoint grass to forest vegetation as shown by the disappearance of the E horizon (albic horizon) (Shoji et al., 1988a).

Tephra-derived Spodosols show many chemical and clay mineralogical properties common for nontephra-derived Spodosols as shown by a comparison of tephra-derived Spodosols from Alaska and nontephra-derived Spodosols from New England (Shoji and Yamada, 1991). For example, both types of soils have albic horizons showing a predominance of expandable 2:1 minerals among the crystalline clay minerals and the presence of pedogenic opal. The spodic horizons of both soils show the existence of allophane, imogolite and Al/Fe-humus complexes, and the predominance of chloritized 2:1 layer silicates among crystalline clay minerals.

Intense podzolization is required for the formation of tephra-derived Spodosols whose parent material is dominated by highly weatherable materials such as volcanic glass. Since the factors favoring podzolization are observed in only a few areas as indicated earlier, the distribution of tephra-derived Spodosols is very limited.

3.3.2. Inceptisols

Transition of Andisols to Inceptisols proceeds with formation of 1:1 layer silicates or increased concentrations of 2:1 layer silicates. Formation of these minerals can consume a significant amount of active Al, resulting in a decrease of acid oxalate extractable Al and P retention, and increased bulk densities.

Weathering sequences containing Andisols, Inceptisols, and Ultisols have been observed for soils developed on andesitic ash at various elevations associated with different climatic conditions in Irazu-Turrialba of Costa Rica. The major clays in Humitropepts and Palehumults are kaolinite and gibbsite (Martini, 1976). Halloysite formation has also been confirmed for Inceptisols graded from Andisols in New Zealand (Parfitt et al., 1983), Cameroon (Delvaux et al., 1989) and the Philippines (Otsuka et al., 1988).

Some light-colored volcanic ash soils from Kitakami, northeastern Japan are nonallophanic and have acid oxalate extractable Al + 1/2 Fe < 2%, P retention < 85%, and bulk density > 0.9 g cm^{-3}. Therefore, they fail to meet the andic soil property criteria. Hence, they are classified as Andic Dystrochrepts. The occurrence of 2 : 1 clay minerals in such soils can consume a considerable amount of active Al to form hydroxy-Al interlayers, resulting in a weakening of andic soil properties. This process can be called an "anti-andic reaction".

3.3.3. Alfisols and Ultisols

Clay illuviation does not occur in soils containing large amounts of noncrystalline materials because such materials are difficult to disperse under field conditions. Therefore, tephra-derived Alfisols and Ultisols will be formed only after transformation of noncrystalline materials to layer lattice clays such as kaolins. Intense formation of kaolins has been observed for Palehumults in Costa Rica (Martini, 1976), for Hapludults in the Philippines (Otsuka et al., 1988) and for Udalfs and Udults in Cameroon (Delvaux et al., 1989).

3.3.4. Mollisols and Vertisols

Mollisols and Vertisols are formed in tephras under ustic or udic marginal to ustic soil moisture regimes and isothermic–isohyperthermic soil temperature regimes. Volcanic ash in El Salvador has weathered to smectite, vermiculite, kaolinite and iron oxides and the soils formed were classified as Mollisols and Vertisols (Yerima et al., 1987). Peralkaline volcanic ash in Kenya has weathered to Mollisols, Vertisols, and Oxisols (Wielemaker and Wakatsuki, 1984; Wakatsuki, 1990).

Transition of Andisols to Mollisols (Plate 12) has been observed in Abashiri, Hokkaido, north Japan where mean annual temperature is 5–6°C and mean annual precipitation is about 800 mm (Shoji et al., 1990a). They have developed in old tephras (12,000–32,000 years old) dominated by noncolored glass and display transformation of allophane-imogolite to halloysite resulting in a notable decrease of acid oxalate extractable Al (<0.6%), bulk density > 0.9 g cm^{-3}, and P retention < 60%. Halloysite formation was also observed for Andic Hapludolls formed from basic volcanic ash in the northwestern highlands of Rwanda where soil temperature is isothermic and the soil moisture regime is udic (Mizota and Chapelle, 1988).

3.3.5. Oxisols

Tephra-derived Oxisols are formed under udic/perudic moisture regimes and an isohyperthermic temperature regime. These soils accumulate a large amount of noncrystalline materials, but they tend to be dominated by low-activity clays such as oxyhydroxides of aluminum or ferric iron. Such Oxisols have bulk densities of 1.1–1.2 g cm^{-3} as indicated for pedons from Western Samoa and Chile (Kimble and Eswaran, 1988).

REFERENCES

Amundson, R. and Jenny, H., 1991. The place of humans in the state factor theory of ecosystems and their soils. Soil Sci., 151: 99–109.

Andisol TU Database, 1992. Database on Andisols from Japan, Alaska, and northwestern U.S.A. prepared by Soil Science Laboratory, Tohoku University (see Appendix).

Antos, A.J. and Zobel, D.B., 1985. Plant form, developmental plasticity, and survival following burial by volcanic ejecta. Can. J. Bot., 63: 2083–2090.

Antos, A. J. and Zobel, D.B., 1986. Seedling establishment in forests affected by tephra from Mount St. Helens. Am. J. Bot., 73: 495–499.

Delvaux, B., Herbillon, A.J. and Vielvoye, L., 1989. Characterization of a weathering sequence of soils derived from volcanic ash in Cameroon-taxonomic, mineralogical and agronomic implications. Geoderma, 45: 375–388.

Duchaufour, P., 1984. Pedologie, Masson, Paris.

FAO/Unesco, 1964. Meeting on the Classification and Correlation of Soils from Volcanic Ash. Tokyo, Japan, 11–27 June, 1964. World Soil Resources Reports.

Flach, K.W., Holzhey, C.S., DeConinck, F. and Bartlett, R.J., 1980. Genesis and Classification of Andepts and Spodosols. In: B.K.G. Theng (Editor), Soils with Variable Charge. New Zealand Society of Soil Science, Lower Hutt, pp. 411–426.

Garcia-Rodeja, E., Silva, B.M. and Macias, F., 1987. Andosols developed from non-volcanic materials in Galicia, NW Spain. J. Soil Sci., 38: 573–591.

Gibbs, H.S., Cowie, J.D. and Pullar, W.A., 1968. Soils of North Island. In: J. Luke (Editor), Soils of New Zealand, part 1. New Zealand Soil Bureau Bulletin 26 (1), pp. 48–67.

Haruki, M., 1988. Destruction and recovery of flora and fauna. In: H. Kadomura, H. Okada and T. Araya (Editors), 1977–82 volcanism and environmental hazards of Usu volcano. Hokkaido Univ. Press, Japan, pp. 165–193 (in Japanese).

Hewitt, A.E. and Witty, J.E., 1988. Andisol transitions with other soils, particularly the Inceptisols. In: D.I. Kinloch, S. Shoji, F.H. Beinroth and H. Eswaran (Editors), Proc. 9th Int. Soil Classification Workshop, Japan, 20 July to 1 August, 1987. Publ. by Jap. Committee for 9th Int. Soil Classification Workshop, for the Soil Management Support Services, Washington D.C., U.S.A., pp. 233–244.

Hunter, C.R., Frazier, B.E. and Busacca, A.J., 1987. Lytell series: a nonvolcanic Andisol. Soil Sci. Soc. Am. J., 51: 376–383.

Inoue, K. and Naruse, T., 1990. Asian long-range eolian dust deposited on soils and paleosols along the Japan Sea coast. Quat. Res., 29: 209–222 (in Japanese, with English summary).

Ito, K. and Haruki, M., 1984. Revegetation surveys of Mt. Usu during six years after the eruption in 1977–1978. In: H. Kadomura (Editor), A Follow-up Survey of Environmental Changes Caused by the 1977–1978 Eruptions of Mt. Usu and Related Events. Graduate School of Environ. Sci., Hokkaido Univ., Sapporo, Japan, pp. 61–84 (in Japanese, with English abstract).

Jenny, H., 1941. Factors of Soil Formation. McGraw-Hill Book Company, Inc., New York and London.

Kanno, I., 1961. Genesis and classification of main genetic soil types in Japan, 1. Introduction and Humic Allophane soils. Bull. Kyushu Agr. Exp. Sta., 7: 13–185 (in Japanese, with English abstract).

Katayama, M., Sasaki, T., Tomioka, E. and Amano, Y., 1983. Soil Survey of Kushiro Subprefecture. Hokkaido Nat. Agr. Exp. Sta., 28 (in Japanese, with English abstract).

Kato, Y., 1983. Formation mechanism of volcanic ash soils. In: N. Yoshinaga (Editor), Volcanic Ash Soils. Hakuyusha, Tokyo, pp. 5–30 (in Japanese).

Kato, Y., Hamada, R. and Sakagami, K., 1988. Genesis of Andisols in Japan. In: D.I. Kinloch, S. Shoji, F.H. Beinroth and H. Eswaran (Editors), Proc. 9th Int. Soil Classification Workshop, Japan, 20 July to 1 August, 1987. Publ. by Jap. Committee for 9th Int. Soil Classification Workshop, for the Soil Management Support Services, Washington D.C., U.S.A. pp. 376–388.

Kawamuro, K. and Osumi, Y., 1988. Melanic and fulvic Andisols in central Japan, Soil pollen and plant opal analyses. In: D.I. Kinloch, S. Shoji, F.H. Beinroth and H. Eswaran (Editors), Proc. 9th Int. Soil

Classification Workshop, Japan, 20 July to 1 August, 1987. Publ. by Jap. Committee for 9th Int. Soil Classification Workshop, for the Soil Management Support Services, Washington D.C., U.S.A., pp. 389–401.

Kawamuro, K. and Torii, A., 1986. Difference in past vegetation between Black soils and Brown Forest soils derived from volcanic ash at Mt. Kurohime, Nagano Pref., Japan. Quat. Res., 25: 81–98 (in Japanese, with English abstract).

Kikuchi, K., 1981. Interpretative classification of the soils in Tokachi District, its mapping and application to practical uses of soil improvement. Rep. Hokkaido Pref. Agr. Exp. Stn., 34: 1–118 (in Japanese, with English abstract).

Kimble J.M. and Eswaran, H., 1988. The Andisol-Oxisol transition. In: D.I. Kinloch, S. Shoji, F.H. Beinroth and H. Eswaran (Editors), Proc. 9th Int. Soil Classification Workshop, Japan, 20 July to 1 August, 1987. Publ. by Jap. Committee for 9th Int. Soil Classification Workshop, for the Soil Management Support Services, Washington D.C., U.S.A., pp. 219–232.

Kirkman, J.H., 1980. Clay mineralogy of a sequence of andesitic tephra beds of western Taranaki, New Zealand. Clay Minerals, 15: 157–163.

Kirkman, J.H. and McHardy, W.J., 1980. A comparative study of the morphology, chemical composition and weathering of rhyolitic and andesitic glass. Clay Minerals, 15: 165–173.

Kondo, R. and Sase, T., 1986. Opal phytoliths, their nature and application. Quat. Res., 25: 31–64 (in Japanese).

Kondo, R., Sase, T. and Kato, Y., 1988. Opal phytolith analysis of Andisols with regard to interpretation of paleovegetation. In: D.I. Kinloch, S. Shoji, F.H. Beinroth and H. Eswaran (Editors), Proc. 9th Int. Soil Classification Workshop, Japan, 20 July to 1 August, 1987. Publ. by Jap. Committee for 9th Int. Soil Classification Workshop, for the Soil Management Support Services, Washington D.C., U.S.A., pp. 520–534.

Kumada, K., Ohta, S., Arai, S., Kitamura, M. and Imai, K., 1985. Changes in soil organic matter and nitrogen fertility during the slash-and-burn and cropping phases in experimental shifting cultivation. Soil Sci. Plant Nutr., 31: 611–623.

Kumada, K., Sato, O., Ohsumi, Y. and Ohta, S., 1967. Humus composition of mountain soils in central Japan with special reference to the distribution of P type humic acid. Soil Sci. Plant. Nutr., 13: 151–158.

Leamy, M.L., Smith, G.D., Colmet-Daage, F. and Otowa, M., 1980. The morphological characteristics of Andisols. In: B.K.G. Theng (Editor), Soils with Variable Charge. New Zealand Soc. Soil Sci., Lower Hutt, New Zealand, pp. 17–34.

Lowe, D.J., 1986. Controls on the rates of weathering and clay mineral genesis in airfall tephras. In: S.M. Colman and D.P. Dethier (Editors), Rates of Chemical Weathering of Rocks and Minerals. Academic Press, Orlando, pp. 265–330.

Martini, J.A., 1976. The evolution of soil properties as it relates to the genesis of volcanic ash soils in Costa Rica. Soil Sci Soc. Am. J., 40: 895–900.

Matsui, T., 1988. Introductory Soil Geography. Tsukiji Shoin, Tokyo (in Japanese).

Midorikawa, B., 1978. Method to determine primary production, In: M. Numata (Editor), Handbook of Grassland Investigation. Univ. of Tokyo Press, Tokyo, pp. 61–77 (in Japanese).

Mizota, C. and Chapelle, J., 1988. Characteristics of some Andepts and andic soils in Rwanda, Central Africa. Geoderma, 41: 193–209.

Mohr, E.C.J., Van Baren, F.A. and Van Schuylenborgh, J., 1972. Andisols. In: Tropical Soils. Mouton-Ichtiar Baru-Van Hoeve, Hague-Paris-Djakarta, pp. 397–418.

Oba, Y. and Nagatsuka, S., 1988. Soil Genesis and Classification. Yokendo, Tokyo (in Japanese).

Ohmori, H. and Yanagimachi, O., 1988. Thermal conditions both of the upper and lower limits of the *Fagus crenata* forest zone, and changes in summer temperature from the latest Pleistocene to the middle Holocene in Japan. Quat. Res., 27: 81–100 (in Japanese, with English summary).

Oike, S., 1972. Holocene tephrochronology in the eastern foothills of the Towada volcano, northeastern Honshu, Japan. Quat. Res., 11: 228–235 (in Japanese, with English abstract).

Ono, T. and Shoji, S., 1978. Genesis of Andosols at Kitakami, Iwate Prefecture, northeast Japan, 2. Relationships between parent materials and soil formation. Quat. Res., 17: 15–23 (in Japanese, with English abstract).

Ono, T., Kobayashi, S. and Shoji, S., 1981. Relationships between land use and properties of Andosols on Geto alluvial fan, Iwate Prefecture. J. Sci. Soil Manure, Jpn. 52: 87–98 (in Japanese).

Ossaka, J., 1982. Activity of volcanoes and clay minerals. Nendo Kagaku (Clay Science), 22: 127–137 (in Japanese, with English abstract).

Otsuka, H., Briones, A.A., Daquiado, N.P. and Evangelio, F.A., 1988. Characteristics and genesis of volcanic ash soils in the Philippines. Tech. Bull. Tropical Agr. Res. Center, Japan, 24: 1–122.

Parfitt, R.L., 1990. Soils formed in tephra in different climatic regions. Trans. International Congress of Soil Sci., 14th, 1990, VII: 134–139.

Parfitt, R.L. and Kimble, J.M., 1989. Conditions for formation of allophane in soils. Soil Sci. Soc. Am. J., 53: 971–977.

Parfitt, R.L. and Saigusa, M., 1985. Allophane and humus-aluminum in Spodosols and Andepts formed from the same volcanic ash beds in New Zealand. Soil Sci., 139: 149–155.

Parfitt, R.L., Russell, M. and Orbell, G.E., 1983. Weathering sequence of soils from volcanic ash involving allophane and halloysite, New Zealand. Geoderma, 29: 41–57.

Ping, C.L., Shoji, S. and Ito, T., 1988. Properties and classification of three volcanic ash-derived pedons from Alueutian Islands and Alaska Peninsula, Alaska. Soil Sci. Soc., Am. J., 52: 455–462.

Ping, C.L., Shoji, S., Ito, T., Takahashi, T. and Moore, J.P., 1989. Characteristics and classification of volcanic-ash-derived soils in Alaska. Soil Sci., 148: 8–28.

Saigusa, M. and Shoji, S., 1986. Surface weathering in Zao tephra dominated by mafic glass. Soil Sci. Plant Nutr., 32: 617–628.

Saigusa, M., Shoji, S. and Kato, T., 1978. Origin and nature of halloysite in Ando soils from Towada tephra, Japan. Geoderma, 20: 115–129.

Saigusa, M., Shoji, S. and Otowa, M., 1991. Clay mineralogy of two Andisols showing a hydrosequence and its relationships to their physical and chemical properties. Pedologist, 35: 21–33 (in Japanese, with English abstract).

Sase, T., 1986a. Plant opal analysis of Andisols in North Island, New Zealand. Pedologist, 30: 2–12 (in Japanese, with English abstract).

Sase, T., 1986b. Opal phytolith analysis of volcanic ash soils from recent Towada ashes. Pedologist, 30: 102–114 (in Japanese, with English abstract).

Sase, T. and Kato, Y., 1976a. The study on phytogenic particles, especially on the plant opals in humic horizons of present and buried volcanic ash soils, 1. The problem on the source of plant opals. Quat. Res., 15: 21–34 (in Japanese, with English abstract).

Sase, T. and Kato, Y., 1976b. The study on phytogenic particles, especially on the plant opals in humic horizons of present and buried volcanic ash soils, 2. The problem on the origin of humus in volcanic ash soils and the assumption of paleoclimate by plant opals. Quat. Res., 15: 66–74 (in Japanese, with English abstract).

Sase, T., Hosono, M., Utsugawa, T. and Aoki, K., 1988. Opal phytolith analysis of present and buried volcanic ash soils at Te Ngae Road tephra section, Rotorua Basin, North Island, New Zealand. Quat. Res., 27: 153–163 (in Japanese, with English abstract).

Sase, T., Kondo, R. and Inoue, K., 1990. Vegetational environments of volcanic ash soils developed at the foot of Mt. Iwate during the last 13,000 yrs., with special reference to opal phytolith analysis. Pedologist, 34: 15–30 (in Japanese, with English abstract).

Shinagawa, A., 1962. Further accumulation of humus on the volcanic ash soils originated from volcano Sakurajima's ashes. Bull. Fac. Agr. Kagoshima Univ., 11: 155–205 (in Japanese, with English abstract).

Shindo, H., Higashi, T. and Matsui, Y., 1986. Comparison of humic acid from charred residues of Susuki (Eulalia, *Miscanthus sinensis* A.) and from the A horizon of volcanic ash soils. Soil Sci. Plant Nutr., 32: 579–586.

Shoji, S., 1983. Mineralogical properties of volcanic ash soils. In: N. Yoshinaga (Editor), Volcanic Ash Soils. Hakuyusha, Tokyo, pp. 31–72 (in Japanese).

Shoji, S., 1986. Mineralogical characteristics, 1. Primary minerals. In: K. Wada (Editor), Ando Soils in Japan. Kyushu University Press, Fukuoka, Japan, pp. 21–40.

Shoji, S. and Fujiwara, T., 1984. Active aluminum and iron in the humus horizons of Andosols from northeastern Japan: Their forms, properties, and significance in clay weathering. Soil Sci., 137: 216–226.

Shoji, S. and Ito, T., 1990. Classification of tephra-derived Spodosols. Soil Sci., 150: 799–815.

Shoji, S. and Ono, T., 1978a. Physical and chemical properties and clay mineralogy of Andosols from Kitakami, Japan. Soil Sci., 126: 297–312.

Shoji, S., and Ono, T., 1978b. Genesis of Andosols at Kitakami, Iwate Prefecture, northeast Japan, 1. Relationships between topography and soil formation. Quat. Res., 16: 247–254 (in Japanese, with English abstract).

Shoji, S. and Saigusa, M., 1977. Amorphous clay materials of Towada Ando soils. Soil Sci. Plant Nutr., 23: 437–455.

Shoji, S. and Yamada, H., 1991. Comparisons of mineralogical properties between tephra-derived Spodosols from Alaska and nontephra-derived Spodosols from New England. Soil Sci., 152: 162–183.

Shoji, S., Kobayashi, S. and Masui, J., 1974. Chemical composition of ferromagnetic minerals in volcanic ash layers with special reference to their origins. J. Japan. Assoc. Min. Petr. Econ. Geol., 69: 110–120.

Shoji, S., Fujiwara, Y., Yamada, I. and Saigusa, M., 1982. Chemistry and clay mineralogy of Ando soils, Brown forest soils, and Podzolic soils formed from recent Towada ashes, northeastern Japan. Soil Sci., 133: 69–86.

Shoji, S., Ito, T., Nakamura, S. and Saigusa, M., 1987. Properties of humus of Andosols from New Zealand, Chili, and Ecuador. Jap. J. Soil Sci. Plant. Nutr., 58: 473–479 (in Japanese).

Shoji, S., Takahashi, T., Ito, T. and Ping, C.L., 1988a. Properties and classification of selected volcanic ash soils from Kenai Peninsula, Alaska. Soil Sci., 145: 395–413.

Shoji, S., Takahashi, T., Saigusa, M., Yamada, I. and Ugolini, F.C., 1988b. Properties of Spodosols and Andisols showing climosequential and biosequential relations in southern Hakkoda, northeastern Japan. Soil Sci., 145: 135–150.

Shoji, S., Hakamada, T. and Tomioka, E., 1990a. Properties and classification of selected volcanic ash soils from Abashiri, northern Japan-Transition of Andisols to Mollisols. Soil Sci. Plant Nutr., 36: 409–423.

Shoji, S., Kurebayashi, T. and Yamada, I., 1990b. Growth and chemical composition of Japanese pampas grass (*Miscanthus sinensis*) with special reference to the formation of dark-colored Andisols in northeastern Japan. Soil Sci. Plant Nutr., 36: 105–120.

Shoji, S., Nanzyo, M., Shirato, Y. and Ito, T., 1993. Chemical kinetics of weathering in young Andisols from northeastern Japan using soil age normalized to $10°C$. Soil Sci., 155: 53–60.

Simonson, R.W. and Rieger, S., 1967. Soils of the Andept suborder in Alaska. Proc. Soil Sci. Soc. Am., 31: 692–699.

Smeck, N.E., Runge, E.C.A. and Mackintosh, E.E., 1983. Dynamics and Genetic Modelling of Soil Systems. In: L.P. Wilding, N.E. Smeck and G.F. Hall (Editors), Pedogenesis and Soil Taxonomy, 1. Concepts and Interactions. Elsevier, Amsterdam-Oxford-New York, Developments in Soil Science 11A, pp. 51–82.

Soil Survey Staff, 1990. Keys to Soil Taxonomy, 4th edition. AID, USDA-SMSS, Technical Monograph, No. 19, Blacksburg, Virginia.

Soil Survey Staff, 1992. Keys to Soil Taxonomy, 5th edition. AID, USDA-SMSS Technical Monograph No. 19, Blacksburg, Virginia.

Takahashi, T., Shoji, S. and Sato, A., 1989. Clayey Spodosols and Andisols showing a biosequential relation from Shimokita Peninsula, northeastern Japan. Soil Sci., 148: 204–218.

Takehara, H., 1964. Classification. In: Volcanic Ash Soils in Japan. Ministry of Agriculture and Forestry, Japanese Government, pp. 135–145.

Tan, K.H., 1984. Andosols. Van Nostrand Reinhold Soil Science Series. Van Nostrand Reinhold, New York, 418 pp.

Tate, K.R., Yamamoto, K., Churchman, G.J., Meinhold, R. and Newman, R.H., 1990. Relationships between the type and carbon chemistry of humic acids from some New Zealand and Japanese soils. Soil Sci. Plant Nutr., 36: 611–621.

Tomioka, H., Ishizuka, M. and Kanazawa, Y., 1990. Forest succession after the eruption of Mt. Usu. IV. Ten-year changes in forest floor vegetation following the eruption. Trans. of Meeting in Hokkaido Branch, Jap. Forestry Soc., 38: 97–100 (in Japanese).

Ugolini, F.C. and Dahlgren, R.A., 1991. Weathering environment and occurrence of imogolite/allophane in selected Andisols and Spodosols. Soil Sci. Soc. Am. J., 55: 1166–1171.

Ugolini, F.C., Minden, R., Dawson, H. and Zachara, J., 1977. An example of soil processes in the *Abies amabilis* zone of central Cascades, Washington. Soil Sci., 124: 291–302.

Ugolini, F.C., Dahlgren, R., Shoji, S. and Ito, T., 1988. An example of andosolization and podzolization as revealed by soil solution studies, Southern Hakkoda, northeastern Japan. Soil Sci., 145: 111–125.

Van Wambeke, A., 1992. Andisols. In: Soils of the Tropics. McGraw-Hill, Inc. New York et al., pp. 207–232.

Wada, K., 1985. The distinctive properties of Andosols. In: B.A. Stewart (Editor), Advances in Soil Sci., 2. Springer-Verlage, New York, Berlin, Heidelberg, Tokyo, pp. 173–229.

Wada, K. and Aomine, S., 1973. Soil development on volcanic materials during Quaternary. Soil Sci., 116: 170–177.

Wakatsuki, T., 1990. Distribution and properties of volcanic ash soils. In: Soil Resources of Tropical Africa. Association for International Cooperation of Agriculture and Forestry, Tokyo, No. 13, pp. 114–123 (in Japanese).

Wielemaker, W.G. and Wakatsuki, T., 1984. Properties, weathering and classification of some soils formed in peralkaline volcanic ash in Kenya. Geoderma, 32: 21–44.

Yamada, Y., 1986. The characterization of humus accumulation in Andosols by ^{14}C dating. Bull. Nat. Inst. Agro.-Env. Sci., 3: 23–86 (in Japanese, with English abstract).

Yamamoto, K., Tate, K.R. and Churchman, G.J., 1989. A comparison of the humic substances from some volcanic ash soils in New Zealand and Japan. Soil Sci. Plant Nutr., 35: 257–270.

Yasuda, Y., 1987. The cold climate of the last glacial age in Japan, A comparison with southern Europe. Quat. Res., 25: 277–294 (in Japanese, with English abstract).

Yerima, B.P.K., Wilding, L.P., Calhoun,F.G. and Hallmark, C.T., 1987. Volcanic ash-influenced Vertisols and associated Mollisols of El Salvador — Physical, chemical, and morphological properties. Soil Sci. Soc. Am. J., 51: 699–708.

Yoshioka, K., 1966. Development and recovery of vegetation since the 1929 eruption of Mt. Komagatake. Hokkaido, 1. Akaikawa pumice flow. Ecol. Rev., 16: 271–292.

Chapter 4

CLASSIFICATION OF VOLCANIC ASH SOILS

S. SHOJI, R. DAHLGREN and M. NANZYO

4.1. INTRODUCTION

In concept, an international system of soil classification should provide an integrated knowledge of the soils of the world, and contribute to an inventory of the world's soil resources. Furthermore, it should afford a scientific basis for, and hasten the agrotechnology transfer between soils with similar properties and physical environmental factors. These conceptual constructs were achieved in the recently completed international classification of volcanic ash soils which resulted in creation of the Andisol order in Soil Taxonomy (Soil Survey Staff, 1992).

The concept of agrotechnology transfer, as related to soils in general, could be effectively applied in developing a global strategy for sustainable use of our soil resource and preservation of environmental quality. Recently, environmental studies have shown that chemicals, such as fertilizers applied to soils may contribute to environmental degradation to a significant extent. From the point of view of the soil scientist, different soils have a widely varying susceptibility to environmental contamination due to the vast range of physical, chemical and mineralogical properties associated with soils on a global scale. Thus, the classification of soils provides a resource-based tool for evaluating and constructing solutions to these problems.

The genesis, unique properties, and factors relating to the productivity of volcanic ash soils were recognized in Japan and New Zealand several decades ago. However, these soils did not receive world-wide recognition among soil scientists until the middle of this century. It was only in 1960 that "volcanic ash soils" were identified for the first time in an international system of soil classification which was proposed by the Soil Survey Staff (1960).

The demand for food production to feed the rapidly increasing world population after World War II, especially in the 1960's to 1970's, necessitated the elaboration of an effective strategy for agrotechnology transfer. This promoted the development of an international classification for volcanic ash soils because their distribution was extensive in many densely populated regions of the developing tropics.

Volcanic ash soils were first included as the Andept suborder of Inceptisols in the Seventh Approximation (Soil Survey Staff, 1960) and this suborder was finally introduced into Soil Taxonomy (Soil Survey Staff, 1975). During this period, the

FAO/Unesco held the 1964 meeting on the classification of volcanic ash soils in Japan (FAO/Unesco, 1964) and created the major soil grouping Andosols for the Soil Map of the World (FAO/Unesco, 1974).

In 1978 Smith proposed the reclassification of Andepts and provided a rationale for development of the new soil order Andisols. His proposal led to the development and establishment of the current international classification of Andisols as described later in this chapter (Soil Survey Staff, 1990, 1992).

Several countries such as Japan and New Zealand have developed their own national classification systems for volcanic ash soils. Each national system organizes knowledge of the distribution, properties, productivity, utilization, and management of soils in its respective jurisdiction. However, the national systems are not in competition with the international systems. Rather, they are complementary and provide the basic information and data needed for the development of the more comprehensive international systems.

This chapter first describes the classification of Andisols in Soil Taxonomy and then briefly outlines the national systems for classification of volcanic ash soils established independently in Japan and New Zealand.

4.2. CLASSIFICATION OF ANDISOLS

The current classification of Andisols introduced into the Keys to Soil Taxonomy (Soil Survey Staff, 1990, 1992) is the most widely used system for the international classification of volcanic ash soils. The development of this classification had its origin in the Andisol proposal put forth by Smith in 1978 and was further developed through the collective experience and efforts of members of the International Committee on the Classification of Andisols (ICOMAND) chaired by M.L. Leamy for ten years from 1978 to 1988. This section summarizes the development and testing of the concept and classification of Andisols from Smith's original proposal (Smith, 1978) to the current Keys to Soil Taxonomy (Soil Survey Staff, 1990, 1992).

4.2.1. The Andisol proposal by G.D. Smith (1978)

The classification of Andepts in Soil Taxonomy (Soil Survey Staff, 1975) was based almost exclusively on data from volcanic ash soils in Hawaii, the western United States, and Alaska. Realizing the limitations of such a narrowly defined classification system for international purposes, Smith (1978) proposed that the Andept suborder be re-evaluated to make it internationally applicable. Smith's proposal has come to be widely known as the Andisol proposal. In this proposal, he pointed out many serious defects in the classification of Andepts as established in 1975 Soil Taxonomy and developed the rationale for the establishment of a new order called Andisols. The proposal outlined several important criteria necessary for the reclassification of volcanic ash soils that are the basis of the current classification of Andisols (Soil Survey Staff, 1992).

(1) Limitations in the classification of Andepts and creation of the Andisol order

Smith (1978) showed that the classification of Andepts presented in Soil Taxonomy (Soil Survey Staff, 1975) had a number of serious limitations:

(1) The definition of the suborder clearly excludes a number of soils that should be included if we consider all of their properties,

(2) Base saturation determined by $NH_4 OAc$ (pH = 7) has been used as a differentia with a limit of 50%, the same limit used for mineral soils that have crystalline clays. The significance of this limit is open to serious questioning because the clays of many Andisols are dominated by amorphous materials displaying pH dependent CEC,

(3) Thixotropy has been used as a differentia, but the decision that a given horizon is or is not thixotropic is very subjective and can not be made uniformly,

(4) The soil moisture regime has not been used as a differentia for Andepts as it has for all other soil orders. Interpretations for a given family cannot be made without the use of climatic regimes,

(5) The darkness of the epipedon is weighted heavily in subgroup definitions, but in warm intertropical areas there seems to be little or no relation between color and carbon content, degree of weathering, or any other property,

(6) For most mineral soils, a fragmental particle-size class is provided, but not for Andepts, and

(7) Inadequate emphasis was given to the unique moisture retention properties of the Andepts.

In order to correct these defects or limitations, Smith proposed that the reclassification of Andepts should consider the following:

(1) To elevate the suborder of Andepts to an order creating a new Andisol order,

(2) To introduce soil moisture and temperature regimes to define suborders of Andisols,

(3) To divide some great groups of Andepts among the suborders, and

(4) To propose definitions of typic subgroups for the great groups.

He proposed the name "Andisols" for the new soil order rather than "Andosols". The rationale was that "Andosol" was used at that time in other classification systems (for example, Andosols in FAO/Unesco Soil Map of the World) and that the connecting vowel *o* was supposed to be restricted to Greek formative elements.

(2) Central concept and properties of Andisols in the Andisol proposal

According to the Andisol proposal, the central concept of Andisols is that of a soil developing in volcanic ash, pumice, cinders, and other volcanic ejecta or in volcaniclastic materials. These soils have an exchange complex that is dominated by X-ray amorphous compounds of Al, Si, and humus, or a matrix dominated by glass, and having one or more diagnostic horizons other than an ochric epipedon.

This central concept consists of two important items: the parent materials are of volcanic origin and the soils are dominated by noncrystalline materials.

However, it does not include the presence of large amounts of dark-colored organic matter that is an essential characteristic for the central concept of Kurobokudo (Andosols) in Japan (see 4.3.2). It also makes no mention of Al/Fe humus complexes that are the major form of active Al and Fe in nonallophanic Andisols (Shoji et al., 1985).

Smith (1978) summarized the important properties of Andisols as follows: abundance of amorphous clays with a low permanent charge and a high pH dependent charge, virtual absence of aluminum toxicity, high phosphate fixation and water retention, high organic carbon content; low bulk density, unique engineering properties, etc. The need to amend these criteria became apparent quickly with reports that aluminum toxicity is rare for allophanic Andisols but common for nonallophanic Andisols with low base saturation (Saigusa et al., 1980; Shoji et al., 1980). These findings lead to the revision of Smith's original proposal and to the development of the existing concept. On the basis of the central concept and important properties, Smith proposed a strict definition for Andisols. This definition significantly contributed to the current definition of Andisols as described later (Soil Survey Staff, 1990, 1992).

(3) Suborders and great groups in the Andisol proposal

As described in Table 4.1, Smith proposed six suborders for Andisols that were based on the moisture and temperature regimes similar to the suborders of Alfisols and several other orders. In addition, a suborder of Tropands was included to parallel Tropepts in the Inceptisol order.

Seven great groups were identified for the six suborders. They were cryic (cold), placic (placic horizon), duric (duripan), hydric (high 1500 kPa water), melanic (very dark and organic rich), vitric (glassy), and haplic (minimum development).

When compared to the current classification of Andisols (Soil Survey Staff, 1990, 1992), it appears that considerable modifications of Smith's original proposal were incorporated following international testing of the classification criteria by the ICO-MAND (Leamy et al., 1988). Revisions to the original proposal include deletion of Tropands, changing Borands to Cryands, and addition of Torrands and Vitrands at the suborder level (see Table 4.4). However, the Andisol proposal has been the basis for the development of the present classification of the Andisol order.

(4) Subgroups and families in the Andisol proposal

Many new subgroups were introduced in the Andisol proposal. The total number of subgroups for the seven great groups amounted to 94. Each subgroup was only included after careful testing by the ICOMAND.

Since it is often difficult to determine particle-size classes used to distinguish families of Andisols, a combination of particle-size and mineralogy classes was proposed in the Andisol proposal. Three categories for the nonfragmental class consisting of ashy, medial, and hydrous were established. These categories are determined from 1500 kPa water retention. Two substitutes for the fragmental class were established and consist of the pumiceous and cindery categories.

TABLE 4.1

Suborders and great groups in the Andisol proposal (Smith, 1978)

Suborder	Great group
Aquands: artificial drainage or an aquic moisture regime	Vitraquands
	Haplaquands
Borands: a frigid or cryic temperature regime	Melanoborands
	Cryoborands
	Placoborands
	Vitriborands
	Haploborands
Xerands: a xeric moisture regime	Durixerands
	Vitrixerands
	Haploxerands
Ustands: an ustic moisture regime or a duripan, or both	Durustands
	Vitrustands
	Haplustands
Tropands: an isomesic or warmer iso-temperature regime or a hyperthermic temperature regime	Plactropands
	Hydrotropands
	Vitritropands
	Haplotropands
Udands: a udic or perudic moisture regime	Placudands
	Hydrudands
	Melanudands
	Vitrudands
	Hapludands

4.2.2. The current classification of Andisols

The classification of Andisols in the 1990 Keys to Soil Taxonomy (Soil Survey Staff, 1990) was developed on the basis of the final report of the ICOMAND (Leamy et al., 1988). However, since the adopted classification system had some remaining problems as pointed out later, it was further revised in the 1992 Keys to Soil Taxonomy (Soil Survey Staff, 1992) according to the proposals of the International Committee on the Classification of Spodosols (ICOMOD) and the International Committee on the Classification of Aquic Soils (ICOMAQ).

(1) Central concept of Andisols and andic soil properties

The central concept of Andisols described by Smith (1978), was revised according to recent advances in mineralogy and chemistry of Andisols. The important unique properties of Andepts or Andisols used to be ascribed to allophane (1975 Soil Taxonomy) or to X-ray amorphous compounds of Al, Si, and humus (the Andisol proposal). However, the discovery of nonallophanic Andisols by Shoji and Ono (1978) indicated that Al/Fe humus complexes also substantially contribute to many of the unique properties of nonallophanic Andisols that are also common to allophanic Andisols (Shoji et al., 1985). The parent materials of such Andisols from northeastern Japan are mixtures of tephra and eolian dusts from the Asian continent (Ito, T. and Shoji, S., 1992, unpublished). In addition, noncrystalline aluminum and iron hydroxides also are conspicuously related to the unique properties of some Andisols, such as Hydrudands in Hawaii (Wada and Wada, 1976; Parfitt et al., 1988). Thus, it is currently recognized that important noncrystalline materials or short-range-order minerals of Andisols include not only allophane, but also imogolite, ferrihydrite, and Al/Fe-humus complexes.

Considering these findings, ICOMAND (Leamy, 1988) established that the central concept of an Andisol is that of a soil developing in volcanic ejecta (such as volcanic ash, pumice, cinders, lava, etc.), and/or in volcaniclastic material, whose colloidal fraction is dominated by noncrystalline materials, and/or Al-humus complexes. ICOMAND also recognized that under some environmental conditions, weathering of primary aluminosilicates in parent materials of nonvolcanic origin may lead to the formation of noncrystalline materials; some of these soils are also included in Andisols. For example, Garcia-Rodeja et al. (1987) described Andisols from Spain that formed from gabbros, amphibolites, and schists. Many of their properties were similar to those of nonallophanic Andisols from Japan (Shoji et al., 1985).

The original definition of andic soil properties was first documented by ICO-MAND (Leamy, 1984) following the Sixth International Soil Classification Workshop on the classification and management of Andisols held in Chile and Ecuador in 1984. The definition was finally established following testing by ICOMAND (Leamy, 1988) and was introduced into the Keys to Soil Taxonomy (Soil Survey Staff, 1990, 1992) as presented in Table 4.2.

Criteria 1 in the definition of andic soil properties is used to classify Andisols that show a high degree of weathering as determined by ammonium oxalate extractable Al plus a half Fe (Al_o + $1/2 Fe_o$). Such soils typically have a high phosphate retention and low bulk density.

Common tephras having rhyolite, dacite and andesite composition are dominated by noncolored volcanic glass. The mean total content of Al + 1/2 Fe in noncolored volcanic glass is 8 percent (Shoji, 1986). Therefore, to meet the andic soil property criteria of Al_o + $1/2 Fe_o$ > 2 percent, 25 g of noncolored volcanic glass per 100 g soil must completely weather to form noncrystalline materials of Al and Fe, assuming the origin of all ammonium oxalate extractable Al and Fe is volcanic glass.

Criteria 2 is applied to classify Andisols which are vitric or have developed from mixed parent materials of tephra and nontephra deposits. Soils with mixed parent materials typically have high bulk densities (>0.9 g cm^{-3}) so that they fail to meet andic soil properties as defined by criteria 1. Therefore, the bulk density requirement is waived in criteria 2 and the phosphate retention >25 percent requirement is included to separate Andisols from Entisols (Shoji et al., 1987b). Criteria 2a is further used to define andic soil properties in soils showing a low degree of weathering or vitric properties. It has two requirements consisting of minimum concentrations for glass content and Al_o + $1/2 Fe_o$.

TABLE 4.2

Definition of andic soil properties in the 1992 Keys to Soil Taxonomy (Soil Survey Staff, 1992)

To be recognized as having andic soil properties, soil materials must contain less than 25 percent (by weight) organic carbon and meet one or both of the following requirements:

1. In the fine-earth fraction, all of the following:

   a. Aluminum plus 1/2 iron percentages (by ammonium oxalate) totaling 2.0 percent or more, and
   b. A bulk density, measured at 33 kPa water retention, of 0.90 g/cm^3 or less, and
   c. A phosphate retention of 85 percent or more; or

2. In the fine-earth fraction, a phosphate retention of 25 percent or more, 30 percent or more particles of 0.02 to 2.0 mm, and one of the following:

   a. Aluminum plus 1/2 iron percentages (by ammonium oxalate) totaling 0.40 or more and, in the 0.02-to-2.0-mm fraction, 30 percent or more volcanic glass; or
   b. Aluminum plus 1/2 iron percentages (by ammonium oxalate) totaling 2.0 or more and, in the 0.02-to-2.0-mm fraction, 5 percent or more volcanic glass; or
   c. Aluminum plus 1/2 iron percentages (by ammonium oxalate) totaling between 0.40 and 2.0 and, in the 0.02-to-2.0-mm fraction, enough volcanic glass so that the glass percentage, when plotted against the value obtained by adding aluminum plus 1/2 iron percentages in the fine-earth fraction, falls within the shaded area of Fig. 4.1.

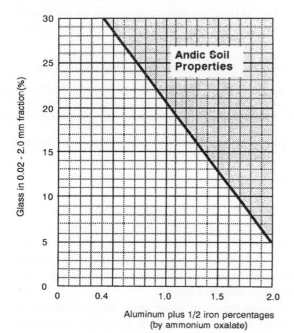

Fig. 4.1. Soils that plot in the shaded area have andic soil properties if the less than 2.0 mm fraction has phosphate retention of more than 25 percent and the 0.02 to 2.0 mm fraction is at least 30 percent of the less than 2.0 mm fraction (Soil Survey Staff, 1990).

The theoretical basis for criteria 2 has a serious conceptual flaw as described below. According to the criterion for the glass content, soils that show low glass contents in the <2 mm fraction may meet andic soil properties. This can be understood by studying the relationship between the glass content and Al_o + $1/2\,Fe_o$ of the bulk soil.

A given soil (<2 mm) has a mixture of volcanic ash and nonvolcanic deposits, Al_o + $1/2\,Fe_o$ of 0.4 percent, and a 0.02 to 2.0 mm fraction that comprises 30 percent of the bulk soil. If the 0.02–2.0 mm fraction contains a minimum of 30 percent noncolored glass, the total glass in the bulk soil is 9 g. On the other hand, Al_o + $1/2\,Fe_o$ of 0.4 g is equivalent to the Al + $1/2\,Fe$ released from 5 g of noncolored glass if all the weathering products are present in ammonium oxalate extractable forms. Thus, the soil can meet andic soil properties if the original glass content of the parent materials is at least 14 g (9 g nonweathered glass plus 5 g weathered glass) per 100 g soil as described below. It is obvious that this glass content is too low.

As described before, noncolored glass can release 2 g of Al_o + $1/2\,Fe_o$ if 25 g of glass completely weathers to form noncrystalline Al and Fe minerals which are extracted by the ammonium oxalate reagent. Fourteen grams of glass can release only 1.12 g of Al_o + $1/2\,Fe_o$. Therefore, a soil with 14 g of glass could never weather to produce the 2 g of Al_o + $1/2\,Fe_o$ required by criteria 1. It is clear that the lower limit of glass content in criteria 2 is too low. Thus, the lower limit of glass content for soils having Al_o + $1/2\,Fe_o$ of 0.4 to 2.0 percent should be raised as presented in Fig. 4.2 and criteria 2 of the andic soil properties should be revised as follows:

2. In the fine-earth fraction, a phosphate retention of 25 percent or more and one of the following:

   a. Aluminum plus 1/2 iron percentages (by ammonium oxalate) totaling 0.40 or more and enough glass so that the glass percentage in the fine-earth fraction, when plotted against the value obtained by adding aluminum plus 1/2 iron percentages in the fine-earth fraction, falls within the shaded area of Fig. 4.2; or
   b. Aluminum plus 1/2 iron percentages (by ammonium oxalate) totaling 2.0 or more and 2 percent or more volcanic glass in the fine-earth fraction.

The previous 2a and 2c are combined into 2a in the proposed revision.

(2) Definition of Andisols and their boundary problems with other soil orders

The definition of Andisols in Keys to Soil Taxonomy (Soil Survey Staff, 1990, 1992) is based on the existence of subhorizons meeting andic soil properties which have a cumulative thickness of more than 35 cm within 60 cm of either the mineral soil surface or the top of an organic layer with andic soil properties (Table 4.3). Andisols are keyed out after Spodosols and before Oxisols.

This definition has an option for organic layers meeting andic soil properties. It is not uncommon that Andisols in Japan have a thick organic layer whose organic carbon content is greater than 18 percent and whose organic matter is highly humified (Andisol TU Database, 1992). Since the most humified organic layer has developed concomitantly with the formation of Al-humus complexes under

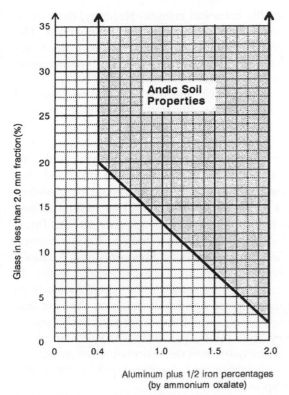

Fig. 4.2. Soils that plot in the shaded area have andic soil properties if the fine-earth fraction has phosphate retention of 25 percent or more (proposed by S. Shoji, 1992, unpublished).

TABLE 4.3

Definitions of Andisols in the 1992 Keys to Soil Taxonomy (Soil Survey Staff, 1992)

Other soils that have andic soil properties in 60 percent or more of the thickness either:

1. Within 60 cm either of the mineral soil surface, or of the top of an organic layer with andic soil properties, whichever is shallower, if there is no lithic or paralithic contact, duripan, or petrocalcic horizon within that depth; or

2. Between either the mineral soil surface, or the top of an organic layer with andic soil properties, whichever is shallower, and a lithic or paralithic contact, duripan, or petrocalcic horizon.

---

well-drained conditions, it is reasonable to differentiate the organic-rich layer with andic soil properties from histic horizons.

Andisols and Spodosols, especially tephra-derived Spodosols, have significant similarities that are ascribed to the presence of noncrystalline materials such as allophane, imogolite, ferrihydrite and Al/Fe-humus complexes. Therefore, it was often difficult to accurately separate the two soil orders according to the

1990 Keys to Soil Taxonomy using spodic chemical criteria. Recently, there have been two different philosophies on classification of tephra-derived Spodosols. The first proposes to exclude tephra-derived Spodosols from the Spodosol order and the second recognizes tephra-derived Spodosols which have the characteristic morphology of Spodosols (see Plate 11) as reviewed by Shoji and Ito (1990).

The original philosophy of Soil Taxonomy was that a soil should be classified based on its properties (Soil Survey Staff, 1975). Recently, however, the genetic background in Soil Taxonomy has been stressed as indicated by the statement of Arnold (1988) that the order category is abstractly defined as soils whose properties are the result of, or reflect a major soil-forming process. In keeping with this philosophy, Shoji and Ito (1990) and Shoji and Yamada (1991) proposed that all soils showing the characteristic morphology of Spodosols should be classified as Spodosols irrespective of the kind of parent material, and that tephra-derived Spodosols be classified as Spodosols if they meet the relevant criteria. The International Committee on the Classification of Spodosols (Rourke, 1991) reviewed various proposals on the classification of tephra-derived Spodosols and Andisols, and proposed a new definition of Andisols. This definition excludes soils having an albic horizon and an underlying horizon meeting spodic color requirements from the Andisol order. Consequently, tephra-derived Spodosols satisfying the albic and spodic requirements are classified as Spodosols even if they meet andic soil property requirements. Finally, the Soil Survey Staff (1992) has established new Spodosol criteria in the 1992 Keys to Soil Taxonomy that stress soil morphological properties, solves the Andisol/Spodosol separation problem and classifies tephra-derived soils with the morphology of Spodosols as Spodosols.

(3) Suborders and great groups of Andisols

The classification of suborders and great groups of Andisols established by the Soil Survey Staff (1992) is shown in Table 4.4. A number of significant revisions were incorporated into the original Andisol proposal (Table 4.1) to formulate the present classification.

Seven suborders are provided for the Andisol order in the current classification. These include five suborders proposed in the original Andisol proposal: Aquands, Xerands, Ustands, and Udands, and Cryands which were originally included as Borands (Smith, 1978). The suborders of Torrands and Vitrands were proposed by ICOMAND (Leamy et al., 1988) and were adopted in the current classification.

The suborder of Tropands that is comparable to Tropepts in 1975 Soil Taxonomy, was proposed in the Andisol proposal by Smith (1978) but this suborder was abolished. A suborder of Allands was proposed by Shoji et al. (1985) to classify nonallophanic Andisols containing large amounts of KCl-extractable Al or toxic Al. However, soils with high KCl-extractable Al are included at the subgroup level due to the effects of management practices on this property. A suborder of Perands was also considered to classify Andisols with a perudic soil moisture regime (Leamy et al., 1987), but it was not adopted in the Keys to Soil Taxonomy (Soil Survey Staff, 1990, 1992).

TABLE 4.4

Suborders and great groups in the 1992 Keys to Soil Taxonomy (Soil Survey Staff, 1992)

Suborder	Great group	Suborder	Great Group
Aquands	Cryaquands	Xerands	Vitrixerands
	Placaquands		Melanoxerands
	Duraquands		Haploxerands
	Vitraquands		
	Melanaquands	Vitrands	Ustivitrands
	Epiaquands		Udivitrands
	Endoaquands		
		Ustands	Durustands
Cryands	Gelicryands		Haplustands
	Melanocryands		
	Fulvicryands	Udands	Placudands
	Hydrocryands		Durudands
	Vitricryands		Melanudands
	Haplocryands		Fulvudands
			Hydrudands
Torrands	Vitritorrands		Hapludands

Aquands. These are Andisols having an aquic moisture regime caused by restricted drainage as imposed by depressed landscape position or the existence of imperme-able layers and are defined as presented in Table 4.5. It is difficult to define some Aquands according to the chroma requirement for mottles because the formation of low chroma mottles in these soils is inhibited by coating of iron minerals with non-crystalline Al hydroxides (Shoji, 1988a). In addition, determination of mottles and the pyridyl test are often obscured by the presence of dark-colored organic matter.

It is common in well-drained Andisols used for wetland rice that migration of iron and manganese from the Ap horizon to the subsurface horizon takes place during irrigation resulting in mottles being found mostly in the horizon below the

TABLE 4.5

Definition of Aquand suborder in the 1992 Keys to Soil Taxonomy (Soil Survey Staff, 1992)

Andisols that have either:

1. A histic epipedon; or
2. In a layer between 40 and 50 cm either from the mineral soil surface or from the top of an organic layer with andic soil properties, whichever is shallower, aquic conditions for some time most years (or artificial drainage) and one or more of the following:

   a. Two percent or more redox concentrations; or
   b. A color value, moist, of 4 or more, and 50 percent or more chroma of 2 or less either in redox depletions on faces of peds, or in the matrix if peds are absent; or
   c. Enough active ferrous iron to give a positive reaction to α, α'-dipyridyl at a time when the soil is not being irrigated.

Ap horizon. Such Andisols should be excluded from the Aquands and should be classified as anthraquic subgroups (Shoji, 1988a).

Aquands are formed in volcanic materials under a wide range of climatic conditions and include seven great groups: cryic, placic, duric, vitric, melanic, epiaquic and endoaquic (Table 4.4). Two of the seven great groups are adopted from the original Andisol proposal (Table 4.1). Distribution and development of Aquands have been described by Shoji and Ping (1992). Cryaquands are extensive in Alaska, U.S.A. and Melanaquands characteristically occur in the lowlands of Japan.

Cryands. Cryands are defined as Andisols having a cryic or pergelic soil temperature regime (Soil Survey Staff, 1992). There are six great groups within the Cryand suborder: gelic, melanic, fulvic, hydric, vitric, and haplic (Table 4.4).

The classification criteria for Cryands was developed primarily from studies of Andisols in Alaska, U.S.A. and in the high mountains in northeastern Japan (Ping et al., 1988, 1989; Shoji et al., 1988a, b). Great groups in the current classification differ considerably from those of Borands proposed in the original Andisol proposal (Table 4.1). The Gelicryand great group is provided to accommodate Cryands that have a mean annual soil temperature of 0°C or lower, indicating the existence of permafrost. This is a very important consideration in land use interpretation in Alaska.

Cryands and Spodosols such as Haplocryods, Humicryods, and Placocryods have commonly developed in the same parent materials under a cryic or pergelic soil temperature regime. They share a great number of important physical and chemical properties. Thus, the separation of these soil orders was rigorously investigated and a proposal to differentiate these soils was proposed by Shoji and Ito (1990).

Torrands. Torrands are Andisols that have an aridic moisture regime (Soil Survey Staff, 1992). Even though tephra is dominated by volcanic glass, which is the most weatherable component in tephra, the rate of its chemical weathering is very low under dry conditions. Thus, only the great group of Vitritorrands is provided within the Torrand suborder, resulting in all Torrands having a vitric great group (Table 4.4).

Xerands. Xerands are Andisols that have a xeric moisture regime (Soil Survey Staff, 1992). The Xerand suborder includes three great groups: vitric, melanic, and haplic (Table 4.4). It is interesting to note that the fulvic great group is not provided in the Xerand suborder. This suggests that humification of soil organic matter is favorable under a xeric soil moisture regime resulting in formation of melanic epipedons as observed in California, U.S.A. (Takahashi and Dahlgren, 1993, unpublished).

Vitrands. Vitrands are Andisols that have 1500-kPa water retention values less than 15 percent on air-dried samples and less than 30 percent on undried samples, throughout one or more horizons with a total thickness of 35 cm or more within 60 cm of either the mineral soil surface or the top of an organic layer with andic soil properties, whichever is shallower (Soil Survey Staff, 1992). The parent material of Vitrands is commonly pumice, cinder, or scoria, and is weakly weathered.

The creation of the Vitrand suborder has caused some problems because it is inconsistent with the concept of the other suborders which are based on soil moisture and temperature regimes. As shown in Table 4.4, the Vitrand suborder is keyed out before the suborders of Ustands and Udands. As a result vitric great groups are not provided in the suborders of Ustands and Udands.

The Vitrand suborder contains two great groups, ustic and udic, indicating the soil moisture regime that typically is provided at the suborder level. These great groups occur extensively in the vicinity of volcanoes with recent strongly explosive eruptions as observed in Japan (Otowa, 1986). Their morphological properties are closely related to the type of explosive eruption, distance from the source volcano, and mode of transportation and sedimentation.

Ustands. Ustands are Andisols that have an ustic soil moisture regime (Soil Survey Staff, 1992). They include two great groups, duric and haplic, which were adapted from the original Andisol proposal (Tables 4.1 and 4.4). Haplustands occur in Hawaii, U.S.A. (Ikawa et al., 1988) and the Philippines (Otsuka et al., 1988).

Udands. Udands are Andisols that have udic soil moisture regimes and are the most widely used Andisols for agriculture. Since they occur in the humid temperate to humid tropics, weathering rates are rapid, forming an abundance of noncrystalline materials. Six great groups are provided for Udands: placic, duric, melanic, fulvic, hydric, and haplic (Table 4.4). Four of these great groups are inherited from the Andisol proposal (Table 4.1).

Melanudands and Fulvudands have attracted much attention during the past several years. As a result, we describe these great groups in detail below. Detailed morphological properties with genetic considerations of these great groups were discussed in Chapter 2.

Melanudands. The traditional concept of Kurobokudo (Andosols) in Japan is soils having a black humus-rich horizon and a friable consistence (Plates 4 and 7). The central concept of melanic Andisols proposed by Smith (1978) originated from such a humus horizon that is the primary characteristic for Kurobokudo (Third Division of Soils, 1982).

The definition of a melanic epipedon and the requirements for Melanudands have been intensively discussed by ICOMAND. Most of this discussion centered around the thickness, color, melanic index parameter to describe humus character-istics, organic carbon content, and depth of the melanic epipedon within the profile (Shoji, 1988b). The revised definition of the melanic epipedon that was finally adopted in Keys to Soil Taxonomy is shown in Table 4.6.

The lower limit for the thickness of a melanic horizon was 30 cm in the original Andisol proposal, but it has been revised to a cumulative thickness of 30 cm or more within a total thickness of 40 cm (Leamy, 1988). Determination of soil color

TABLE 4.6

Definition of the melanic epipedon in the 1992 Keys to Soil Taxonomy (Soil Survey Staff, 1992)

The melanic epipedon is a thick black horizon at or near the soil surface which contains high concentrations of organic carbon, usually associated with short-range-order minerals or aluminum–humus complexes. The intense black color is attributed to the accumulation of organic matter from which "Type A" humic acids are extracted. This organic matter is thought to result from large amounts of root residues supplied by a gramineous vegetation, and can be distinguished from organic matter formed under forest vegetation by the melanic index (Honna et al., 1988).

The suite of secondary minerals is usually dominated by allophane, and the soil material has a low bulk density and a high anion adsorption capacity.

The melanic epipedon has both of the following:

1. An upper boundary at, or within 30 cm of, either the mineral soil surface or the upper boundary of an organic layer with andic soil properties (defined below), whichever is shallower; and
2. In layers with a cumulative thickness of 30 cm or more within a total thickness of 40 cm, all of the following:

    a. Andic soil properties throughout; and
    b. A color value, moist, and chroma (Munsell designations) of 2 or less throughout, and a melanic index of 1.70 or less throughout; and
    c. Six percent or more organic carbon as a weighted average, and 4 percent or more organic carbon in all layers.

---

using the Munsell color book varies considerably with the soil conditions and is fairly subjective. Thus, Shoji (1988b) proposed that the melanic epipedon should have both Munsell color values and chromas, moist, of 2 or less throughout and a melanic index of 1.70 or less throughout. This proposal was introduced into the current definition of the melanic epipedon (Table 4.6). A melanic index of 1.70 or less was proposed by Honna et al. (1988). It indicates an abundance of A-type humic acid in the soil organic matter that is responsible for the characteristic black color.

Smith (1978) first proposed a minimum requirement of 8 percent organic carbon for Melanudands. Since his proposal, organic carbon requirements of 8 percent or more and 6 percent or more were debated and tested. Shoji (1988b) emphasized that the contribution of organic matter to soil properties (bulk density, liquid limit, plastic limit, and 1500 kPa water retention) is notable when the organic carbon content exceeds 6 percent. Furthermore, the Ap horizons of Kurobokudo in Japan show a bimodal frequency of organic carbon content when the width of the class is set at one percent organic carbon. The first mode appears at 4–5 percent and the second mode at 6–7 percent, with a minimum in the range 5–6 percent (see section 6.3.1 of Chapter 6). Thus, he proposed that the minimum requirement for organic carbon should be 6 percent as a weighted average to reflect the typical distribution of organic carbon in Kurobokudo. This proposal was adopted into the current definition of the melanic epipedon (Table 4.6). The Andisol TU Database (1992) shows that Andisols from Japan, Alaska, and northwestern

United States are reasonably separated into two groups according to this criterion.

From an agronomic point of view, not only the melanic epipedon, but also the fulvic horizon are especially important when they occur in the upper part of the soil profile. Thus, Shoji (1988b) proposed that all or the greater part of the melanic epipedon should occur within the upper 50 cm of the soil profile. This proposal was modified by ICOMAND (Leamy, 1988) as an upper boundary of the melanic epipedon at or within 30 cm of the surface and finally adopted into Keys to Soil Taxonomy (Soil Survey Staff, 1990, 1992) (Table 4.6).

Melanudands are extensively formed in *Miscanthus sinensis* (Japanese pampas grass) and *Sasa* ecosystems in Japan. They also have been identified in Indonesia (Nanzyo et al., 1992) and the Philippines (Otsuka et al., 1988).

Fulvudands. The Fulvudand great group was first proposed by Otowa (1985) in order to classify Andisols with thick dark brown epipedons (Plate 5) that were not as dark as melanic epipedons. The fulvic requirements were evaluated and tested along with the melanic requirements by ICOMAND.

Fulvudands have been identified in Chile, Indonesia, Japan, New Zealand, etc. (Nanzyo et al., 1992; Otowa, 1985; Shoji et al., 1987b; Shoji et al., 1988b). In Japan, Fulvudands commonly form under forest ecosystems and often show a biosequential relationship with Melanudands (Kawamuro and Osumi, 1988; Shoji et al., 1988a). Although the two great groups have thick, organic matter rich horizons, they differ remarkably in their humus characteristics reflecting differences primarily due to vegetation (Shoji et al., 1988b) (Plates 5, 6, 7 and 8). Fulvudands show high fulvic acid to humic acid ratios and commonly have organic matter dominated by P-type humic acid with the lowest degree of humification (Honna et al., 1988; Shoji et al., 1988b). The higher melanic indices (>1.70) of Fulvudands are attributable to the unique absorption spectrum of fulvic acid (Honna et al., 1988).

Andisols satisfying all melanic requirements except soil color and melanic index should be classified into the fulvic great group. The current definition specifies that the color value and chroma, moist, must be 3 or less (Soil Survey Staff, 1992); however, there are some Andisols having high organic carbon but colors lighter than this requirement. Thus, the current color requirement for fulvic epipedons is unnecessary.

(4) Subgroups of Andisols

Smith (1978) proposed 20 subgroups in the original Andisol proposal. ICOMAND (Leamy, 1987) proposed many subgroups, some of which evolved directly from the original Andisol proposal. ICOMAND also proposed the principles for subgroup keying within each great group: the subgroups should be ranked according to the relative importance of each characteristic in the particular context and multiple subgroup recognition should be employed. Accordingly, the subgroups were further revised (Leamy et al., 1988) and were finally adopted into Keys to Soil Taxonomy. For example, subgroups of both Melanudands and Hapludands are

TABLE 4.7

Subgroups of Melanudands and Hapludands in the 1992 Keys to Soil Taxonomy (Soil Survey Staff, 1992)

Great Group	Subgroup	Great Group	Subgroup
Melanudands	Lithic	Hapludands	Lithic
	Anthraquic		Petroferric
	Alic Aquic		Anthraquic
	Alic Pachic		Aquic Duric
	Alic Thaptic		Duric
	Alic		Alic
	Aquic		Aquic
	Acrudoxic Vitric		Acrudoxic Hydric
	Acrudoxic Hydric		Acrudoxic Thaptic
	Acrudoxic		Acrudoxic Ultic
	Pachic Vitric		Acrudoxic
	Eutric Vitric		Vitric
	Vitric		Hydric Thaptic
	Hydric Pachic		Hydric
	Pachic		Eutric Thaptic
	Eutric Hydric		Thaptic
	Hydric		Eutric
	Thaptic		Oxic
	Ultic		Ultic
	Typic		Alfic
			Typic

keyed out as shown in Table 4.7. This table shows that a large number of subgroups have been provided with multiple subgroup recognition. The next section discusses attributes of selected subgroups.

Alic subgroups. An alic (allic) subgroup was first proposed in the Andisol proposal to distinguish Andisols with relatively large amounts of KCl-extractable Al (nonallophanic Andisols) from common Andisols. Shoji et al. (1985) showed that nonallophanic Andisols, occupying considerable areas of Japan, have unique properties such as low bulk density, high oxalate-extractable Al and Fe, large phosphate retention, etc. which are common to allophanic Andisols. In contrast to allophanic Andisols, nonallophanic soils exhibit an abundance of 2:1 layer silicates, Al-humus complexes as the primary form of active Al, very strong acidity (pH < 5.0), high Al saturation and strong Al toxicity.

Based on the dissimilarities between nonallophanic and allophanic Andisols and also the agronomic significance of nonallophanic Andisols, Shoji et al. (1985) proposed the creation of an Alland suborder. However, this proposal was later withdrawn because the exchangeable Al was not considered to be the result of a major soil-forming process and its levels can be changed by farming practices. Finally, the alic subgroups have been provided in great groups of several suborders such as Aquands, Cryands and Udands to indicate the presence of potentially toxic Al (Soil Survey Staff, 1990, 1992).

Alic Andisols have 1 M KCl-extractable Al of more than 2.0 cmol$_c$ kg^{-1} in the fine-earth fraction throughout a layer 10 cm or more thick between a depth of 25 and 50 cm (Soil Survey Staff, 1990). This level of exchangeable Al is potentially toxic enough to severely reduce the root growth of common crops (Saigusa et al., 1980). The layer underlying the Ap or surface horizon (25 to 50 cm) is employed to determine the level of exchangeable Al because its soil properties are difficult to change by farming practices.

Aquic and anthraquic subgroups. The aquic subgroups are identified by the presence of aquic features in some subhorizon between 50 and 100 cm from the mineral soil surface or, from the upper boundary of an organic layer with andic soil properties. On the other hand, well drained to excessively drained Andisols have their properties changed after puddling of the surface soil and continuous flooding for rice cultivation. Changes include degradation of soil structure, development of a plowpan, loss of iron and manganese from the Ap horizon, and accumulation of iron and manganese (mottles and nodules) in the subsurface horizon. These features are characteristic for the anthraquic subgroup designation (Eswaran, 1989; Mitsuchi, 1989).

Soils with anthraquic features do not have the limitations for soil management and plant growth shown by Aquands or Wet Kurobokudo (Shoji, 1988a). Thus, anthraquic subgroups are provided for Melanudands and Hapludands (Soil Survey Staff, 1992). These great groups are extensively used for wetland rice in Japan and display the characteristic anthraquic features.

Humic subgroups. The mollic and umbric subgroups were provided in the 1990 Keys to Soil Taxonomy according to the criteria of mollic and umbric epipedons, respectively. A mollic epipedon is a surface horizon that has strong structure, color values darker than 3.5 moist and 5.5 dry, and chromas less than 3.5 moist. In addition organic carbon content must be at least 0.6 percent when mixed to a depth of 18 cm and base saturation must be 50 percent or more as determined by the NH$_4$ OAc (pH = 7) method. The umbric epipedon is a surface horizon like the mollic epipedon except that its base saturation is less than 50 percent (Soil Survey Staff, 1990).

The criteria for both mollic and umbric subgroups include base saturation by the NH$_4$ OAc (pH = 7) method that was pointed out by Smith (1978) as a defect in the classification of Andepts (Andisols) because Andepts have variable charge mineralogy. Thus, the mollic and umbric subgroups were combined into the new humic subgroup in the 1992 Keys to Soil Taxonomy which solves the base saturation problem (Soil Survey Staff, 1992). The establishment of the humic subgroup enhances the consistency of subgroup placement in Andisols.

Pachic subgroups. Pachic subgroups of Andisols have more than 6.0 percent organic carbon and meet the color requirements of the mollic epipedon throughout at least 50 cm of the upper 60 cm, excluding any overlying layers that do not have andic soil properties (Soil Survey Staff, 1990, 1992). Since the range of mollic colors includes the color requirements for both melanic epipedons and fulvic surface horizons, the pachic subgroups are Andisols with a thick melanic epipedon or fulvic surface horizon.

Thaptic subgroups. Many Andisols have a multisequum profile, reflecting repeated tephra deposition and subsequent soil formation. The thaptic subgroups are useful for describing the morphological properties of such Andisols. Thaptic subgroups of Andisols are defined as soils that have between 25 and 100 cm (between 40 and 100 cm for melanic and fulvic great groups) from either the mineral soil surface, or the top of an organic layer with andic soil properties, whichever is shallower, a layer 10 cm or more thick with more than 3.0 percent organic carbon and colors of a mollic epipedon throughout, underlying one or more horizons with a total thickness of 10 cm or more that have a color value, moist, 1 unit or more higher and an organic-carbon content 1 percent or more (absolute) lower (Soil Survey Staff, 1992).

(5) Families of Andisols

Particle-size analysis is difficult to apply to Andisols because the soil material commonly consists of stable aggregates cemented by noncrystalline materials and organic matter. Therefore, modifiers that combined particle-size and mineralogical properties are used to distinguish families of Andisols. The current modifiers evolved from Smith's proposal to define classes based on a combination of particle-size and mineralogy (Smith, 1978). They were revised by ICOMAND (Leamy et al., 1988) and were subsequently adopted into Keys to Soil Taxonomy as shown in Table 4.8 (Soil Survey Staff, 1992).

The substitutes for the fragmental classes are pumiceous and cindery. These classes have insufficient fine-earth material to fill 10 percent of the total volume. The substitutes for the nonfragmental classes consist of nine classes. These classes have sufficient fine-earth material to fill 10 percent or more of the total volume. The ashy, medial, and hydrous classes determined using 1500 kPa water retention are comparable with sandy, loamy and clayey textural classes, respectively, for nonandic mineral soils.

4.3. CLASSIFICATION OF ANDOSOLS IN THE FAO/UNESCO SOIL MAP OF THE WORLD.

The Andosol major soil grouping was created by FAO/Unesco (1974) in order to prepare the Soil Map of the World at a scale of 1 : 5,000,000. The definition of this grouping was drawn from that of the Andept suborder in Soil Taxonomy (Soil Survey Staff, 1975). Andosols were classified into four soil units: mollic (Tm), humic (Th), ochric (To) and vitric (Tv) Andosols. It is noted that wet Andosols were not provided for in this classification.

TABLE 4.8

Modifiers that replace names of particle-size classes in the 1992 Keys to Soil Taxonomy (Soil Survey Staff, 1992)

1. Substitutes for the fragmental particle-size class:

These classes have a fine-earth component of less than 10 percent of the total volume.

Pumiceous — In the whole soil, more than 60 percent (by weight) volcanic ash, cinders, lapilli, pumice and pumice-like fragments with diameters of more than 1 mm; in the fraction coarser than 2.0 mm, two thirds or more (by volume) pumice or pumice-like fragments.

Cindery — In the whole soil, more than 60 percent (by weight) volcanic ash, cinders, lapilli, pumice and pumice-like fragments with diameters of more than 1 mm; in the fraction coarser than 2.0 mm, less than two thirds (by volume) pumice and pumice-like fragments.

2. Substitutes for the non-fragmental particle-size classes:

These classes have a fine-earth component of 10 percent or more of the total volume.

Ashy — Less than 35 percent (by volume) rock fragments; a fine-earth fraction which contains 30 percent or more (by weight) particles between 0.02 and 2.0 mm in diameter and which has either:

a. Andic soil properties, and a water content at 1500 kPa tension of less than 30 percent on undried samples and less than 12 percent on dried samples; or
b. No andic soil properties, and a total of 30 percent or more of the 0.02-to-2.0-mm fraction (by grain count) consisting of volcanic glass, glass aggregates, glass coated grains, and other vitric volcaniclastics.

Ashy–pumiceous — 35 percent or more (by volume) rock fragments, of which two thirds or more (by volume) are pumice or pumice-like fragments; an ashy fine-earth fraction.

Ashy–skeletal — 35 percent or more (by volume) rock fragments, of which less than two-thirds (by volume) are pumice and pumice-like fragments; an ashy fine-earth fraction.

Medial — A fine-earth fraction which has andic soil properties, and which has a water content at 1500 kPa tension of 12 percent or more on air-dried samples and of 30 to 100 percent on undried samples; less than 35 percent (by volume) rock fragments.

Medial–pumiceous — 35 percent or more (by volume) rock fragments, of which two thirds or more (by volume) are pumice or pumice-like fragments; a medial fine-earth fraction.

Medial–skeletal — 35 percent or more (by volume) rock fragments, of which less than two thirds (by volume) are pumice or pumice-like fragments; a medial fine-earth fraction.

Hydrous — A fine-earth fraction which has andic soil properties, and which has a water content at 1500 kPa tension of 100 percent or more on undried samples; less than 35 percent (by volume) rock fragments.

Hydrous–pumiceous — 35 percent or more (by volume) rock fragments, of which two thirds or more (by volume) are pumice or pumice-like fragments; a hydrous fine-earth fraction.

Hydrous–skeletal — 35 percent or more (by volume) rock fragments, of which less than two thirds (by volume) are pumice or pumice-like fragments; a hydrous fine-earth fraction.

FAO/Unesco (1989) revised the classification of Andosols (FAO/Unesco, 1974) by introducing the andic soil properties criteria. As presented in Table 4.9, six soil units are provided for in the current classification. The Gelic Andosols having permafrost and Gleyic Andosols showing gleyic properties were created. The humic soil unit was renamed the umbric soil unit. The haplic soil unit was provided

TABLE 4.9

Classification of Andosols in the FAO/Unesco Soil Map of the World (FAO/Unesco, 1989)

Major Soil Grouping	Soil Units
ANDOSOLS (AN) Soils showing andic properties to a depth of 35 cm or more from the surface and having a mollic or an umbric A horizon possibly overlying a cambic B horizon, or an ochric A horizon and a cambic B horizon; having no other diagnostic horizons; lacking gleyic properties within 50 cm of the surface; lacking the characteristics which are diagnostic for Vertisols; lacking salic properties.	
	Gelic Andosols (ANi) Andosols having permafrost within 200 cm of the surface.
	Gleyic Andosols (ANg) Andosols showing gleyic properties within 100 cm of the surface; lacking permafrost within 200 cm of the surface.
	Vitric Andosols (ANz) Andosols lacking a smeary consistence or having a texture which is coarser than silt loam on the weighted average for all horizons within 100 cm of the surface, or both; lacking gleyic properties within 100 cm of the surface; lacking permafrost within 200 cm of the surface.
	Mollic Andosols (ANm) Andosols having a mollic A horizon; having a smeary consistence, and having a texture which is silt loam or finer on the weighted average for all horizons within 100 cm of the surface; lacking gleyic properties within 100 cm of the surface; lacking permafrost within 200 cm of the surface.
	Umbric Andosols (ANu) Andosols having an umbric A horizon; having a smeary consistence and having a texture which is silt loam or finer on the weighted average for all horizons within 100 cm of the surface; lacking gleyic properties within 100 cm of the surface; lacking permafrost within 200 cm of the surface.
	Haplic Andosols (ANh) Andosols having an ochric A horizon and a cambic B horizon; having a smeary consistence and having a texture which is silt loam or finer on the weighted average for all horizons within 100 cm of the surface; lacking gleyic properties within 100 cm of the surface; lacking permafrost within 200 cm of the surface.

instead of the ochric soil unit. This classification of Andosols has two significant defects: the criteria for soil units require base saturation and particle-size data whose determinations are difficult to reproduce as described earlier.

4.4. NATIONAL CLASSIFICATIONS OF VOLCANIC ASH SOILS

4.4.1. *Soil classification of agricultural land in Hokkaido, Japan*

In Hokkaido, Japan, the classification and mapping of volcanic ash soils were historically based on the origin and stratigraphy of tephras, because young volcanic

·ash soils are extensive on this island. However, determination of the source volcanoes and identification of tephra layers are not always possible. Furthermore, such traditional classification was not adequately based on the attributes of the soils or the major soil-forming processes. This classification system also failed to address the agronomic aspects of the soil. Thus, soil scientists of Hokkaido have proposed the "Soil Classification of Agricultural Land in Hokkaido" (Hokkaido Soil Classification Committee, 1979) which includes an improved classification of volcanic ash soils as shown in Table 4.10.

Volcanic ash soils in Hokkaido are primarily classified into Regosols and Andosols (Hokkaido Soil Classification Committee, 1979). Some volcanic ash soils are also identified as Brown forest soils (Shoji et al., 1990). The major category of Regosols has two groups namely, Volcanogenous Regosols and Gleyic Volcanogenous Regosols. These groups have developed primarily in pumiceous sand and gravel and are the least weathered resulting in low agricultural productivity. Each of these groups is further classified into four subgroups according to the existence of regosolic horizons (typic), stratified regosolic horizons (stratic), and the kind of underlying nonandic soil.

The Andosol category has seven groups which are classified according to the soil moisture regime and diagnostic horizons for volcanic ash soils. They are further divided into three to six subgroups according to diagnostic horizons, horizon sequences, and kind of underlying nonandic soil (Table 4.10).

TABLE 4.10

Classification of volcanic ash soils in the Hokkaido system (Hokkaido Soil Classification Committee, 1979)

Major Group	Group	Subgroup
Regosols	Volcanogenous Regosols: that have formed from pumiceous sand and gravel.	Typic Volcanogenous Regosols Stratic Volcanogenous Regosols Volcanogenous Regosols Burying Brown Forest soils Volcanogenous Regosols Burying Lowland soils
	Gleyic Volcanogenous Regosols: that are Volcanogenous Regosols having characteristics associated with wetness within 50 cm of the surface.	Typic Gleyic Volcanogenous Regosols Stratic Gleyic Volcanogenous Regosols Gleyic Volcanogenous Regosols Burying Lowland soils Gleyic Volcanogenous Regosols Burying Peat soils
Andosols	Regosolic Andosols: that have a rego-andosolic horizon thicker than 25 cm within 50 cm of the surface.	Typic Regosolic Andosols Stratic Regosolic Andosols Regosolic Andosols Burying Brown Forest soils

TABLE 4.10 (continued)

Major Group	Group	Subgroup
Andosols		Regosolic Andosols Burying Lowland soils
	Gleyic Regosolic Andosols: that are Regosolic Andosols having characteristics associated with wetness within 50 cm of the surface.	Stratic Gleyic Regosolic Andosols Gleyic Regosolic Andosols Burying Pseudogleys Gleyic Regosolic Andosols Burying Lowland soils
	Brown Andosols: that have brown fluffy andosolic or loamy andosolic horizons thicker than 25 cm within 50 cm of the surface.	Fluffy Brown Andosols Loamy Brown Andosols Stratic Fluffy Andosols Stratic Loamy Andosols Fluffy Brown Andosols Burying Brown Forest soils Fluffy Brown Andosols Burying Lowland soils
	Ordinary Andosols: that have dark-colored fluffy andosolic or loamy andosolic horizons thicker than 25 cm within 50 cm of the surface.	Fluffy Ordinary Andosols Loamy Ordinary Andosols Stratic Fluffy Ordinary Andosols Stratic Loamy Ordinary Andosols Fluffy Ordinary Andosols Burying Brown Forest soils Fluffy Ordinary Andosols Burying Lowland soils
	Gleyic Ordinary Andosols: that are Ordinary Andosols having characteristics associated with wetness within 50 cm of the surface.	Fluffy Gleyic Ordinary Andosols Loamy Gleyic Ordinary Andosols Stratic Fluffy Gleyic Ordinary Andosols Fluffy Ordinary Andosols Burying Pseudogleys Fluffy Gleyic Ordinary Andosols Burying Lowland soils Fluffy Gleyic Ordinary Andosols Burying Peat soils
	Cumulic Andosols: that have an upper boundary of a black fluffy andosolic or loamy andosolic horizon thicker than 30 cm within 25 cm of the surface. Average humus content of black horizons is over 12 percent.	Typic Cumulic Andosols Cumulic Andosols Burying Brown Forest soils Cumulic Andosols Burying Lowland soils
	Gleyic Cumulic Andosols: that are cumulic Andosols having characteristics associated with wetness within 50 cm of the surface.	Typic Gleyic Cumulic Andosols Gleyic Cumulic Andosols Burying Pseudogleys Gleyic Cumulic Andosols Burying Lowland soils

The classification of volcanic ash soils established by the Hokkaido Soil Classification Committee (1979) is correlated with that of Andisols in the 1990 Keys to Soil Taxonomy as follows:

Groups in the 1979 Hokkaido system	Great groups in the 1990 Keys to Soil Taxonomy
Volcanogenous Regosols	Udipsamments
Gleyic Volcanogenous Regosols	Psammaquents
Regosolic Andosols	Udivitrands
Gleyic Regosolic Andisols	Vitraquands
Brown Andosols	Hapludands
Ordinary Andosols	Hapludands and Melanudands
Gleyic Ordinary Andosols	Haplaquands and Melanaquands
Cumulic Andosols	Melanudands
Gleyic Cumulic Andosols	Melanaquands

4.4.2. Classification of cultivated soils in Japan

Volcanic ash soils are identified in the soil groups of Andosols (Kurobokudo), Wet Andosols (Wet Kurobokudo) and Gleyed Andosols (Gleyed Kurobokudo) in the Classification of Cultivated Soils in Japan as presented in Table 4.11 (The Third Division of Soils, 1982).

The central concept of Andosols (Kurobokudo) are soils that have developed in coarse to fine pyroclastic materials and are characterized by a high content of vitric materials and/or allophane. Most of them have black to dark-colored A horizons over brown to yellowish brown B horizons and are characterized by low bulk density (usually < 0.85 g cm^{-3}), high cation exchange capacity as determined by NH$_4$ OAc (pH $= 7$) method (usually > 30 cmol$_c$ kg^{-1}), and high phosphate adsorption coefficients (>1500 mg P$_2$O$_5$/100 g).

Andosols occur mainly on terraces and on gentle footslopes of volcanoes and are mostly used as upland crop fields, pastures, orchards, tea gardens, etc. They are classified into five soil-series groups according to differences in the thickness of humus layers and humus content (Table 4.11).

Wet Andosols (Wet Kurobokudo) are hydromorphic Andosols characterized by iron mottles or manganese concretions in their profiles. These characteristics developed mainly under the influence of irrigation for rice cultivation and others due to groundwater or stagnant water. The parent materials are pyroclastic materials or their secondary deposits. Wet Andosols are mainly distributed on alluvial lowlands and in depressions on terraces and hills. They are used mainly as paddy rice fields and partly for upland crops. Wet Andosols are also divided into five soil-series groups according to the same criterion as those of Andosols (Table 4.11).

Gleyed Andosols (Gleyed Kurobokudo) are hydromorphic Andosols that have developed under permanently or nearly permanently water-saturated conditions and are characterized by the occurrence of gley horizons in the profiles. Thus,

TABLE 4.11

Classification of volcanic ash soils in the national system for cultivated soils in Japan (The Third Division of Soils, 1982)

Soil-group	Soil-series Group
Andosols	Thick High-humic Andosols [1] Thick Humic Andosols [2] High-humic Andosols [3] Humic Andosols [4] Light-colored Andosols [5]
Wet Andosols	Thick High-humic Wet Andosols [1] Thick Humic Wet Andosols [2] High-humic Wet Andosols [3] Humic Wet Andosols [4] Light-colored Wet Andosols [5]
Gleyed Andosols	High-humic Gleyed Andosols [1] Humic Gleyed Andosols [2] Light-colored Gleyed Andosols [5]

[1] The humus horizons have a cumulative thickness of 50 cm or more from the surface and have 10 percent or more humus as a weighted average.
[2] The humus horizons have a cumulative thickness of 50 cm or more from the surface and have 5 to 10 percent humus as a weighted average.
[3] The humus horizons have a cumulative thickness of less than 50 to 25 cm from the surface and have 10 percent or more humus as a weighted average.
[4] The humus horizons have a cumulative thickness of less than 50 to 25 cm from the surface and have 5 to 10 percent humus as a weighted average.
[5] The humus horizons have a cumulative thickness of less than 25 cm from the surface and have less than 5 percent humus as a weighted average.

Gleyed Andosols are comparable with Aquands in the 1992 Keys to Soil Taxonomy. Gleyed Andosols occur on alluvial lowlands and in depressions on terraces and are primarily used as paddy rice fields. They are divided into three soil-series groups according to differences in the humus content (Table 4.11).

From the foregoing, it is noticeable that the thickness and organic carbon content of humus horizons are weighted most heavily in the classification of soil-series groups.

4.4.3. New Zealand soil classification (Version 2.0)

The New Zealand soil classification system was based on studies of specific groups of the New Zealand Genetic Soil Classification and has maintained possible continuity with successful parts of this classification system (Hewitt, 1989). In this classification system, as presented in Table 4.12, volcanic ash soils are identified mainly in the soil order of Allophanic Soils, Pumice Soils, Granular Soils, and Recent Soils. Some volcanic ash soils are included in the order of Podzols.

TABLE 4.12

Classification of volcanic ash soils in the New Zealand system (Version 2.0) (Hewitt, 1989)

Order	Group	Subgroup
Allophanic Soils	Perch-gley	ironstone, lithic, typic
	Gley	peaty, typic
	Impeded	mottled-ironstone, mottled–lithic, lithic, mottled–ultic, ultic, typic
	Orthic	bouldery, mottled, vitric–acidic, vitric, red, acidic, typic
Pumice Soils	Perch-gley	duric, typic
	Ground-water-gley	typic
	Impeded	mottled–welded, welded, typic
	Orthic	mottled, podzolised, allophanic, buried–allophanic, immature, typic
Granular Soils	Perch-gley	oxidic, acidic, typic
	Melanic	allophanic, typic
	Oxidic	allophanic, acidic, mottled–acidic, mottled, typic
	Orthic	allophanic, mottled–acidic, mottled, acidic, typic
Recent Soils	Tephric	gleyed, buried–pumice, buried–allophanic, typic
Podzols	Orthic	sandy, pumice, pumice-mantled–allophanic, allophanic–clayey, allophanic, stony, typic

Allophanic Soils are keyed out after Podzols and are defined as soils that have a layer or layers of allophanic soil material having a total thickness of 35 cm or more within 60 cm of the mineral soil surface. However, the properties of allophanic soil material are not always due to allophane. Imogolite and ferrihydrite are also recognized as important clays.

Allophanic Soils are divided into four groups according to the water permeability and characteristics associated with wetness. Each group is further divided into several subgroups (Table 4.12).

Pumice Soils are keyed out after Allophanic Soils and are soils that have both

(1) A layer of vitric material extending from the mineral soil surface to 25 cm or more, or 35 cm or more thick occurring within 60 cm of the mineral soil surface, and

(2) A B horizon 10 cm or more thick.

They are correlated mainly with the Vitrands in the 1992 Keys to Soil Taxonomy. Pumice Soils have properties dominated by a pumiceous and glassy skeleton which has a low clay content; allophane and imogolite typically occur as coatings on the skeleton. Pumice Soils develop in sandy or pumiceous tephra of 700–3500 yr B.P.

in New Zealand. Pumice Soils are also divided into four groups and each group is divided into one to six subgroups.

Granular Soils are a grouping of strongly weathered volcanic ash soils generally older than 50,000 years. They contrast with Allophanic and Pumice Soils by having a clay mineralogical assemblage dominated by kaolin minerals and associated vermiculite and hydrous-interlayer-vermiculite. Some Granular Soils also have been derived from basaltic and andesitic rocks.

Granular Soils are keyed out after Oxidic Soils and are defined as soils that have a clayey, moderately or strongly pedal cutanic horizon and that have either

(1) A reductimorphic horizon within 15 cm of the base of the A horizon, or within 30 cm of the mineral soil surface; or

(2) Matrix color value more than 4 and polyhedral aggregates 2 cm or less in size in most of the B horizons within 60 cm of the mineral soil surface.

Most Granular Soils are correlated with Ultisols but a few with the Alfisols in Soil Taxonomy. They are divided into four groups and each group into two to five subgroups (Table 4.12).

Volcanic ash soils belonging to Recent Soils are identified in the Tephric group that have developed in tephra deposits from the mineral soil surface to a depth of 30 cm or more. The group is divided into four subgroups (Table 4.12).

Tephra-derived Spodosols are provided in two subgroups of the Orthic group of Podzols in the New Zealand system (Table 4.12).

REFERENCES

Andisol TU Database, 1992. Database on Andisols from Japan, Alaska, and northwestern U.S.A. prepared by Soil Science Laboratory, Tohoku University. (see Appendix 2)

Arnold, R.W., 1988. The worldwide distribution of Andisols and the need for an Andisol order in Soil Taxonomy. In: D.I. Kinloch, S. Shoji, F.H. Beinroth and H. Eswaran (Editors), Proc. 9th Int. Soil Classification Workshop, Japan, 20 July to 1 August, 1987. Publ. by Jap. Committee for 9th Int. Soil Classification Workshop, for the Soil Management Support Services, Washington D.C., U.S.A., pp. 5–12.

Eswaran, H., 1989. Classification of rice growing soils. In: J. Bay-Petersen (Editor), Classification and management of rice growing soils. Proc. 5th Int. Soil Management Workshop, FFTC Book series No.39. Taipei, Taiwan, pp. 28–32.

FAO/Unesco, 1964. Meeting on the Classification and Correlation of Soils from Volcanic Ash, Tokyo, Japan, 11–27 June, 1964. World Soil Resources Reports.

FAO/Unesco, 1974. FAO-Unesco Soil Map of the World 1:5,000,000, Volume I: Legend. Unesco, Paris.

FAO/Unesco, 1989. FAO-Unesco Soil Map of the World, Revised Legend. World Resources Report 60, FAO, Rome. Wageningen.

Garcia-Rodeja, E., Silva, B.M. and Macias, F., 1987. Andosols developed from nonvolcanic materials in Galicia, NW Spain. J. Soil Sci., 38: 573–591.

Hewitt, A.E., 1989. New Zealand soil classification (version 2.0). DSIR Division of Land and Soil Sciences Technical Record DN 2. DSIR, New Zealand.

Hokkaido Soil Classification Committee, 1979. Soil classification of agricultural land in Hokkaido, 2nd approximation. Misc. Publ. Hokkaido Nat. Agr. Exp. Sta., 17: 1–89 (in Japanese, with English abstract).

Honna, T., Yamamoto, S. and Matsui, K., 1988. A simple procedure to determine melanic index useful to separation of melanic and fulvic Andisols. Pedologist, 32: 69–78.

Ikawa, H., Sato, H. and Chu, A.E., 1988. Properties of Eutrandepts and Vitrandepts of Hawaii with ustic soil moisture regime. In: D.I. Kinloch, S. Shoji, F.H. Beinroth and H. Eswaran (Editors), Proc. 9th Int. Soil Classification Workshop, Japan, 20 July to 1 August, 1987. Publ. by Jap. Committee for 9th Int. Soil Classification Workshop, for the Soil Management Support Services, Washington D.C., U.S.A., pp. 456–462.

Kawamuro, K. and Osumi, Y., 1988. Melanic and fulvic Andisols in central Japan: soil pollen and plant opal analyses. In: D.I. Kinloch, S. Shoji, F.H. Beinroth and H. Eswaran (Editors), Proc. 9th Int. Soil Classification Workshop, Japan, 20 July to 1 August, 1987. Publ. by Jap. Committee for 9th Int. Soil Classification Workshop, for the Soil Management Support Services, Washington D.C., U.S.A., pp. 389–401.

Leamy, M.L., 1984. International Committee on the Classification of Andisols (ICOMAND) Circular letter No.6. New Zealand Soil Bureau, DSIR, Lower Hutt.

Leamy, M.L., 1987. International Committee on the Classification of Andisols (ICOMAND) Circular letter No.9. New Zealand Soil Bureau, DSIR, Lower Hutt.

Leamy, M.L., 1988. International Committee on the Classification of Andisols (ICOMAND) Circular letter No.10. New Zealand Soil Bureau, DSIR, Lower Hutt.

Leamy, M.L., Clayden, B., Parfitt, R.L., Kinloch, D.I. and Childs, C.W., 1988. Final proposal of the International Committee on the Classification of Andisols (ICOMAND). New Zealand Soil Bureau, DSIR, Lower Hutt.

Mitsuchi, M., 1989. Rice soils in Japan. In: J. Bay-Petersen (Editor), Classification and management of rice growing soils. Proc. 5th Int. Soil Management Workshop. FFTC Book series No. 39. Taipei Taiwan, pp. 33–40.

Nanzyo, M., Shoji, S. and Sudjadi, M., 1992. Properties and classification of Andisols from West Java, Indonesia. In preparation for contribution.

Otowa, M., 1985. Criteria of great groups of Andisols. In: F.H. Beinroth, W. Luzio L., F. Maldonado P., and H. Eswaran (Editors), Proc. 6th Int. Soil Classification Workshop, Chile and Ecuador. Part 1: Papers. Sociedad Chilena de la Ciencia del Suelo, Santiago, Chile, pp. 191–208.

Otowa, M., 1986. Morphology and Classification. In: K. Wada (Editor), Ando Soils in Japan. Kyushu University Press, Fukuoka, Japan, pp. 3–20.

Otsuka, H., Briones, A.A., Daquiado, N.P. and Evangelio, F.A., 1988. Characteristics and genesis of volcanic ash soils in the Philippines. Tech. Bull. Trop. Agr. Res. Center, Japan, 24: 1–122.

Parfitt, R.L., Childs, C.W. and Eden, D.N., 1988. Ferrihydrite and allophane in four Andepts from Hawaii and implications for their classification. Geoderma, 41: 223–241.

Ping, C.L., Shoji, S. and Ito, T., 1988. Properties and classification of three volcanic ash-derived pedons from Aleutian Islands and Alaska Peninsula, Alaska. Soil Sci. Soc. Am. J., 52: 455–462.

Ping, C.L., Shoji, S., Ito, T., Takahashi, T. and Moore, J.P., 1989. Characteristics and classification of volcanic-ash-derived soils in Alaska. Soil Sci., 148: 8–28.

Rourke, R., 1991. International Committee on the Classification of Spodosols. Circular letter No. 10. Maine.

Saigusa, M., Shoji, S. and Takahashi, T., 1980. Plant root growth in acid Andosols from northeastern Japan, 2. Exchange acidity Y_1 as a realistic measure of aluminum toxicity potential. Soil Sci., 130: 242–250.

Shoji, S., 1986. Primary Minerals. In: K. Wada (Editor), Ando Soils in Japan. Kyushu University Press, Fukuoka, Japan, pp. 21–40.

Shoji, S., 1988a. Wet Andisols. In: Classification and management of rice-growing soils. 5th Int. Soil Management Workshop, 1. FFTC for the Asian and Pacific Region. Taipei, Taiwan, pp. 11.1–11.13.

Shoji, S., 1988b. Separation of melanic and fulvic Andisols. Soil Sci. Plant Nutr., 34: 303–306.

Shoji, S. and Ito, T., 1990. Classification of tephra-derived Spodosols. Soil Sci., 150: 799–815.

Shoji, S. and Ono, T., 1978. Physical and chemical properties and clay mineralogy of Andosols from Kitakami, Japan. Soil Sci., 126: 297–312.

Shoji, S. and Ping, C.L., 1992. Wet Andisols. Proc. 8th International Soil Correlation Meeting, Louisiana, and Texas October 6–20, 1990, pp. 230–234.

Shoji, S. and Yamada, H., 1991. Comparisons of mineralogical properties between tephra-derived Spodosols from Alaska and nontephra-derived Spodosols from New England. Soil Sci., 152: 162–183.

Shoji, S., Saigusa, M. and Takahashi, T., 1980. Plant root growth in acid Andosols from northeastern Japan, 1. Soil properties and root growth of burdock, barley, and orchard grass. Soil Sci., 130: 124–131.

Shoji, S., Ito, T., Saigusa, M. and Yamada, I., 1985. The case for recognizing a suborder of nonallophanic Andisols (Allands). In: F.H. Beinroth, W. Luzio L., F. Maldonado P., and H. Eswaran (Editors), Proc. 6th Int. Soil Classification Workshop, Chile and Ecuador, Part 1: Papers. Sociedad Chilena de la Ciencia del Suelo, Santiago, Chile, pp. 175–190.

Shoji, S., Ito, T., Nakamura, S. and Saigusa, M., 1987a. Properties of humus of Andosols from New Zealand, Chile and Ecuador. Jap. J. Soil Sci. Plant Nutr., 58: 473–479 (in Japanese).

Shoji, S., Ito, T. and Saigusa, M., 1987b. Andisol–Entisol transition problem. Pedologist, 31: 171–175.

Shoji, S., Takahashi, T., Ito, T. and Ping, C.L., 1988a. Properties and classification of selected volcanic ash soils from Kenai Peninsula, Alaska. Soil Sci., 145: 395–413.

Shoji, S., Takahashi, T., Saigusa, M., Yamada, I. and Ugolini, F.C., 1988b. Properties of Spodosols and Andisols showing climosequential and biosequential relations in southern Hakkoda, northeastern Japan. Soil Sci., 145: 135–150.

Shoji, S., Hakamada, T. and Tomioka, E., 1990. Properties and classification of selected volcanic ash soils from Abashiri, Northern Japan-Transition of Andisols to Mollisols. Soil Sci. Plant Nutr., 36: 409–423.

Smith, G.D., 1978. A preliminary proposal for the reclassification of Andepts and some andic subgroups (the Andisol proposal, 1978). New Zealand Soil Bureau Record, DSIR, Lower Hutt, 96.

Soil Survey Staff, 1960. Soil Classification, a comprehensive system — 7th approximation. USDA, Washington D.C., USGPO.

Soil Survey Staff, 1975. Soil Taxonomy. A basic system of soil classification for making and interpreting soil survey. Agr. Hdbk. No.436. USDA, Washington D.C., USGPO.

Soil Survey Staff, 1990. Keys to Soil Taxonomy, 4th edition. AID, USDA-SMSS, Technical Monograph, No. 19, Blacksburg, Virginia.

Soil Survey Staff, 1992. Keys to Soil Taxonomy, 5th edition. AID, USDA-SMSS Technical Monograph No. 19, Blacksburg, Virginia.

The Third Division of Soils, 1982. Classification of cultivated soils in Japan. Department of Soils and Fertilizers, National Institute of Agricultural Sciences (Japan).

Wada, K. and Wada, S., 1976. Clay mineralogy of the B horizons of two Hydrandepts, a Torrox and a Humitropept in Hawaii. Geoderma, 16: 139–157.

Chapter 5

MINERALOGICAL CHARACTERISTICS OF VOLCANIC ASH SOILS

R. DAHLGREN, S. SHOJI and M. NANZYO

5.1. INTRODUCTION

A distinctive feature of soils derived from volcanic materials is the occurrence of a unique clay-size mineral assemblage dominated by noncrystalline components. Noncrystalline materials common in volcanic ash soils include: allophane, imogolite, opaline silica, and ferrihydrite. These materials can not be defined as "crystalline minerals" because they do not have a fixed chemical composition or a regular three-dimensional structural framework. However, some of these materials do produce broad X-ray diffraction patterns indicating some ordering in their atomic arrangement. Nomenclature, such as short-range-order and paracrystalline minerals have been used to describe imogolite, since this phase does possess a regular atomic arrangement along its tube unit, yet there is some randomness involved in the arrangement of the tube units to form threads. For allophane, both X-ray diffraction and electron diffraction data indicate no repeat of structural units in any of the three dimensions. Therefore, "noncrystalline" appears to be more appropriate than "short-range-order" for describing allophane; however, several previous definitions of allophane have described this phase as having short-range-order (e.g. van Olphen, 1971; Wada, 1989). Since the term "short-range-order" has been used to describe materials with a wide range of crystallinity, it would be confusing to use this term to describe materials such as allophane and ferrihydrite that show no repeat of structural units in any of the three dimensions. Therefore, to facilitate the discussion of materials composing the colloidal fraction of volcanic ash soils, we have chosen to use only two terms, "crystalline" and "noncrystalline" in this text.

Preferential formation of noncrystalline materials results in part from rapid weathering of volcanic glass, which shows the least resistance to chemical weathering. Rapid weathering releases elements faster than crystalline minerals can form. As a result, soil solutions become over-saturated with respect to several poorly ordered solid-phase materials. Rapid precipitation kinetics favor formation of these metastable solid-phases. With increased weathering, the glass fraction is depleted and soil solution activities of weathering products, such as silicon, are decreased. This leads to transformation of metastable, poorly ordered solid-phase materials to more stable, crystalline minerals. A second factor contributing to the relative absence of crystalline minerals in the clay-size fraction is that volcanic ash parent materials typically contain few precursor minerals (e.g. chlorite, mica) that can weather directly to crystalline 2:1 and 1:1 layer silicates. This deficiency of layer silicate clays may further impede formation of crystalline clay minerals due to the lack of a template that can serve to catalyze precipitation of crystalline layer silicates from solution. Therefore, the formation of noncrystalline materials can be directly linked to the unique properties of volcanic ash parent material.

The mineralogical composition of the colloidal fraction of soils derived from volcanic materials varies widely depending on (1) chemical, mineralogical, and physical properties of the parent material, (2) post-depositional weathering environment, and (3) the stage of soil formation. The mineralogy, chemical composition, and texture of the parent material largely determine the rate of chemical weathering, the amount and distribution of reactants for synthesis of secondary materials, and the pH through its influence on the base status of the soil. The post-depositional weathering environment is determined by interactions between soil temperature, rainfall and leaching, drainage effects, organic matter accumulations, additions of overburden ash deposits, and the pH of the weathering environment. The stage of soil formation relates to the length of time that weathering has proceeded in a given deposit. As weathering proceeds in a well-drained profile, the soil profile transforms from a Si-rich to Si-depleted environment as weathering reactions consume the glassy components of the parent material and Si is leached from the profile. All these factors and conditions interact to determine the soil forming factors that control soil genesis and the formation and transformation of clay minerals.

The purpose of this chapter is to outline our present knowledge of the nature, properties, and genesis of clay minerals in soils derived from volcanic materials. Specifically, this chapter discusses the significance of volcano-derived ejecta as a parent material, the major clay mineral constituents, weathering of volcanic materials, and clay mineral formation and transformation. Reviews dealing with the subject matter of this chapter have been published by Wada and Harward (1974), Fieldes and Claridge (1975), Nagasawa (1978), Wada (1978, 1980, 1985, 1989), Shoji (1983, 1986), Yoshinaga (1986, 1988), Lowe (1986), Mizota and van Reeuwijk (1989), and others.

5.2. VOLCANIC ASH AS A PARENT MATERIAL

Volcanic ash, in the context of this section, refers collectively to volcanic ejecta or tephra, including both pyroclastic fall and flow materials such as volcanic ash, pumice, and scoria (or cinders). All such materials are dominated by volcanic glass which shows the least resistance to chemical weathering and is thus of great importance as a parent material. The fine particle-size, glassy nature of the particles, and high porosity and permeability of volcanic ash enhance weathering and interaction in the soil environment. It is this rapid weathering that contributes to the formation of a unique colloidal assemblage and the physical and chemical properties that define the Andisol soil order. This section examines the mineralogical characteristics of volcanic ash as a parent material. For a detailed discussion of mineralogical and lithological properties of volcanic ash, readers should refer to the review by Shoji (1986).

5.2.1. Classification and chemical composition of volcanic ash

The characterization and classification of volcanic ash has commonly been made on the basis of the mineralogical composition. However, there may be significant contradictions between the classification of volcanic ash based on mineralogical properties as compared to chemical composition (Shoji et al., 1975; Yamada et al., 1975). Because of the difficulty in classifying volcanic ash by mineralogic properties, Shoji et al. (1975) proposed a classification into one of five rock types based on total silica content: rhyolite, dacite, andesite, basaltic andesite, and basalt (Table 5.1). Volcanic ashes comprising the parent material of the Andisol soil order show a complete range of chemical compositions with the exception of the ultra-basic rock type (Neall, 1985).

Classification based on silica concentrations is substantiated because the silica content of volcanic ash is closely correlated with the concentrations of all major elements (except K) (Table 5.2) and many minor elements (Kobayashi and Shoji, 1976; Shoji et al., 1975; Shoji et al., 1980). Classification based on chemical composition provides an effective criteria for volcanic ash, but may have limitations in some cases because dispersal of air-borne volcanic ash results in sorting of particles by size and density. This sorting typically leads to differences in the chemical composition and classification of volcanic ash as a function of distance from the volcanic vent. Classification based on mineralogical composition is also affected because heavy minerals are preferentially deposited nearest the volcanic vent.

The total chemical analysis for various volcanic ash samples shows that there is a strong relationship between SiO_2 content and the concentrations of most chemical elements (Table 5.2). Statistically significant relationships were shown for all elements, except for K in volcanic ash samples from Japan. A comparison of regression constants between tephras from Japan and New Zealand indicates that

TABLE 5.1

Classification of volcanic ash based on total silica content (Shoji et al., 1975).

Rock type	Rock	Total SiO_2 %
Felsic (acidic)	Rhyolite	100–70
	Dacite	70–62
Intermediate	Andesite	62–58
	Basaltic andesite	58–53.5
Mafic (basic)	Basalt	53.5–45

TABLE 5.2

Relationship between total SiO_2 content and selected elements in volcanic ash from Japan (n = 26) and New Zealand (n = 9) (Yamada, 1988).

Element	Japan Tephra		New Zealand Tephra	
Al_2O_3 %	$y = -0.242x + 30.90$	$r = -0.905$ [b]	$y = -0.247x + 31.09$	$r = -0.925$ [b]
Fe_2O_3 %	$y = -0.380x + 35.24$	$r = -0.962$ [b]	$y = -0.343x + 26.23$	$r = -0.976$ [b]
MgO %	$y = -0.255x + 18.38$	$r = -0.926$ [b]	$y = -0.218x + 16.05$	$r = -0.842$ [b]
CaO %	$y = -0.321x + 25.86$	$r = -0.967$ [b]	$y = -0.360x + 27.40$	$r = -0.956$ [b]
Na_2O %	$y = 0.097x - 2.58$	$r = 0.869$ [b]	$y = 0.346x + 0.346$	$r = 0.814$ [b]
K_2O %	$y = 0.032x - 0.82$	$r = 0.351$	$y = 0.096x - 3.936$	$r = 0.927$ [b]
TiO_2 %	$y = -0.014x + 1.54$	$r = -0.546$ [a]	$y = -0.029x + 2.31$	$r = -0.996$ [b]

Level of statistical significance: [a] = 95%; [b] = 99%

elemental relationships with the SiO_2 content are similar geographically, with the exception of K_2O and Na_2O. Concentrations of the alkali elements K and Na vary strongly depending on the source of the magma from which the volcanic ash originates. In Japan, the chemical composition of the magma source changes as a function of distance from the Japan trench. The three volcanic zones which parallel the Japan trench are designated as the tholeiite zone, high-alumina basalt zone, and alkali basalt zone, in order of increasing distance from the trench (Kuno, 1960). The K concentration of the magma and resulting volcanic ash shows a distinct increase with distance from the Japan trench. It is this difference in the source of magma that results in the poor correlation for SiO_2% versus K_2O% in the tephras of Japan. Shoji et al. (1975) found that the concentration of V in ferromagnetic minerals is negatively related not only with the concentration of Zn in ferromagnetic minerals, but also with SiO_2 content of volcanic ash. Thus, they showed that the rock types of volcanic ash are determined by the V–Zn relationship of ferromagnetic minerals.

5.2.2. Mineralogy of volcanic ash

The primary mineral composition of volcanic ash is typically characterized by first dividing minerals into light (specific gravity [SG] < 2.8–3.0) and heavy (SG > 2.8–3.0) mineral categories. Light minerals dominate in volcanic ash with an abundance mostly ranging between 70 and 95% (Shoji, 1986; Yoshinaga, 1988). Within the light mineral category, the relative abundance generally follows: noncolored volcanic glass $\gg$ plagioclase feldspars $\gg$ silica minerals (quartz, cristobalite and tridymite) $\simeq$ mica. The distribution of plagioclase and alkali feldspars is dependent on the chemical composition of the magma and is therefore quite variable. In general, the plagioclase feldspars are most prevalent and form a solid-solution series ranging from albite ($NaSi_3AlO_8$) to anorthite ($CaSi_2Al_2O_8$). Of the plagioclases, andesine ($Ab_{70}An_{30}$–$Ab_{50}An_{50}$) and labradorite ($Ab_{50}An_{50}$–$Ab_{30}An_{70}$) are the most common (Yamada et al., 1975; Dethier et al., 1981).

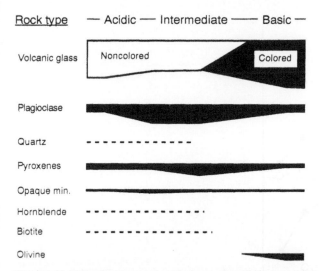

Fig. 5.1. Relationship between rock types of tephritic materials and their mineralogy (adapted from Shoji, 1983; copyright Hakuyusha, with permission).

Heavy minerals comprise only a small fraction of volcanic ash having a felsic or intermediate chemical composition. The relative abundance of minerals within the heavy mineral category follows: hypersthene $\simeq$ opaque minerals > augite $\simeq$ hornblende. The occurrence of other ferromagnesian minerals is related to the volcanic zones (magma sources). In volcanic ash with a basalt and basaltic andesite composition, colored volcanic glass becomes the dominant component in the heavy mineral fraction and olivine commonly occurs as a characteristic mineral (Fig. 5.1).

As one would expect, the mineralogic composition of volcanic ash varies according to rock types (Figs. 5.1 and 5.2). The content of crystalline minerals is highest in andesitic ashes and lowest in rhyolitic and basaltic ashes. Ash having the composition of rhyolite, dacite or andesite is dominated by noncolored volcanic glass with lesser amounts of plagioclase, pyroxenes, and ferromagnetic minerals. In contrast, volcanic ash of basalt and basaltic andesite composition is dominated by colored volcanic glass accompanied by plagioclase, olivine, pyroxenes, and ferromagnetic minerals.

The mineralogic composition of volcanic ash varies widely as a function of particle-size as shown in Fig. 5.2. Crystalline minerals are most common in the size range 100–500 μm. Plagioclase shows a relatively uniform distribution throughout the silt and sand size fractions. In contrast, the heavy mineral content shows a pronounced decrease with decreasing particle size, until heavy minerals are virtually absent in the size fractions less than 50 μm. Volcanic glass increases in relative proportion to plagioclase and heavy minerals as the particle size decreases.

The phenocrysts in tephra usually have a glassy coating that varies in thickness according to the kind of mineral and the explosiveness of the volcanic eruption.

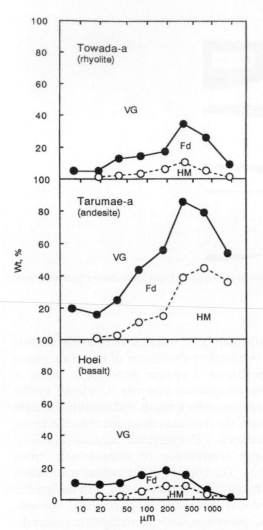

Fig. 5.2. Mineral composition of each size-fraction of three tephritic materials having the compositions of rhyolite, dacite and basalt. *VG* = volcanic glass, *Fd* = feldspars, *HM* = heavy minerals (adapted from Yamada and Shoji, 1975).

Often times, the ferromagnesian minerals hypersthene and hornblende are more highly glass-coated than the other minerals (Meyer, 1971). The amount of glass coating is rather small when the eruption is violent, since the glass crusts are abraded from the phenocrysts during the eruption.

Volcanic ash may also contain appreciable quantities of minerals that originate from previously altered materials associated with the volcano. These minerals, often called accessory, accidental or exotic minerals, are formed by weathering and/or hydrothermal alteration of materials comprising the cone of the volcano.

The accessory minerals along with the minerals formed from solidification of the magma are mixed and deposited together during a volcanic eruption. Accessory minerals attributed to previously altered materials often include: opal, cristobalite, kaolinite, allophane, halloysite, smectite, and interstratified layer silicates (Ossaka, 1982).

Following deposition of volcanic ash, the ash deposits are subject to further additions from eolian sources. Eolian inputs may be responsible for additions of quartz, mica, and other 1.4 nm minerals to Andisols (e.g. Inoue, 1981; Mizota, 1983; Mizota and Matsuhisa, 1985). The occurrence of quartz is almost ubiquitous in the silt and clay fractions of Andisols in Japan. Quartz concentrations decrease with increasing depth in the soil profile and are found even in ash of basaltic composition. These observations indicate that at least a portion of the quartz and possibly other 2:1 layer silicates originate from eolian sources.

5.2.3. Classification of volcanic glasses

Volcanic glass may be viewed as a metastable supercooled liquid which results from a reduction in nucleation and crystallization rates due to the rapid cooling of the magma. The chemical composition of volcanic glass is determined by the composition of the magma. Because cooling of the magma occurs so fast, there is little fractionation of elements during the solidification process. Volcanic glass is typically divided into colored and noncolored categories based on refractive indices of >1.52 and <1.52, respectively. As shown in Fig. 5.1, the content and type of volcanic glasses are closely related to the rock type of the volcanic ash. Noncolored glass is found in volcanic ashes having a rock composition of rhyolitic, dacitic, and andesitic, while colored glass is found in basaltic andesite and basaltic rock types. The relationship between total silica content of volcanic ash and the refractive index of glass is shown in Fig. 5.3. Noncolored glasses have refractive indices ranging between 1.49 and 1.51, while colored glasses have refractive indices greater than 1.52.

The straight line in Fig. 5.3 indicates the regression equation relating the refractive index and silica content of the glasses produced by fusion of various volcanic rocks (Kittleman, 1963). The glasses produced by fusion yield a highly significant linear relationship that differs from the relationship between total silica content of the volcanic ash and the refractive index of the glass component. This deviation results from the difference in the distribution of volcanic glass and crystalline minerals as a function of rock type. The deviation was largest for andesitic ash which contains the largest content of crystalline minerals (~50%) and lowest for rhyolitic and basaltic ash that contain few crystalline minerals (~10%) (Fig. 5.2). Thus, there is only a small difference in chemical composition between the bulk ash and the volcanic glass component in rhyolitic and basaltic ash.

Striking differences occur between the chemical composition of noncolored and colored glasses (Yamada and Shoji, 1983). Noncolored and colored glasses

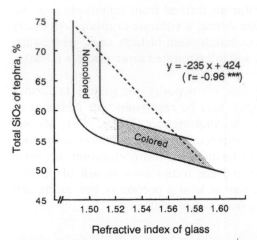

Fig. 5.3. Relationship between the refractive indices of volcanic glasses from various tephritic materials and the total silica content of the corresponding volcanic glass (adapted from Kobayashi et al., 1976).

display total silica contents of 74–78% and 62–54%, respectively. Colored glass contains higher concentrations of the cations Al, Fe, Ti, Mg, and Ca, and lower concentrations of K. Volcanic glasses show close correlations between their silica content and the contents of other elements, with the exception of Na and K (Shoji, 1986; Yamada, 1988).

Volcanic glass exhibits a wide variation in its morphology as a result of the magma viscosity (determined by the chemical composition and gas content of the magma) and the type of volcanic eruption. Yamada and Shoji (1983) divided volcanic glass particles (100 to 200 μm size range) into four categories based on morphological characteristics: sponge-like, fibrous, curved platy, and berry-like (Fig. 5.4). Sponge-like glass particles are highly vesicular suggesting that these glasses were produced by violent explosions of highly viscous magma. Fibrous glass particles contain elongated vesicles. Curved platy particles are produced by the pulverization of glasses having relatively large vesicles. Berry-like glass particles show angular blocky or subangular blocky forms with low vesicularity and contain crystallites of plagioclase. Noncolored volcanic glass consists mainly of sponge-like particles which are siliceous and highly vesicular. In contrast, colored volcanic glass in basaltic andesite and basaltic ashes contains only berry-like particles.

5.2.4. Alteration of volcanic ash

Volcanic ash contains a number of components that are highly weatherable in the soil environment. Among the various minerals, volcanic glass is the most weatherable component as a result of its amorphous nature. Morphologically, volcanic glass weathers through several stages (Ruxton, 1988). Initially, the clear glass becomes discolored in shades of yellow-brown and then becomes semi-

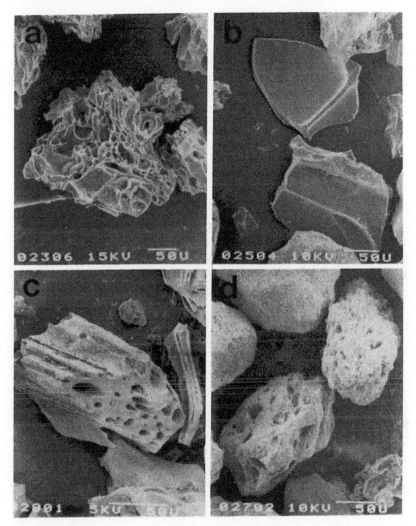

Fig. 5.4. Electron micrographs of four morphological types of volcanic glass. Reference lines represent 50 μm. a. Sponge-like; b. Curved platy; c. Fiberous; d. Berry-like (courtesy of I. Yamada).

opaque. With increased weathering, the grains become opaque and often develop rounded or nodular form. The nodular sheath surrounding the fresh interiors appears similar to the incongruent weathering of some primary minerals in laboratory experiments (Wollast and Chou, 1984).

Weathering of volcanic glass can be described in terms of a combination of parabolic and linear kinetics, reflecting the hydration of glass and the formation of clay minerals, respectively (Hodder et al., 1990). Laboratory glass weathering experiments have consistently shown rapid hydration of glass and release of cations into solution according to parabolic kinetics (White, 1983; White and Claassen,

1980). The initial rapid reactions involve an interchange of aqueous hydrogen ions for cations situated both on and near the surface of the glass particle. This results in formation of a cation-depleted leached zone at the glass surface that is composed of structurally bonded silica and aluminum. For extended periods of reaction, the release of Na and K continues to display parabolic kinetics while the release of Al and Si are observed to display "linear kinetics" (approximately first order reactions). A coupled weathering model, involving surface dissolution with concurrent diffusion of cations, produces a mass balance between the aqueous and glass phases. Dissolution kinetics are controlled by ion transport to and from sites within the glass. Laboratory experiments indicate that the parabolic diffusion rate of a chemical species from the solid is a nonlinear function of its aqueous concentration. Therefore, weathering rates are strongly influenced by the rate of leaching and removal of weathering products.

Field investigations of glass weathering indicate two types of volcanic glass alteration during the early stages of weathering in noncolored glass (Yamada and Shoji, 1983). The first type of alteration was characterized by a loss of Na and resorption of K by the glass phase under strongly acidic conditions (pH < 5) in nonallophanic surface soils. The alteration involved an exchange reaction between Na^+ and H^+ plus K^+, without any significant change in the glass structure. It was determined that the Na^+–H^+ exchange reaction occurred first followed by the H^+–K^+ exchange reaction. Such a resorption mechanism is probably controlled by changing surface potentials related to changes in surface cation concentrations. Potassium resorption becomes progressively less with increasing pH and is not apparent at neutral pH values (White and Claassen, 1980).

The second type of alteration was characterized by a gain in Na and/or Ca concentrations and a decrease in K concentration. This process occurred in soil horizons that contained allophanic materials, such as weakly acidic surface soils, subsurface soils, and buried soils. The cation enrichment/depletion proceeds concomitantly with desilication of the glass structure. Aluminum and iron concentrations were increased by negative enrichment due to loss of silica.

Weathering rates are very sensitive to the characteristics of the glass and to the nature of the weathering environment. Studies of long-term weathering from ash deposits of various ages indicate half-lives for volcanic glass ranging from 1650 to 5000 years (Ruxton, 1988). In the soil environment, the weathering rate of glass is regulated by the chemical composition of glass, surface area (controlled by grain size and the vesicular nature of particles), soil temperature, leaching potential (amount of water percolating through the ash), thickness of ash being weathered, and pH and complexing ligand concentrations in percolating soil solutions.

Differences in the chemical composition of glass result in colored glass being more susceptible to weathering than noncolored glass. Colored glass has a much greater concentration of cations, such as Al, Fe, Ca, and Mg, which leads to substitution of these cations for silica in the glass structure. Cation substitution weakens the bonding characteristics of the glass because there are fewer silica–silica linkages. In silica-rich glasses, the silicate tetrahedra are highly condensed because individual silica tetrahedra are linked together largely through oxygen ions and these strong bonds are not easily broken by chemical weathering. Cations are also susceptible to rapid removal from both the surface and within the glass by ion exchange reactions with hydrogen ions. Removal of cations weakens the glass and increases the porosity of the glass which makes the glass more susceptible to dissolution. Laboratory dissolution studies showed that elemental release rates were 1.5 times greater for colored glass (basaltic andesite composition) than for noncolored glass (rhyolite composition) (Shoji et al., 1993). Among noncolored glass, andesitic glass with a smaller Si : Al atomic ratio and a more porous structure showed greater weathering than rhyolitic glass (Kirkman and McHardy, 1980).

The chemical composition and gas content of magma along with the nature of the volcanic eruption determine the particle-size and vesicularity of the glass. Decreasing particle-size and increasing vesicularity result in a greater surface area exposed to chemical weathering. The viscosity and gas content of the magma combine to determine the amount of vesicles found in glass. The rate constant for dissolution of glass might be expected to be independent of the total surface area. However, in the absence of continuous leaching, a doubling of the surface area exposed to weathering increased total dissolution by a factor of approximately 1.4 times (White and Claassen, 1980). The nonlinear increase in weathering results from higher aqueous concentrations in the soil solution that increases the concentration of species at the glass surface and decreases the diffusion gradient within the glass.

Leaching potential and the thickness of ash deposits both affect weathering rates through their control on solute concentrations. As solute concentrations increase, mass transfer of weathering products is decreased as the diffusional driving force for ion exchange and transport processes are reduced, with a subsequent decrease in the weathering rate. Volcanic ash from the 1980 eruption of Mt. St. Helens was artificially added in thicknesses of 15 and 5 cm to forest soils in the Cascades of Washington State. Soil solution studies indicated that the weathering rate in the 5 cm thickness of ash was approximately twice that found in the 15 cm ash thickness (Dahlgren and Ugolini, 1989a). The greater weathering rates in the 5 cm thickness of ash were attributed to the efficiency at which weathering products were removed; the number of pore volumes of leachate being three times greater in the 5 cm addition.

The coupled weathering model for glass, involving surface dissolution with concurrent diffusion is highly sensitive to changes in soil temperature. Laboratory dissolution studies by Shoji et al. (1993) for both colored and noncolored glass showed that weathering rates increased by a factor of 1.4–1.5 for each 10°C increase in temperature. The temperature dependency of weathering rates determined in the laboratory was substantiated by comparison to weathering rates determined for Andisols in Japan showing a range of soil temperatures.

The primary reactions in the dissolution of volcanic glass involve the consumption of hydrogen ions. Therefore, the weathering rates are dependent on the hydrogen ion concentration in the soil solution. White and Claassen (1980)

determined that the H-ion dependency for glass weathering based on Na dissolution rates was $[H^+]^{0.49}$. A stronger pH dependence for Ca and Mg is expected
since exchangeable divalent cation equilibria are exponentially related to H ion
concentration. In contrast, no ion-exchange reactions occur between H ions and
uncharged silica resulting in the rate constant for silica being largely independent
of pH. Weathering rates are also influenced by soluble organic compounds that act
as proton donors and complex soluble metals such as Al. Ligands that form complexes with Al typically enhance weathering rates of aluminum bearing minerals
(Stumm et al., 1985). The mechanism is ligand exchange with OH on the surface
and subsequent detachment of the metal-ligand group. The dissolution rate in the
presence of complexing ligands is proportional to the concentration of the surface
complex (Stumm et al., 1985).

In relation to volcanic glass, the other primary minerals found in volcanic
ash have the following approximate stability with regard to chemical weathering
(Aomine and Wada, 1962; Loughnan, 1969; Mitchell, 1975; Shoji et al., 1974;
Yamada et al., 1978):

Colored volcanic glass < noncolored glass = olivine < plagioclase
  < augite < hypersthene < hornblende < ferromagnetic minerals

Olivine is considered to be as susceptible to chemical weathering as volcanic
glass; however, it occurs in only small amounts in volcanic ash. Of the ferromagnesian minerals, augite is relatively susceptible to chemical weathering while
hornblende is very stable and observed to be almost fresh or slightly etched in
5000 years old soils in northeastern Japan (Yamada et al., 1978). Plagioclase
minerals are the second most abundant component of volcanic ash next to glass
and are therefore an extremely important primary mineral in volcanic soils. The
plagioclase minerals are highly weatherable among the crystalline primary minerals
with those of calcic composition being more susceptible to weathering than those
of sodic composition. The weathering rates of these crystalline primary minerals
are often altered due to their encapsulation in a protective coating of glass.

5.3. COLLOIDAL CONSTITUENTS COMMON TO VOLCANIC ASH SOILS

The unique physical and chemical properties of Andisols and other soils
derived or influenced by volcanic parent materials result from active (highly
reactive) solid-phase pools of Al and Fe. Active Al and Fe in the clay-size fraction
of soils derived from volcanic ash consist of allophane, imogolite, and ferrihydrite
on the one hand, or Al/Fe-humus complexes often together with opaline silica on
the other. These two groups of noncrystalline constituents may occur together, but
there is an inverse relationship because the two groups have opposing conditions
favoring their formation (Shoji et al., 1982; Shoji and Fujiwara, 1984; Saigusa
and Shoji, 1986; Mizota and van Reeuwijk, 1989; Parfitt and Kimble, 1989).
Metal-humus complexes are dominant at pH values less than 5, while allophane

and imogolite become predominant at higher pH values. Complexation of Al by humus lowers Al activities in humus-rich horizons and has been shown to inhibit formation of allophane and imogolite, with the excess Si precipitating in the form of opaline silica in young Andisols. Inhibition of allophane and imogolite formation by incorporation of Al into Al-humus complexes is an example of the "anti-allophanic effect" that was first observed in young Andisols from Hokkaido, Japan (Shoji and Masui, 1972). The anti-allophanic effect is a dominant process in nonallophanic Andisols where pH values less than 5.0 and high organic matter concentrations favor the formation of Al-humus complexes.

Halloysite is another common constituent in many volcanic soils and displays a wide range of structural disorder. An inverse relationship has been shown for the abundance of halloysite compared to Al-rich allophane and imogolite (Saigusa et al., 1978; Parfitt et al., 1983; Parfitt et al., 1984; Parfitt and Wilson, 1985; Singleton et al., 1989). Halloysite is found as the dominant clay mineral in Si-rich environments, while Al-rich allophane and imogolite are found at comparatively lower Si concentrations. In addition to these poorly ordered constituents, kaolinite, gibbsite, and various 2:1 or hydroxy-Al interlayered 2:1 layer silicates (chloritized 2:1 layer silicates) are commonly found in soils derived from volcanic materials. The following sections briefly describe the major colloidal constituents commonly found in soils derived from volcanic ash parent material.

5.3.1. Allophane and "allophane-like constituents"

Allophane is a group name given to a series of naturally occurring, noncrystalline, hydrous aluminosilicates with a widely varying chemical composition (van Olphen, 1971). Synthetic experiments indicate that allophane, along with imogolite, is formed in nature by coprecipitation of monomeric or condensed Si anions with hydroxy-Al (Farmer et al., 1979a; Farmer, 1982; Wada et al., 1979; Wada and Wada, 1980). Allophane consists of hollow, irregularly spherical particles with outside diameters of 3.5 to 5 nm and a wall thickness of 0.7–1 nm (Fig. 5.5). Specific surface area measurements range from 581 m^2 g^{-1} by nitrogen at 77 K (Hall et al., 1985; Vandickelen et al., 1980) to 700–1100 m^2 g^{-1} by adsorption of ethylene glycol monoethyl ether (Egashira and Aomine, 1974). Allophane has no definite chemical composition and displays a range in Al and Si concentrations from an Al:Si atomic ratio of 1:1 to 2:1. Allophanes with Al:Si atomic ratios less than 1:1 and greater than 2:1 may be present in nature; but distinct, individual phases have not been isolated and characterized.

Based on the varying ratio of Al:Si, allophanes can be operationally divided into two end-members, Al-rich (Al:Si = 2:1) and Si-rich (Al:Si = 1:1) allophanes. Al-rich allophane (also termed proto-imogolite allophanes or imogolite-like allophanes) is related to imogolite by having the same local atomic arrangement and chemical composition (Al:Si = 2:1); however, they differ morphologically. Al-rich allophanes are considered to be made up of fragments having the same local atomic arrangement and chemical composition as the imogolite

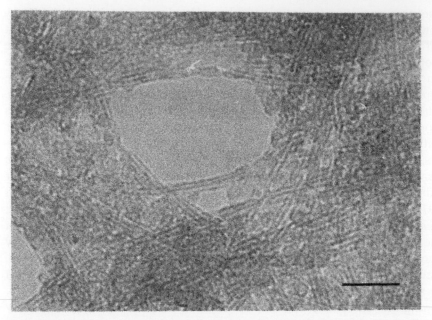

Fig. 5.5. Electron micrograph of allophane (spherules) and imogolite (threads) (clay specimen of Towada Andisol). The reference line represents 20 nm (courtesy of M. Saigusa).

structure over a short range (Parfitt and Henmi, 1980). The infrared spectra of imogolite and allophane have many similar features indicating a similarity in their atomic arrangement; however, the absorption bands for allophane are generally broader due to the greater degree of disorder in the allophane structure. Differences exist in the 800–1200 cm^{-1} absorption region and in the intensity of the 348 cm^{-1} absorption maxima (Wada, 1989). Allophanes show a single absorption maximum, associated with Si(Al)O vibrations, whose frequency increases from approximately 975 to 1020 cm^{-1} as the Al/Si atomic ratio increases from 1:1 to 2:1. In contrast, imogolite shows two absorption maxima at approximately 940 and 1000 cm^{-1}; the band near 1000 cm^{-1} being due to the tubular morphology of imogolite (Parfitt and Hemni, 1980). Imogolite displays a well-defined IR absorption band at 348 cm^{-1} that is either missing or substantially weaker for the allophane structures (Farmer et al., 1977).

Si-rich allophanes with Al:Si atomic ratios of approximately 1:1 have polymerized silicate together with some orthosilicate groups. Both infrared and NMR spectroscopy indicate that the portion of polymerized Si increases as the Al:Si atomic ratio decreases (Parfitt et al., 1980; Goodman et al., 1985; Shimizu et al., 1988). The Al occurs mainly in octahedral sites with some present in tetrahedral sites (Kirkman, 1975; Goodman et al., 1985; Shimizu et al., 1988). The octahedral Al layer provides the structural framework for Si-rich allophanes and links to silicate polymers (Parfitt et al., 1980).

Al-rich allophane (Al:Si $\simeq$ 2:1) is the most abundant type of allophane found in soils (Parfitt and Kimble, 1989; Andisol TU Database, 1992). Allophane with an Al:Si atomic ratio of 1:1 is not commonly found in Andisols and Spodosols, especially in soils with udic soil moisture regimes. Because of their lesser abundance, Si-rich allophanes are less well understood than their Al-rich allophane counterpart.

Allophanes with Al:Si ratios less than 1 to as high as 4 have been proposed based on the results of selective dissolution (Parfitt and Kimble, 1989; Farmer and Russell, 1990). However, natural soil allophanes with Al:Si chemical compositions other than 1 or 2 have not been positively identified. Soil allophanes with Al:Si atomic ratios between 1 and 2 have been shown to have structural properties of both Al-rich and Si-rich allophanes. This may indicate that these allophanes are simply mixtures of Al-rich and Si-rich allophanes mixed together in various proportions (Parfitt et al., 1980). These mixtures could consist of distinctly different phases and/or mixtures of structures within the allophane particles.

Allophanes with Al:Si ratios greater than 2 may be explained by the non-specificity of selective dissolution techniques or to replacement of orthosilicate tetrahedrons by hydroxyls in the allophane structure. Acid oxalate is the reagent of choice for dissolving allophane and imogolite from soils. The Al:Si ratio is typically estimated by using the formula $(Al_o - Al_p)/Si_o$, where Al_o and Si_o are acid oxalate extractable Al and Si, respectively, and Al_p is pyrophosphate-extractable Al (Parfitt and Wilson, 1985). In addition to allophane and imogolite, oxalate also dissolves Al-humus complexes, some gibbsite, some hydroxy-Al interlayers, Al-substituted ferrihydrite, and incongruently removes Al from poorly crystalline halloysite (Wada and Kakuto, 1985; Parfitt and Childs, 1985). Aluminum extracted by pyrophosphate reagent is typically subtracted from acid oxalate extractable Al to correct for removal of Al-humus complexes. However, the many other Al-bearing phases partially attacked by acid oxalate lead to an overestimation of the Al:Si atomic ratio. This may account for Al:Si ratios in excess of 2, especially when Si_o concentrations are low leading to division by a small number. Dahlgren and Ugolini (1991) were able to improve the estimate of the Al:Si atomic ratio in imogolite-rich Spodosols by substituting dithionite-citrate-extractable Al (Al_d) for Al_p as follows: $(Al_o - Al_d)/Si_o$. Dithionite-citrate did not appreciably attack imogolite, but it did partially remove hydroxy-Al interlayers and Al-substituted ferrihydrite similar to acid oxalate. Alternatively, allophanes with Al:Si ratios greater than 2 may exist as natural phases. The hydroxyl ligand may replace the orthosilicate tetrahedra leading to a Si-depleted phase.

Allophane-like constituents are defined as noncrystalline aluminosilicates that are dissolved by dithionite-citrate and hot 2% Na_2CO_3 (Wada and Greenland, 1970). Allophane-like structures have not been isolated and their existence is only inferred from difference infrared spectra, using clay specimens before and after chemical treatment with dithionite-citrate or Na_2CO_3. Features of the infrared spectrum indicate similarity to both allophane and imogolite. Allophane-like constituents may simply be due to incongruent dissolution of allophane and/or imogolite by selective dissolution techniques (Parfitt et al., 1980). Alternatively, they may be a polymeric hydroxy-aluminosilicate cation as suggested from synthesis of these cations in dilute solutions containing orthosilicic acid and hydroxy-Al ions (Wada and Wada, 1980). Whether the allophane-like constituents are a distinct phase or just a product of incongruent dissolution requires further study.

The structure of allophane remains a topic of considerable debate. X-ray fluorescence spectroscopy has shown that allophanes may contain Al in both 4- and 6-fold coordination and that the content of Al in 4-fold coordination increases as the Si content of the allophane increases. That allophane has a sheet structure related to kaolinite has been proposed by Udagawa et al. (1969), Brindley and Fancher (1969), Okada et al. (1975), and Wada and Wada (1977). Wada and Wada (1977) suggested that the wall of the allophane sphere consists of a defective kaolin structure containing on octahedral (Al^{VI}) sheet and one tetrahedral (Si, Al^{IV}) sheet. They proposed that both the Al-rich and Si-rich end members contain tetrahedral sheets with Al substituting for Si in one-third of the cation sites. The depletion of Al in the Si-rich allophane (Al : Si = 1) resulted from two-thirds of the possible octahedral sites being vacant, suggesting that the tetrahedral sheet provides the framework of the structure.

Parfitt and Henmi (1980) showed that some Al-rich allophanes from New Zealand were composed of distorted imogolite structures. These samples contained all their Al in 6-fold coordination and the octahedral (gibbsite-like) sheet provided the structural framework. The gibbsitic sheet forms the outer surface with O_3SiOH groups bonded on the inside surface. There is evidence that pores exist in the allophane spherule with diameters ranging between 0.3 and 2.0 nm (Paterson, 1977). These pores appear to originate from omission of two Si atoms and two Al atoms from every six units cells as shown in Fig. 5.6.

Si-rich allophanes investigated by Farmer et al. (1979a) suggested a structure with properties similar to halloysite. These Si-rich allophanes contained only a small fraction of Al^{IV} in contrast to those examined by Wada and Wada (1977). Thus, it was concluded that the Al was primarily in octahedral sites, and the octahedral sheet provides the structural framework for the wall of the allophane spherule.

5.3.2. Imogolite

Imogolite was first described by Yoshinaga and Aomine (1962) in a soil derived from the glassy volcanic ash, known as "imogo". Imogolite commonly has been found in association with allophane and is similar to allophane in many of its chemical properties. Imogolite is a short-range-order hydrous aluminosilicate mineral with a distinct tubular morphology that may extend several microns in length (Fig. 5.5). It consists of bundles of well-defined fine tubes with inner and outer diameters of 1.0 and 2.0 nm, respectively. The external surface of the tube is composed of a curved gibbsite-like structure with an orthosilicate (O_3SiOH) group coordinated through oxygen with three aluminum atoms in the interior as shown in Fig. 5.7

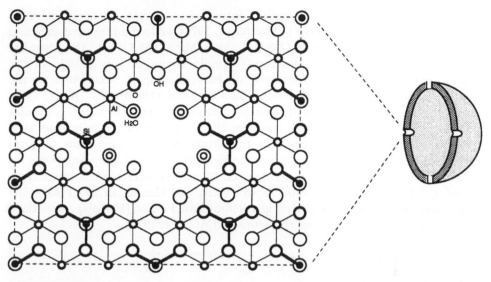

Fig. 5.6. Schematic of the structure of allophane and a micropore on its surface (prepared from Parfitt and Henmi, 1980).

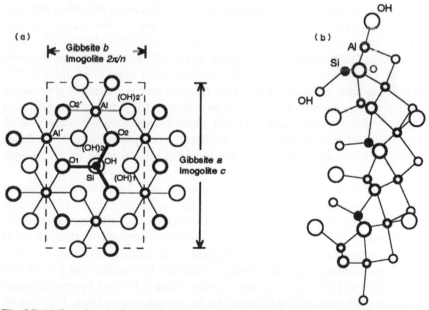

Fig. 5.7. (a) Postulated relationship between the structural unit of imogolite and that of gibbsite. SiOH groups that would lie at the cell corners in imogolite are omitted from the diagram. (b) Curling of the gibbsite (hydroxide) sheet induced by contraction of one surface to accommodate SiO$_3$OH tetrahedra; projection along the imogolite c-axis (prepared from Cradwick et al., 1972 and K. Wada, 1989, with permission).

(Cradwick et al., 1972). The best empirical formula for naturally occurring imogolite is $1.1 SiO_2 Al_2O_3 2.3 - 2.8 H_2O(+)$ (Yoshinaga, 1970) while the proposed structure indicates a formula of $(OH)_3Al_2O_3SiOH$ [$SiO_2 \cdot Al_2O_3 \cdot 2 H_2O(+)$]. The chemical composition of the proposed structural formula compares very closely to the formula obtained by chemical analysis of natural samples. Aluminum in the imogolite structure exists only in 6-fold coordination (Henmi and Wada, 1976; Shimizu et al., 1988). The atomic arrangement in the imogolite tubes is regular along the axis (long-range-order); however, the diameters of the tubes can vary. Some randomness is also involved in the arrangement of tube units to form thread-like bundles. Surface area measurements for imogolite vary depending on the method and experimental pretreatments. Measured surface areas range from 700 m^2 g^{-1} by adsorption of water vapor (Wada and Henmi, 1972) to 900–1100 m^2 g^{-1} determined by the ethylene glycol monoethyl ether (EGME) method on freeze-dried imogolite specimens (Egashira and Aomine, 1974).

5.3.3. Opaline silica

Two types of opaline silica are common in young volcanic ash soils: pedogenic opaline silica (commonly called laminar opaline silica) and biogenic opaline silica (plant opal and diatoms). These types of opaline silica can be easily distinguished according to their morphological properties. Laminar opaline silica appears as a common constituent of the clay-size fraction in surface horizons of young volcanic ash soils (Shoji and Masui, 1969, 1971). They occur only in the 0.2–5 μm size fraction and are most abundant in the 0.4 to 2 μm range. Morphologically, they appear as extremely thin particles with circular, elliptical, rectangular, and rhombic shapes (Fig. 5.8). Of the four shapes, circular and elliptical types predominate; the elliptical type being most common in the fine fraction and the circular type being dominant in the coarse fraction. Particles show a fine-grained, uneven surface and appear to be very porous suggesting that the particles are actually composed of extremely fine silica spheres. Many of the weathered silica particles reveal a series of rings in cross section suggesting that these particles have formed in a step-wise fashion.

Opaline silica is found more abundantly in younger soils (<4000 years) than in older soils and in humus-rich A horizons rather than underlying B and C horizons. Therefore, opaline silica is a product of the early stages of weathering of volcanic ash, and its formation is favored in the surface horizons where the activity of Al is suppressed by the formation of Al-humus complexes. The formation of opaline silica and that of allophane/imogolite is also antagonistic due to competition for soluble silica. Shoji and Masui (1971) inferred that opaline silica is formed by precipitation from soil solutions over-saturated with silica due to surface evaporation. Freezing of soils has also been shown to lead to formation of opaline silica by concentration of soluble silica (Wada and Nagasato, 1983; Ping et al. 1988). Therefore, the formation of opaline silica is closely correlated with climatic conditions (Shoji and Masui, 1971).

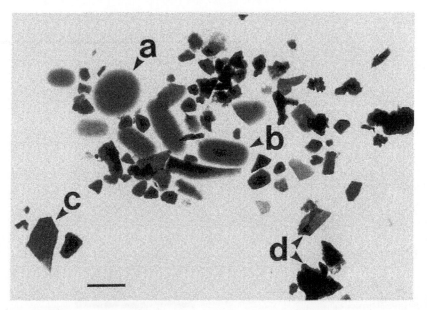

Fig. 5.8. Electron micrograph of laminar (pedogenic) opaline silica (clay specimen of an Andic Entisol from New Berry ash). The reference line represents 2.0 μm. a = circular; b = elliptical; c = diatom; d = volcanic glass.

Opaline silica is a transitory component in young volcanic ash soils and does not have a very significant influence on soil properties. However, its presence does signify high silica activities and its presence may buffer soil solution silica activities at high levels during the early stages of weathering. Other forms of opaline silica such as plant opal and diatoms have been found in soils derived from volcanic materials (Kato, 1983; Mizota et al., 1982; Kondo et al., 1988). Plant opal occurs mainly in A and buried A horizons and is derived mostly from grass and Sasa species, indicating the importance of these vegetation types on silica biocycling. Plant opal particles are thick and large in size with unique morphological forms dictated by the shape of the plant vacuoles in which they precipitate. Because the shape of plant opal is species specific, they have been extensively used to determine paleovegetation when they occur in buried horizons of Andisols (Kondo et al., 1988).

5.3.4. Halloysite

Halloysite is a common constituent in volcanic-derived soils and occurs as the dominant clay mineral in many Si-rich environments (Parfitt and Wilson, 1985). Halloysite is a 1:1 aluminosilicate mineral which is characterized by a diversity of morphology. In rare cases, halloysite has been shown to incorporate appreciable concentrations (13% Fe_2O_3) of iron in the octahedral position (Wada and Mizota, 1982). Generally it occurs with a tubular and spheroidal morphology (Fig. 5.9)

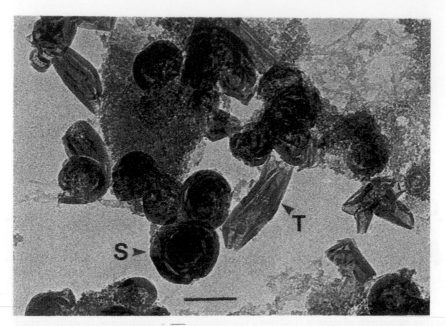

Fig. 5.9. Electron micrograph of halloysite (clay specimen of Towada Andisols). The reference line represents 0.1 μm. S = spheroidal halloysite; T = tubular halloysite (courtesy of M. Saigusa).

in Andisols, but lath-shaped, platy, and crumpled shapes have also been reported (Nagasawa, 1978; Saigusa et al., 1978; Wada and Mizota, 1982; Nagasawa and Noro, 1987). Most halloysites are hydrated (1.0 nm basal spacing) by two interlayer water molecules per formula unit; however, they are susceptible to dehydration (0.7 nm basal spacing) under climatic conditions showing a distinct seasonal moisture deficit (Takahashi et al., 1993).

Halloysite exhibits a wide range of structural disorder due to random stacking of structural layers. Wada and Kakuto (1985) described a poorly-crystalline form of halloysite that they termed embryonic halloysite. These materials showed no X-ray diffraction peaks but exhibited infrared absorption bands characteristic of halloysite. Embryonic halloysites varied in morphology from thin-platy flakes to elongated, rolled thick plates, and spherical particles. Embryonic halloysite was shown to dissolve congruently in hot sodium citrate solution and incongruently (Al/Si ratio = 2.5–3) by acid oxalate treatment (Wada and Kakuto, 1985). Similar embryonic halloysites have also been found in paddy soils derived from volcanic ash in Japan (Wada et al., 1985).

5.3.5. Noncrystalline iron oxides

Iron in soils derived from volcanic materials is present mostly in the form of noncrystalline oxyhydroxides and partly as Fe-humus complexes (Parfitt and

Childs, 1983; Yoshinaga, 1986; Parfitt et al., 1988; Childs et al., 1990, 1991). The greater stability of Fe in oxides as compared with that in humus complexes favors the formation of oxyhydroxides (Wada and Higashi, 1976). The dominant noncrystalline oxyhydroxide is believed to be ferrihydrite, a short-range-order Fe hydroxide mineral with a bulk composition of $5 Fe_2O_3.9 H_2O$ (Schwertmann and Taylor, 1989). A fibrous form of poorly crystalline goethite was also reported in two Andisols formed in volcanic ejecta in Japan (Nakai and Yoshinaga, 1980).

Ferrihydrite is regarded as a highly reactive material due to its hydroxylated surface and high specific surface area (typically 220 to 560 m^2/g). Ferrihydrite is thermodynamically metastable, and with time converts to stable Fe-oxides, usually to goethite under a temperate or cool humid climate, and to hematite under a warmer, dryer climate. Ferrihydrite resembles hematite structurally, except that some Fe positions are vacant and some O and OH groups are replaced by water molecules (Towe and Bradley, 1967). Ferrihydrite appears as individual spherical particles ranging in size between 2 and 5 nm. These particles become highly aggregated forming aggregates 100–300 nm in diameter. Poorly ordered phases have a smooth surface while more crystalline phases tend to have a rough textured surface (Carlson and Schwertmann, 1981). Concentrations of ferrihydrite can be estimated from acid oxalate extractable Fe (Fe_o) concentrations using a factor of 1.7 to convert Fe_o values to ferrihydrite concentrations (Parfitt and Childs, 1988; Childs et al., 1991). Due to the highly reactive nature of the ferrihydrite surface, relatively large quantities of silica (2–6%), organics and phosphates are chemisorbed. This sorption has been shown to impede the crystallization of ferrihydrite to more crystalline goethite and hematite (Schwertmann, 1988; Vempati and Loeppert, 1989).

5.3.6. 2:1 and 2:1:1 layer silicates and their intergrades

Layer silicate minerals of the 2:1 and 2:1:1 type and their intergrades are often found in soils derived from volcanic ejecta. These layer silicates are the dominant clay mineral in the nonallophanic group of Andisols. Nonallophanic Andisols are characterized by $pH(H_2O)$ values less than 5.0, a clay fraction dominated by 2:1 type minerals, and a virtual absence of allophane/imogolite. The origin or formation of 2:1 and 2:1:1 layer silicates in soils derived from volcanic materials has been a topic of considerable interest. Several hypotheses have been proposed to explain their occurrence: (1) alteration products from mafic minerals such as pyroxenes, amphiboles, and micas in the parent material (Kawasaki and Aomine, 1966; Mizota, 1976), (2) formation from amorphous materials as products of an advanced stage of weathering (Masui et al., 1966), (3) hydrothermal alteration products formed in the crater prior to the eruption and deposited with the volcanic ejecta (Kondo et al., 1979; Ossaka, 1982; LaManna and Ugolini, 1987), (4) solid-state transformation of volcanic glass to a 2:1 illite type mineral by K retention (Shoji et al., 1981; Yamada and Shoji, 1982), and (5) inherited from deposition of eolian materials, such as loess (Inoue, 1981; Mizota, 1982, 1983; Mizota and Inoue, 1988; Inoue and Naruse, 1990). No consensus has been reached as to the origin of 2:1 type minerals in volcanic soils. It is certainly possible that all of the above hypotheses operate to differing degrees in volcanic soils but that the dominant mechanism differs from soil to soil depending on the pedologic environment.

Recent oxygen isotope studies of fine-grained quartz has shown that the eolian transport process is important in formation of 2:1 layer silicates in Japan. There is a strong correlation between the content of 2:1 layer silicates and that of fine-grained quartz in many nonallophanic Andisols. These soils are characterized by an abundance of 2:1 layer silicates in the surface horizons with the largest concentrations occurring in soils with the highest snowfall on the windward side (western) of Japan. Direct studies of eolian inputs in Japan range from 0.5 to 1.0 g cm^2 per 1000 years (Mizota and Inoue, 1988). Estimated deposition rates for the last glacial age were significantly higher, ranging from 1.9 to 3.2 g cm^2 per 1000 years. These eolian deposits contain fine-grained quartz and mica as dominant mineral constituents with lesser concentrations of smectite and vermiculite. During the early stages of weathering, smectite, vermiculite, and mica are present as the dominant 2:1 layer silicates. As weathering advances, hydroxy-Al interlayering occurs in the smectite and vermiculite.

5.3.7. Aluminum-humus complexes

Aluminum-humus complexes are the dominant form of active Al in humus horizons of Andisols. Pyrophosphate-extractable Al (Al_p) concentrations approach 3 percent in humus-rich Andisols (see Chapter 6). In contrast, Fe-humus complexes are very low even in humus-rich horizons because iron has a greater stability as Fe oxyhydroxides compared to humus complexes. It is suggested that the accumulation and stabilization of humus in Andisols is in part due to the formation of Al-humus complexes. The formation of Al-humus complexes renders the humus highly resistant to microbial attack. As a result, the mean residence time of organic carbon in Andisols is much greater than the turnover times of organic carbon in Mollisols and Spodosol Bh horizons (Inoue and Higashi, 1988). This process of organic matter stabilization plays a major role in the formation of fulvic and melanic epipedons in Andisols. The humus continues to accumulate as weathering releases Al to form complexes with humus rendering the humus recalcitrant to decomposition.

Interactions between humic substances and Al have been described as ion-exchange, surface adsorption, chelation, peptization and coagulation reactions (Schnitzer, 1978). Metal-humus complexes are believed to form primarily by the interaction of metals with carboxylic functional groups. Humic and fulvic acid fractions from soils of the world have a mean carboxylic group content of 3.6 and 8.2 mole kg^{-1}, respectively (Schnitzer, 1977). These values compare to carboxyl contents of approximately 5 mole kg^{-1} for A-type humic acids (the dominant type of organic carbon in melanic epipedons) extracted from Andisols (Yonebayashi and Hattori, 1988). The stability of Al in humus complexes is determined by the stability constants of all potential complexing sites, the pH of the solution, the concentration and speciation of the aqueous Al, and the concentration of competing aqueous species. Young humus has a low affinity for complexing Al that increases as the humus becomes more highly humified with the development of reactive functional groups. Aluminum-humus complexes preferentially form at pH values less than 4.9 to 5.0 (Shoji and Fujiwara, 1984). Above pH 5, the aqueous concentrations of Al are reduced by the reduced solubility of Al and the hydroxyl ligand (OH) competes with humus to form complexes with Al.

The complexing ability of humic substances is usually evaluated by measuring the ratio: $(Al + Fe)_p/C_p$. Typically ratios range between 0.1–0.2 for most Andisols (Inoue and Higashi, 1988). Theoretical calculations based on functional group analysis of humic substances indicate that a maximum ratio of 0.12 is expected based on complexation of monomeric hydroxy-Al with humus (Higashi et al., 1981). Ratios greater than 0.12 may indicate the presence of polymerized Al associated with the humus or the nonspecificity of pyrophosphate for extracting Al-humus complexes. The presence of Al polymers is supported by the fact that only small amounts of Al are extracted with 1 M KCl from Andisols containing a large amount of Al-humus complexes (Shoji and Ono, 1978; Saigusa et al., 1980). Blocking of ionized carboxyl groups by hydroxy-Al was also inferred from the study of charge characteristics of humus-rich A horizons (see Chapter 6).

5.3.8. Other minerals

Kaolinite may be an important constituent of volcanic soils and is associated with acidic soils and advanced stages of weathering. Some evidence suggests that poorly crystalline kaolinite results from transformation of halloysite as it becomes dehydrated in surface horizons of soils with distinct moisture deficits. Gibbsite has often been reported in soils derived from volcanic ejecta (Yoshinaga, 1986). It is often associated with allophane and imogolite, and in certain cases with vermiculite-chlorite intergrades, and kaolinite. Gibbsite is thought to form by desilication of amorphous materials and hence, is often associated with soils in the advanced stages of weathering (Wada, 1989).

5.4. FORMATION AND TRANSFORMATION OF COLLOIDAL CONSTITUENTS

Soils derived from volcanic ash typically have their colloidal fraction dominated by one of several components: (1) Al-humus complexes with or without 2:1 layer silicate minerals and/or hydroxy-Al interlayered 2:1 minerals, (2) halloysite, or (3) allophane/imogolite. Several formation/transformation sequences have been proposed to explain the occurrence of these distinct colloidal assemblages (e.g. Figure 5.10 and 5.11). While these transformation sequences consider time or climate as the master variable controlling mineralogic properties, these master variables translate literally to factors that regulate soil solution activities or availability of Si and Al. For example, during the early stages of volcanic ash weathering, soluble Si

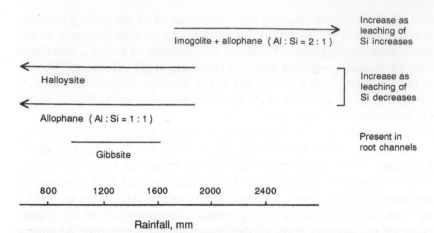

Fig. 5.10. Weathering sequence proposed for rhyolitic tephras in New Zealand with ages between 2,000 and 100,000 years. The abundance of clay minerals increases or decreases according to the degree of leaching of Si that is a function of rainfall (prepared from Parfitt et al., 1983; copyright Elsevier, with permission).

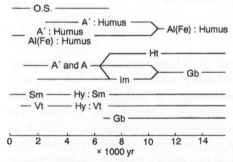

Fig. 5.11. Formation and transformation of clay minerals and their organic complexes in soils derived from volcanic ash in udic, temperate climatic zones. Horizontal bars approximate duration of the respective constituents. A = allophane; A' = allophane-like constituents, $Al(Fe):Humus$ = aluminum/ iron humus complexes; Ch = chlorite; Gb = gibbsite; Ht = halloysite; Im = imogolite; Sm = smectite; $Hy:Sm$ and $Hy:Vt$ = hydroxy-Al interlayered smectite and vermiculite; $O.S.$ = opaline silica; Vt = vermiculite (prepared from Wada, 1989).

activities are maintained at higher levels than during advanced weathering stages due to the abundant supply of highly weatherable materials, such as volcanic glass. As weathering proceeds, the concentrations of highly weatherable materials are depleted and soil solution Si activities decrease. The amount of rainfall is also an important factor controlling soluble Si activities through its regulation of leaching processes. In the following section, the pedogenic factors contributing to the formation and transformation of colloidal components in soils derived from volcanic ash are examined. The discussion will focus on processes that regulate availability of aqueous elemental constituents for mineral synthesis.

5.4.1. Mineral stability relationships

The relationship between aqueous activities and solid-phase mineral constituents can be displayed graphically in the form of stability diagrams. Figure 5.12 represents the solubility relationship of aluminosilicate minerals commonly found in soils derived from volcanic ash in terms of pH and Al^{3+} on the vertical axis and H_4SiO_4 activities on the horizontal axis. The free energy data and equilibrium constants used to construct this diagram (Carlson et al., 1976; Farmer et al., 1979b; Robie et al., 1978) are listed in Appendices 3 and 4. Appropriate thermodynamic data for allophane have not yet been obtained either experimentally or theoretically, and therefore it is not possible to examine stability relationships for allophane. Each line represents the solubility of the particular mineral phase under equilibrium conditions at 25°C.

Although thermodynamic equilibrium modelling is theoretically sound, several pervasive points of error exist. First, reported free energy data are incomplete and a wide range of values can exist for a given mineral phase of interest. Many minerals formed in the weathering environment are complicated because their structures permit a wide range in chemical composition, degree of crystalline disordering, and particle-size distribution. Therefore, the choice of free energy data is very critical and caution should be employed in extrapolating published values to naturally occurring minerals. Secondly, relationships are based on equilibrium conditions, typically at 25°C. Thus, soil solutions must have an adequate residence time to approach equilibrium. Thirdly, although weathering reactions proceed toward the lowest free energy state, metastable products are common and thus kinetic factors are as important as equilibria in determining which minerals actually form. In other words, these diagrams indicate whether a given reaction *could* occur, but kinetic factors determine whether the reaction *will* occur. Metastable minerals found in volcanic ash soils have been shown to persist for substantial periods of time during soil development (Stevens and Vucetich, 1985; Wada, 1989). Consequently,

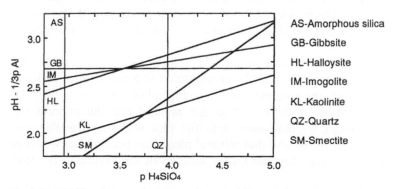

Fig. 5.12. Stability diagram depicting mineral stability relationships of common minerals found in volcanic ash soils at 25°C.

interpretations based on stability diagrams should be viewed with care because by choosing different constants, remarkably different conclusions may be obtained.

In spite of the limitations and assumptions associated with a thermodynamic equilibrium approach, this approach provides valuable insights into the understanding of mineral formation/transformation pathways. Considerable information can be extracted from stability diagrams. For example, if a solubility line for a particular mineral lies below that of another mineral at a given H_4SiO_4 activity, it is the more stable mineral phase because it is less soluble. The order of stability at a H_4SiO_4 activity of 10^{-3} is: smectite > kaolinite > halloysite > imogolite > gibbsite. At a H_4SiO_4 activity of 10^{-4}, kaolinite replaces smectite as the stable mineral phase, and imogolite and gibbsite become more stable than halloysite. While kaolinite is the stable mineral phase at H_4SiO_4 activities between $10^{-5.2}$ and $10^{-3.7}$, other aluminosilicate minerals may form as metastable phases, if the kinetics associated with their formation are favored. Metastable minerals generally exist because their formation is kinetically favored relative to that of the most stable phase. They will ultimately convert to the stable mineral or be completely solubilized over time. Because of kinetic constraints, metastable mineral phases may persist for long periods of time in the soil environment before they are transformed to more stable mineral phases or completely dissolved.

Solubility lines for quartz and amorphous silica appear as vertical line on the stability diagram (Fig. 5.12). This indicates that the solubility of these components does not vary as a function of Al activities. The solubility of silica minerals decreases as a function of increasing packing density of the silica tetrahedra and long-range order. Quartz has a solubility of approximately $10^{-3.95}$ M at 25°C, while amorphous silica has a solubility approximately one order of magnitude greater ($10^{-2.95}$ M). While no published values for the solubility of laminar opaline silica occurring in volcanic ash soils are available, published solubility values for biogenic and pedogenic opal are in the range $10^{-4.75}$ to $10^{-3.30}$ and $10^{-3.97}$ to $10^{-3.05}$, respectively (Drees et al., 1989). Soil solutions of E horizons containing opaline silica from a tephritic Spodosol in the Cascade Range of Washington show H_4SiO_4 concentrations up to $10^{-3.2}$ M (Dahlgren, unpublished data). Therefore, modeling the solubility of laminar opaline silica from volcanic ash soils using the solubility data for amorphous silica may result in an overestimation of soil solution H_4SiO_4 activities.

Perhaps the most useful information obtained from a stability diagram is whether a particular mineral is stable or can form by precipitation at a given set of aqueous activities. If, in a given soil, the aqueous activities plot below a solubility line, the soil solution is undersaturated with respect to that mineral and the mineral will dissolve. On the other hand, if the aqueous activities plot above a particular line, the soil solution is supersaturated and that mineral phase can precipitate. Whether that mineral precipitates or not, depends on the kinetics of the particular reaction involved.

The point where solubility lines cross in a diagram represents the set of aqueous activities where both minerals are in equilibrium with each other and are equally stable. Halloysite, imogolite, and gibbsite are predicted to co-exist with equal stability at H_4SiO_4 activities of approximately $10^{-3.6}$. If no kinetic limitations exist to the formation of imogolite and halloysite, the diagram predicts preferential formation of halloysite at H_4SiO_4 activities greater than $10^{-3.6}$ and formation of imogolite at H_4SiO_4 activities less than this value.

Based on Fig. 5.12, we can make a number of observations. First, gibbsite, kaolinite, or smectite are the stable mineral phases to which the soil will alter with time. Smectite is the stable phase only at high H_4SiO_4 activities ($>10^{-3.7}$), kaolinite at intermediate H_4SiO_4 activities ($10^{-3.7}$–$10^{-5.2}$) and gibbsite at low H_4SiO_4 activities ($<10^{-5.2}$). Whether these minerals are actually present in soils with H_4SiO_4 activity within these ranges depends on the kinetic factors regulating their formation and dissolution. Imogolite and halloysite may form as metastable phases, but solutions will be highly supersaturated with respect to kaolinite, gibbsite, or smectite when these minerals form. The presence of opaline silica, approximated by the amorphous silica line, indicates that high aqueous H_4SiO_4 levels exist at certain times of the year in soils that contain this constituent. In soils containing opaline silica, the H_4SiO_4 activities are high enough to support the stability and formation of smectite.

5.4.2. Formation/transformation of colloidal constituents

(1) Al- and Fe-humus complexes

Formation of metal humus complexes is favored in pedogenic environments rich in organic matter and having a pH < 5.0 (Shoji et al., 1982; Shoji and Fujiwara, 1984; Parfitt and Saigusa, 1985; Shoji et al., 1985). When concentrations of organic acids are low, the pH and weathering are dominated by carbonic acid ($pK_a = 6.3$). At typical soil CO_2 partial pressures (<10–$20\times$ atmospheric), the minimum soil solution pH obtainable by carbonic acid is within the approximate range 5.1–5.5. To get pH values lower than 5.0 requires the presence of a stronger proton donor, which in the case of many soils is organic acids (Ugolini and Sletten, 1991). Organic acids have acidic functional groups with pK_a's ranging from approximately 2.6 to 6.6 (Gregor and Powell, 1988).

The low pH values (pH < 5) at which metal humus complexes are found to be the dominant colloidal phase indicate that organic acids play a major role as proton donors in this environment. Low pH values further indicate that the base supplying capacity of the soil cannot keep pace with the production of organic acids resulting in a portion of the acidic functional groups remaining unneutralized. Therefore, the dissociated ligands of the organic acids are available to form metal-humus complexes. Under these conditions, humus effectively competes for Al leaving little or no Al available for coprecipitation with silica to form aluminosilicate minerals (Dahlgren and Ugolini, 1989a, b, c). This is an example of an anti-allophanic process that inhibits formation of allophane and imogolite. Because Fe has a greater stability in oxides as compared to humus complexes, concentrations of Fe-humus complexes are typically very low in most soils.

Organic matter has a finite capacity to complex metals based on its functional group content. Various studies using sodium pyrophosphate to co-extract metals and organic matter indicate that the maximum complexing capacity occurs at $(Al_p + Fe_p)/C_p$ values ranging from 0.1 to 0.2 (Mokma and Buurman, 1982; Higashi, 1983). Theoretical calculations based on the functional group content of humic substances indicate that if the complexes are composed of monomeric hydroxy Al and Fe ions with humus, that humus complexes with metal/C atomic ratios greater than 0.12 would not be possible (Higashi et al., 1981). Once the capacity of the organic matter to complex Al is saturated, any additional Al released by weathering is available for synthesis of aluminosilicate minerals. With regard to formation of imogolite, horizons with $(Al_p + Fe_p)/C_p$ greater than approximately 0.1–0.2 were shown to contain imogolite, while soil horizons with ratios less than this range contained no imogolite (Parfitt and Saigusa, 1985; Dahlgren and Ugolini, 1991).

(2) Opaline silica

Opaline silica is commonly found in organic-rich A-horizons of young soils forming on volcanic ash. Formation of opaline silica is favored by rapid weathering of glass-rich parent material, a pronounced seasonal period of desiccation or freezing to concentrate solutes, and low Al activities which prevent formation of aluminosilicate minerals. Opaline silica is most often associated with the presence of high concentrations of organic matter (e.g. A horizons of Andisols) or soluble organic acids (e.g. E horizons of Spodosols). Complexation of Al by humus ties up the Al leaving Si to form opaline silica rather than aluminosilicate minerals. High concentrations of soluble Si are necessary to obtain the supersaturation required to precipitate opaline silica. Laboratory solubility studies, while not specifically done on opaline silica from volcanic ash soils, show a wide range of solubility for biogenic and pedogenic opal ranging from $10^{-4.75}$ to $10^{-3.05}$ (Drees et al., 1989). The low solubility of literature reported values relative to amorphous silica (solubility = $10^{-2.95}$ M), may be due to chemisorbed Fe and Al on the silica surface. It appears that the high level of supersaturation for soluble H_4SiO_4 is achieved by evapotranspiration from surface horizons during pronounced dry periods (Shoji and Masui, 1971), or possibly by freezing of the soil solution (Ping et al., 1988). High soluble silica levels must be maintained for opaline silica to remain stable after it has formed. Saito and Shoji (1984) showed that opaline silica dissolves very rapidly in undersaturated soil solutions and follows zero-order reaction kinetics. The maximum concentration of opaline silica occurs in soils <500 years old and few particles remain in soils over 4000–7000 years old. Therefore, the presence of opaline silica is indicative of a silica-rich soil environment in which aqueous Al activities are suppressed by formation of Al-humus complexes.

(3) Allophane and imogolite

Allophane and imogolite are believed to form primarily by precipitation from weathering solutions (Wada, 1989). Because allophane and imogolite are metastable phases, the presumed rapid precipitation kinetics of these noncrystalline materials is thought to favor their synthesis relative to stable crystalline minerals, such as kaolinite. In the soil environment, there is general agreement that imogolite/allophane formation is favored by: $pH(H_2O) = 5-7$, low content of complexing organic compounds, base-rich volcanic ash, certain types of vegetation, the absence of $2:1$ layer silicates or the complete filling of the interlayer position with hydroxy-Al polymers.

Numerous studies have shown that allophane and imogolite form in soils that have $pH(H_2O)$ values between 4.9 and 7 (Shoji et al., 1982; Shoji and Fujiwara, 1984; Parfitt and Kimble, 1989; Ugolini and Dahlgren, 1991). Soil solution studies from Andisols and Spodosols indicate that a large portion of the allophane and imogolite forms by in situ weathering due to carbonic acid in the lower B horizons (Ugolini and Dahlgren, 1987; Ugolini et al., 1988; Dahlgren et al., 1991; Ugolini and Dahlgren, 1991). Organic acids, which depress the pH and complex soluble Al, are removed from solution by precipitation–adsorption reactions in the upper B horizon. After the removal of organic acids, the solution pH increases and allows the dissociation of H_2CO_3 at pH values greater than 5.0. The HCO_3 anion does not chelate Al, which allows the Al to undergo hydrolysis and polymerization reactions in the pH range of carbonic acid weathering (pH 5–7). Soluble Si leached from the upper profile and released by H_2CO_3 weathering of primary minerals in the lower B horizons may then combine with the polymerized Al to form allophane and/or imogolite.

The availability of Al appears to be a critical factor for the synthesis of allophane and imogolite. Both humic substances and $2:1$ layer silicates compete for soluble Al and may lower Al activities below that necessary for the formation of allophane and imogolite (anti-allophanic effect). Humic substances appear to affect synthesis by lowering Al activities and hindering the hydrolysis and polymerization of Al. An example of humic substances lowering Al activities occurs in the A horizons of Andisols and E horizons of Spodosols. Solid-phase and soluble humic substances bind Al rendering it unavailable for co-precipitation with Si. The soluble Si concentrations increase by surface evapotranspiration resulting in precipitation of opaline Si in the A horizons of Andisols and E horizons of Spodosols. This indicates that the Si concentrations of the soil solution would be high enough for the formation of allophane and imogolite, if Al levels were not limiting. Therefore, the limiting factor in allophane/imogolite formation in humus-rich soil environments is Al availability rather than Si availability.

The presence of humic substances does not preclude the formation of allophane and imogolite. Parfitt and Saigusa (1985) showed that allophane and imogolite were present in soil horizons where the $(Al_p + Fe_p)/C_p$ ratio was greater than 0.1. They inferred that a value of 0.1 represented saturation of the humus complexing sites with metals. Therefore, Al available in excess of this ratio would be available for synthesis of allophane and imogolite. In addition to lowering aqueous Al activities, low molecular weight organic acids and soluble humic substances have been shown to perturb the interaction of hydroxy-Al ions with orthosilicic acid and thus hinder the formation of allophane and imogolite (Inoue and Huang, 1984, 1985, 1990).

Organic ligands of different complexing affinities have varying effects on the formation of allophane and imogolite and the nature of precipitation products. The formation of imogolite was completely inhibited at concentrations >30–50 mg/L fulvic acid and >300 mg/L humic acid (Inoue and Huang, 1990). The strong competition of humic substances with orthosilicic acid for the coordination sites of Al is considered to account for the perturbation effects that humic substances have on the formation of allophane and imogolite.

In addition to humic substances, 2:1 layer silicates may also act as a sink for Al released by weathering. Soil pH values in the range of 5–6 have been shown to be ideal for polymerization of hydroxy-Al polymers (Rich, 1968). These hydroxy-Al polymers are reported to have a small positive charge and are readily incorporated into the negatively charged interlayer position of 2:1 layer silicates (Jackson, 1962). Nonallophanic Andisols usually have a considerable amount of 2:1 minerals showing a wide range in the degree of hydroxy-Al interlayer filling. Preferential incorporation of Al into hydroxy-Al polymers may be responsible in part for the absence of allophane and imogolite in nonallophanic Andisols.

Once the interlayer position of 2:1 layer silicates are filled with hydroxy-Al polymers, they no longer serve as a sink for Al and any additional Al may react with Si to form aluminosilicate materials. The Bs horizon of Spodosols formed in volcanic ash and glacial drift in the Cascade Range of Washington provides an example of the co-existence of hydroxy-Al interlayered 2:1 layer silicates and imogolite (Dahlgren and Ugolini, 1989c). In this horizon, the interlayer position of the 2:1 layer silicates were completely filled by hydroxy-Al polymers. The simultaneous equilibrium between the soil solution, interlayer hydroxy-Al (or gibbsite), and imogolite has been postulated by Farmer (1987). For the purpose of equilibrium modeling, hydroxy-Al polymers may be approximated by gibbsite since they have approximately the same chemical composition and solubility as synthetic gibbsite (Dahlgren and Ugolini, 1989c). Combining dissolution equations for interlayer hydroxy-Al and imogolite at 5°C (mean temperature of studied soil) yields the following reaction:

$$2\,Al(OH)_3 + H_4SiO_4 = Al_2SiO_3(OH)_4 + 3\,H_2O \qquad \log K = 4.1$$
(interlayer Al)   (imogolite)

This reaction indicates that when the two mineral phases and solution are equilibrated, the H_4SiO_4 activity in solution will be $10^{-4.1}$ (79 μM). Over the five year period that data were collected from this study, H_4SiO_4 activities averaged $10^{-4.09}$ ($n = 81$) and $10^{-4.0}$ ($n = 63$) in the Bs and C horizons, respectively (Dahlgren and Ugolini, 1989c). In another study, soluble H_4SiO_4 levels in allophanic Bw horizons of two Andisols from Japan were monitored for a two year period. This study also showed an apparent buffering of Si in the range 75–80 μM as proposed by the above reaction (Dahlgren et al., 1991).

The importance of silica activity has been documented as the major factor responsible for preferential formation of either imogolite and Al-rich allophane, Si-rich allophane, or halloysite in soils derived from volcanic ash (Parfitt et al., 1984;

Parfitt and Wilson, 1985; Singleton et al., 1989; Parfitt and Kimble, 1989; Parfitt, 1990). Silica-rich allophane and halloysite tend to form in environments where Si in soil solution is high (>250–350 μM) and Al-rich allophanes and imogolite form in acid soils where the Si in soil solution is relatively low (<100–250 μM). Allophane occurs less frequently in soils with ustic, xeric, and aridic soil moisture regimes because there is less leaching; the allophane that does occur is usually a Si-rich allophane. In these drier environments, halloysite is typically found as the dominant aluminosilicate in volcanic ash soils.

The soil micro-environment has also been shown to affect preferential formation of aluminosilicates due to its presumed influence on Si activities. Silica-rich allophane and imogolite have been shown to co-exist in the same horizon of pumice beds in the Kitakami district of Japan (Wada and Matsubara, 1968). Mineralogic investigations show that Si-rich allophane forms preferentially in the Si-rich environment within the pumice grains, whereas imogolite formed exclusively as gel films in the interstices between pumice grains.

The composition of the parent volcanic ash affects allophane and imogolite synthesis through its control on weathering rates and pH (Saigusa and Shoji, 1986). Allophane and imogolite are found to be most abundant in soils formed on volcanic ash containing colored glass. Colored glass has a weathering rate that is approximately 1.5 times greater than for noncolored glass (Shoji et al., 1993). Rapid weathering rates release elements faster than crystalline minerals can form, leading to supersaturation of soil solutions with respect to metastable materials such as allophane and imogolite. The rapid precipitation kinetics of poorly crystalline phases favors formation of allophane and imogolite compared to their crystalline mineral counterparts. Base-rich parent material also favors allophane and imogolite formation by releasing large quantities of bases which maintain a favorable pH condition (5–7) for allophane/imogolite synthesis.

The effect of vegetation on allophane and imogolite has not been extensively investigated. Grass species have been shown to promote the process of andosolization while forest vegetation tends to foster podzolization (Shoji et al., 1988a, b; Ugolini et al., 1988). Japanese pampas grass (*Miscanthus sinensis*) appears to have a major influence on soil forming processes leading to formation of allophane and imogolite. Pampas grass is very effective in biocycling of base cations that helps maintain a base-rich status and higher pH values. In addition, pampas grass annually produces large amounts of dry matter rich in silicon (approximately 40 g/kg SiO_2) that is subjected to rapid fragmentation and decomposition (Shoji et al., 1990). As a result, pampas grass can biocycle large amounts of Si, contributing to the high Si concentrations in the surface horizons of Andisols (Dahlgren et al., 1991). In contrast, coniferous forest species produce base poor/polyphenol rich litter that contributes to production of organic acids and soil acidification. Low pH (pH < 5) and complexing organic acids inhibit allophane and imogolite formation, especially in the surface mineral horizons. Changes in vegetation have been shown to rapidly affect soil properties. Shoji et al. (1988b) showed a reversal of podzolization to andosolization related to a recent change in vegetation from forest to grass in Alaska. In contrast, conversion of pampas grass to oak vegetation resulted in a lowering of the pH and increased concentrations of soluble organic acids within 50 years in a Japanese Andisol.

(4) Halloysite

Formation of halloysite is favored by a pronounced dry season and a silica-rich environment. Halloysite may crystallize directly from the dissolution products of primary minerals, alteration of feldspars, or as a transformation product of allophane and imogolite via solution and resilication. Parfitt et al. (1982) reported that Al-rich allophane cannot alter directly to halloysite since the Si tetrahedra occur on the inside surface of the hollow allophane spheres as isolated O_3SiOH groups with their apices directed away from the Al octahedral units. This is the reverse of their orientation in halloysite necessitating the allophane to pass through a solution phase (dissolution) with concomitant Si-enrichment and polymerization of the silica sheet.

Wada and Inoue (1974) measured the zero point of silica adsorption for different clay constituents to provide an estimate of equilibrium silica concentrations in soil solutions. They found monomeric silica concentrations as follows: Al-rich allophane and imogolite (Al:Si = 2:1), 180–190 μM; Si-rich allophane (Al:Si = 1:1), 350 μM; and halloysite, 425 μM. These values support the relative stability relationships between minerals shown in Fig. 5.12, and the general mineralogic/equilibrium silica levels determined by Singleton et al. (1989) and Parfitt and Wilson (1985) for the differential formation of imogolite (<250–350 μM H_4SiO_4) versus halloysite (>250–350 μM H_4SiO_4). Silica concentrations required for synthesis of Al-rich allophane and imogolite are clearly lower than for Si-rich allophane and halloysite. This provides a clear explanation for preferential formation of these two groups of materials. However, the silica levels favoring formation of Si-rich allophane and halloysite are very similar and do not provide a distinct explanation for differential formation of these two minerals.

Halloysite is found as the dominant mineral in volcanic ash soils where precipitation is generally less than approximately 1500 mm (Parfitt et al. 1983; Lowe, 1986; Mizota and van Reeuwijk, 1989; Takahashi et al., 1993). Also associated with the lower amounts of precipitation was a pronounced seasonal dry period. During the dry season, it may be assumed that the soil solutions becomes highly concentrated in dissolved components allowing a seasonal resilication of clay materials. However, the high silica concentrations would appear to equally favor the formation of Si-rich allophane and halloysite. Therefore, it appears that the seasonal desiccation is an important distinction between formation of these minerals. Fieldes (1966) showed a strong correlation between the occurrence of allophane and a high, seasonally well-distributed rainfall pattern. On the other hand, halloysite occurred in regions of low precipitation with a pronounced dry season. He proposed that halloysite formation was enhanced by the summer dry period which led to crystallization and Si-O-Al cross-linkage of random aluminosilicate gels during the period of dehydration. This theory is not in conflict with the evidence of Parfitt and Wilson (1985) who found halloysite forming directly from ash rather than as a recrystallization product of allophane and imogolite. In fact, formation from soluble weathering products may proceed through a progression of: solution → sol → gel → poorly crystalline → crystalline. Evidence supporting this dehydration process is found in a toposequence in andesitic volcanic ash in the Sierra Nevada of Mexico (Miehlich, 1984). This study showed that as the climate became drier, the amount of allophane decreased with a concomitant increase in halloysite content; halloysite crystallization was also enhanced as the degree of drying increased.

While seasonal drying of a soil profile may explain the occurrence of halloysite in many incidences, it does not explain the preferential formation of halloysite in buried soil profiles. Thick deposits (>2 m) of volcanic ash provide a silica-rich environment in the buried profile and has been shown to favor the formation of halloysite by resilication processes (Aomine and Wada, 1962; Shoji and Saigusa, 1977; Saigusa et al., 1978; Violante and Wilson, 1983; Wada, 1989). Poor profile drainage also leads to a silica-rich environment that favors formation of both halloysite and Si-rich allophane (also smectites) (Aomine and Wada, 1962; Wada and Harward, 1974; Dudas and Harward, 1975a; Violante and Wilson, 1983; Lowe, 1986; Wada, 1989). In the Waikato region in New Zealand, well drained and poorly drained conditions exist on a rhyolitic volcanogenic and tephra-rich alluvium. The clay fraction of the well-drained site is dominated by allophane, whereas that of the poorly drained site is dominated by halloysite (Lowe, 1986). Under the conditions of deep burial and poorly drained profiles, the preferential formation of halloysite versus Si-rich allophane is not related to a dehydration processes, but rather appears to be a function of the very high silica concentrations.

(5) 2:1 layer silicates and hydroxy-Al interlayered 2:1 minerals

The genesis of 2:1 and hydroxy-Al interlayered 2:1 layer silicates in soils derived from volcanic ash is controversial (Wada, 1980; Mizota, 1983; Mizota and Matsuhisa, 1985). A pedogenic origin for 2:1 layer silicates may occur by a solid-state reaction related to the mobilities of SiO_2, MgO, and K_2O in noncolored volcanic glass (Shoji et al., 1981, 1982; Yamada and Shoji, 1982; Shoji and Fujiwara, 1984). Shoji et al. (1981) proposed the following weathering sequence:

$$\text{glass (noncolored)} \nearrow \text{K-enriched glass} \longrightarrow \text{illite} \longrightarrow 2:1 \text{ layer silicates}$$
$$\searrow \text{amorphous hydrous oxides of Al, Fe, and Si}$$

The preferential retention of K in volcanic glass has been demonstrated by Lipman et al. (1968) and Yamada and Shoji (1982). The feasibility of such solid-state alterations was recently substantiated by Tazaki and Fyfe (1988) with the use of high-resolution TEM. They observed rearrangements in natural and synthetic glass frameworks as 0.33 nm domains and 1.0 and 1.4 nm clay precursors.

Recent studies of oxygen isotope analysis of silt-sized quartz particles isolated from soils in various countries provide evidence for long-distance eolian transport of fine-grained quartz (Mizota, 1982, 1983; Mizota and Matsuhisa, 1985). Quartz, possibly together with 2:1 layer silicates and their precursors, are derived from continental dusts and carried by tropospheric winds. In Japan, the distribution of 2:1 layer silicates and fine-grained quartz shows a distribution most abundant on the windward regions close to the possible source in China and with the greatest abundance in regions of high snowfall (Mizota and Inoue, 1988). Eolian additions are also supported by a strongly decreasing concentration of quartz and 2:1 layer silicates with depth in the soil profile, and by the presence of significant concentrations of quartz in soils formed on basaltic parent materials.

Transformations involving feldspars and ferromagnesian minerals (principally micas, amphiboles, and pyroxenes) are well documented (Cortes and Franzmeier, 1972; Mizota, 1976; Tazaki, 1986; Tazaki and Fyfe, 1987a, b). Other potential sources of 2:1 layer silicates include accessory minerals incorporated into the volcanic ash during the eruption (Kirkman, 1976; Dudas and Harward, 1975b; Ossaka, 1982; Pevear et al., 1982), and mixing of underlying soil material into the ash deposits by processes such as windthrow of trees (Dudas and Harward, 1975b). As indicated in Fig. 5.12, synthesis of smectites should be favorable in surface horizons of Andisols where the presence of opaline silica indicates a Si-rich environment. Smectite precipitation is possible even at lower aqueous Al activities when H_4SiO_4 activities are high. Silicic acid activities may also reach high levels within the vesicular structure of the ash particles due to the inability of the percolating waters to flush these pores (Dudas and Harward, 1975a). Therefore, neoformation of smectites is possible, based on solubility data for smectites, if the precipitation kinetics are favorable.

It is plausible that all of the above mentioned processes may contribute to the occurrence of 2:1 layer silicates in soils derived from volcanic ash. The contribution of eolian sources appears irrefutable in some regions (e.g. windward region of Japan Sea); however, this does not negate the possibility that other processes are contributing to the origin of 2:1 layer silicates in these soils as well.

(6) Ferrihydrite

Formation of noncrystalline or poorly crystalline iron oxyhydroxides is favored compared to crystalline forms in soils derived from volcanic ash. Among the various iron oxyhydroxides, Childs et al. (1991) showed that ferrihydrite was the dominant noncrystalline iron phase in a study of iron oxyhydroxide mineralogy in volcanic ash soils of Japan. The high degree of disorder may result from either rapid formation and/or hindrance of the crystallization process. Ferrihydrite formation has been shown to occur preferentially when Fe^{2+}-containing waters are oxidized very quickly or in the presence of constituents which impede crystal nucleation and growth (Schwertmann and Taylor, 1989).

The weathering conditions of volcanic ash, especially the high rate of weathering and the high level of silicate in the soil solutions, contribute to ferrihydrite formation. The rapid weathering rates of glass and olivines (when present) release large amounts of Fe to the soil solution. In the pH range typical of many volcanic ash soils (pH 4–7), Fe is very insoluble and quickly precipitates. As with allophane and imogolite, the formation of ferrihydrite appears to result from favorable reaction kinetics due to the noncrystalline nature of its structure. In other words, release of Fe from primary mineral sources occurs more rapidly than crystalline minerals, such as goethite and hematite, can precipitate. This leads to supersaturation of the soil solution to the point where metastable phases like ferrihydrite can form. The solubility product of ferrihydrite is approximately 10^{-38} compared to 10^{-42} for goethite (Schwertmann, 1985). Therefore, considerable supersaturation relative to goethite is required for ferrihydrite precipitation.

Laboratory experiments have shown than crystal nucleation and growth of Fe oxyhydroxides are severely inhibited by low concentrations of complexing organics, silicate, or phosphate (Schwertmann, 1985). These components have a strong affinity for sorption to oxyhydroxide surfaces leading to a blocking of the surface sites necessary for continued crystal growth. The effectiveness of organics in suppressing crystallization depends on how strongly the organic adsorbs on the iron precipitate and how strongly it complexes with Fe^{3+} in solution (Cornell and Schwertmann, 1979). The inhibiting effect of organics in soils has been demonstrated in the formation of ferrihydrite in Spodosol B horizons and placic horizons (Campbell and Schwertmann, 1984). In volcanic ash soils, an especially important constituent known to inhibit crystallization is silicate. As stated previously, rapid weathering of volcanic ash leads to potentially very high concentrations of H_4SiO_4 in soil solutions. Ferrihydrite precipitates from Si-rich waters have been shown to contain 2–6 percent Si strongly adsorbed to the ferrihydrite surface (Carlson and Schwertmann, 1981).

A cool and moist climate also appears to kinetically favor formation of metastable ferrihydrite relative to the more stable phases, goethite and hematite. These conditions interfere with the dehydration/reprecipitation (crystallization) processes believed necessary for the transformation of ferrihydrite to hematite or goethite. Another factor that has been shown to strongly suppress goethite formation is high aqueous Al activities. At pH 5–7, less than 1 mole percent of Al was found to be sufficient to completely suppress goethite formation in iron precipitates at 70°C (Schwertmann et al., 1979).

5.4.3. Summary

Weathering rates and clay mineral genesis in volcanic ash soils are determined primarily by the interaction of macro- and micro-environmental factors with the physicochemical composition and mineralogy of the ash. The most important environmental factors are those that affect the concentration of aqueous H_4SiO_4 and the availability of aluminum and iron. The role of pH and agents (organic acids) that hinder the coprecipitation of Al with Si also play critical roles. With regard to Fe, the rate of release of Fe from primary minerals, the presence of constituents that hinder the crystallization process (organics and silicate), and soil moisture and temperature regimes appear to regulate the synthesis of Fe oxyhydroxides. The leaching regime and vegetation, through its regulation of the organic C cycle, are important controlling factors that are in turn dependent upon climate, drainage, and the thickness of the ash deposit.

As volcanic ash weathers, the easily weatherable primary minerals and volcanic glass become depleted and the clay mineralogy changes in response. Soluble H_4SiO_4 concentrations are strongly affected as the degree of weathering increases. Therefore, metastable materials that depend on a Si-rich environment dissolve or transform to more stable mineral phases such as kaolinite or gibbsite. Incorporation of Al and Fe into humus complexes will also reach a steady-state between incorporation into new complexes and degradation of old metal-humus complexes. Degradation of humus will promote the transformation of ferrihydrite to goethite due to removal of the crystallization inhibiting humus. These processes are dramatically affected by weathering rejuvenation due to subsequent ash-falls, depth and composition of overburden, temperature and moisture regimes, and biological factors. Transformation of noncrystalline materials and poorly ordered mineral phases, that contribute to the high concentrations of active Al and Fe, eventually proceeds to the point where the volcanic ash soils pass from Andisols to intergrade soils, and eventually to orders, such as Spodosols, Alfisols, Ultisols, and Oxisols.

REFERENCES

Andisol TU Database, 1992. Database on Andisols from Japan, Alaska, and northwestern U.S.A. Prepared by Soil Science Laboratory, Tohoku University (see Appendix 2).

Aomine, S. and Wada, K., 1962. Differential weathering of volcanic ash and pumice, resulting in formation of hydrated halloysite. Am. Miner., 47: 1024–1048.

Brindley, G.W. and Fancher, D., 1969. Kaolinite defect structures; possible relation to allophanes. Proc. Int. Clay Conf. 1969, 2: 29–34.

Campbell, A.S. and Schwertmann, U., 1984. Iron oxide mineralogy of placic horizons. J. Soil Sci., 35: 569–582.

Carlson, L. and Schwertmann, U., 1981. Natural ferrihydrites in surface deposits from Finland and their association with silica. Geochim. Cosmochim. Acta, 45: 421–429.

Carlson, C.D., Kittrick, J.A., Dixon, J.B. and McKee, T.R., 1976. Stability of soil smectite from a Houston black clay. Clays Clay Miner., 24: 151–155.

Childs, C.W., Palmer, R.W.P. and Ross, C.W. 1990., Thick iron oxide pans in soils of Taranaki, New Zealand. Aust. J. Soil Res., 28: 245–257.

Childs, C.W., Matsue, N. and Yoshinaga, N., 1991. Ferrihydrite in volcanic ash soils of Japan. Soil Sci. Plant Nutr., 37: 299–311.

Cornell, R.M. and Schwertmann, U., 1979. Influence of organic anions on the crystallization of ferrihydrite. Clays Clay Min., 27: 402–410.

Cortes, A. and Franzmeier, D.P., 1972. Weathering of primary minerals in volcanic ash-derived soils of the central Cordillera of Colombia. Geoderma, 8: 165–176.

Cradwick, P.D.G., Farmer, V.C., Russell, J.D., Masson, C.R., Wada, K. and Yoshinaga, N., 1972. Imogolite, a hydrated aluminum silicate of tubular structure. Nature (London) Phys. Sci., 240: 187–189.

Dahlgren, R.A. and Ugolini, F.C., 1989a. Effects of tephra addition on soil processes in Spodosols in the Cascade Range, Washington, U.S.A. Geoderma, 45: 331–355.

Dahlgren, R.A. and Ugolini, F.C., 1989b. Aluminum fractionation of soil solutions from unperturbed and tephra-treated Spodosols, Cascade Range, Washington, USA. Soil Sci. Soc. Am. J., 53: 559–566.

Dahlgren, R.A. and Ugolini, F.C., 1989c. Formation and stability of imogolite in a tephritic Spodosol, Cascade Range, Washington, U.S.A. Geochim. Cosmochim. Acta, 53: 1897–1904.

Dahlgren, R.A. and Ugolini, F.C., 1991. Distribution and characterization of short-range-order minerals in Spodosols from the Washington Cascades. Geoderma, 48: 391–413.

Dahlgren, R.A., Ugolini, F.C., Shoji, S., Ito, T. and Sletten, R.S., 1991. Soil-forming processes in Alic Melanudands under Japanese pampas grass and oak. Soil Sci. Soc. Am. J., 55: 1049–1056.

Dethier, D.P., Pevear, D.R. and Frank, D., 1981. Alteration of new volcanic deposits. In: R.W. Lipman and D.R. Mullineaux (Editors), The 1980 Eruptions of Mount St. Helens, Washington. U.S. Geological Survey Professional Paper 1250, pp. 649–665.

Drees, L.R., Wilding, L.P., Smeck, N.E. and Senkayi, A.L., 1989. Silica in soils: Quartz and disordered silica polymorphs. In: J.B. Dixon and S.B. Weed (Editors), Minerals in Soil Environments, 2nd ed. Soil Science Society of America. Madison, WI, pp. 913–974.

Dudas, M.J. and Harward, M.E., 1975a. Weathering and authigenic halloysite in soil developed in Mazama ash. Soil Sci. Soc. Am. Proc., 39: 561–566.

Dudas, M.J. and Harward, M.E., 1975b. Inherited and detrital 2 : 1 type phyllosilicates in soils developed from Mazama ash. Soil Sci. Soc. Am. Proc., 39: 571–577.

Egashira, K. and Aomine, S., 1974. Effects of drying and heating on the surface area of allophane and imogolite. Clay Sci., 4: 231–242.

Farmer, V.C., 1982. Significance of the presence of allophane and imogolite in podzol Bs horizons for podzol formation mechanism: a review. Soil Sci. Plant Nutr., 28: 571–578.

Farmer, V.C., 1987. The role of inorganic species in the transport of aluminum in Podzols. In: D. Righi and A. Chauvel (Editors), Podzols et Podzolization. Assoc. Fr. Estude Sol, Plaisir, France., pp. 187–194.

Farmer, V.C. and Russell, J.D., 1990. The structure and genesis of allophanes and imogolite: their distribution in nonvolcanic soils. In: M.F. DeBoot, M.H.B. Hayes and A. Herbillon (Editors), Soil Colloids and their Association in Soil Aggregates. Proc. NATO Advanced Studies Workshop, Ghent. 1985., Plenum, New York, pp. 165–178.

Farmer, V.C., Fraser, A.R., Russell, J.D. and Yoshinaga, N., 1977. Recognition of imogolite structures in allophanic clays by infrared spectroscopy. Clay Miner., 12: 55–57.

Farmer, V.C., Fraser, A.R. and Tait, J.M., 1979a. Characterization of the chemical structures of natural and synthetic aluminosilicate gels and sols by infrared spectroscopy. Geochim. Cosmochim. Acta, 43: 1417–1420.

Farmer, V.C., Smith, B.F.L. and Tait, J.M., 1979b. The stability, free energy and heat of formation of imogolite. Clay Miner., 14: 103–107.

Fieldes, M., 1966. The nature of allophane in soils. Part 1. Significance of structural randomness in pedogenesis. N.Z. J. Sci., 9: 599–607.

Fieldes, M. and Claridge, G.G.C., 1975. Allophane. In: J.E. Gieseking (Editor), Soil Components. Springer-Verlag. Heidelberg, 2: 351–393.

Goodman, B.A., Russell, J.D., Montez, B., Oldfield, E. and Kirkpatrick, R.J., 1985. Structural studies of imogolite and allophane by aluminum-27 and silicon-29 nuclear magnetic resonance spectroscopy. Phys. Chem. Miner., 12: 342–346.

Gregor, J.E. and Powell, H.K.P., 1988. Protonation reactions of fulvic acids. J. Soil Sci., 39: 243–252.

Hall, P.L., Churchman, G.J. and Theng, B.K.G., 1985. Size distribution of allophane unit particles in aqueous suspensions. Clays Clay Miner., 33: 345–349.

Henmi, T. and Wada, K., 1976. Morphology and composition of allophane. Am. Miner., 61: 379–390.

Higashi, T., 1983. Characterization of Al/Fe-humus complexes in Dystrandepts through comparison with synthetic forms. Geoderma, 31: 277–288.

Higashi, T., DeConinck, F. and Gelaude, F., 1981. Characterization of some spodic horizons of the

Campine (Belgium) with dithionite-citrate, pyrophosphate and sodium hydroxide-tetraborate. Geoderma, 25: 285–292.

Hodder, A.P.W., Green, B.E. and Lowe, D.J., 1990. A two-stage model for the formation of clay minerals from tephra-derived volcanic glass. Clay Miner., 25: 313–327.

Inoue, K., 1981. Implications of eolian dusts to 14A minerals in the volcanic ash soils in Japan. Pedologist, 25: 97–118 (in Japanese).

Inoue, K. and Higashi, T. 1988. Al- and Fe-humus complexes in Andisols. In: D.I. Kinloch, S. Shoji, F.H. Beinroth and H. Eswaran (Editors), Proc. of the 9th Int. Soil Classification Workshop, Japan. 20 July to 1 August, 1987. Publ. by Jap. Committee for 9th Int. Soil Classification Workshop, for the Soil Management Support Services, Washington, D.C., U.S.A., pp. 81–96.

Inoue, K. and Huang, P.M., 1984. Influence of citric acid on the natural formation of imogolite. Nature., 308: 58–60.

Inoue, K. and Huang, P.M., 1985. Influence of citric acid on the formation of short-range ordered aluminosilicates. Clays Clay Miner., 33: 312–322.

Inoue, K. and Huang, P.M., 1990. Perturbation of imogolite formation by humic substances. Soil Sci. Soc. Am. J., 54: 1490–1497.

Inoue, K. and Naruse, T., 1990. Asian long-range eolian dust deposited on soils and paleosols along the Japan Sea coast. Quat. Res., 29: 209–222 (in Japanese, with English abstract).

Jackson, M.L., 1962. Interlayering of expansible layer silicates in soil chemical weathering. Clays Clay Miner. 11: 29–46.

Kato, Y., 1983. Formation mechanism of volcanic ash soils. In: N. Yoshinaga (Editor), Kazanbaido. Hakuyusha, Tokyo, pp. 5–30 (in Japanese).

Kawasaki, H. and Aomine, S., 1966. So-called 14A clay minerals in some Ando soils. Soil Sci. Plant Nutr., 12: 144–150.

Kirkman, J.H., 1975. Clay mineralogy of some tephra beds of Rotorua area, North Island, New Zealand. Clay Miner., 10: 437–449.

Kirkman, J.H., 1976. Clay mineralogy of Rotomahana sandy loam soil, North Island, New Zealand. N.Z. J. Geol. Geophys., 19: 35–41.

Kirkman, J.H. and McHardy, W.J., 1980. A comparative study of the morphology, chemical composition and weathering of rhyolitic and andesitic glass. Clay Miner., 15: 165–173.

Kittleman, L.R., Jr., 1963. Glass-bead silica determination for a suite of volcanic rocks from the Owyhee Plateau, Oregon. Geol. Soc. Am. Bull., 74: 1405–1409.

Kobayashi, S. and Shoji, S., 1976. Distribution of copper and zinc in volcanic ashes. Soil Sci. Plant Nutr., 22: 401–408.

Kobayashi, S., Shoji, S., Yamada, I. and Masui, J., 1976. Chemical and mineralogical studies on volcanic ashes, III. Some mineralogical and chemical properties of volcanic glasses with special reference to the rock types of volcanic ashes. Soil Sci. Plant Nutr., 22: 7–13.

Kondo, Y., Fujitani, T., Katsui, Y. and Niida, K., 1979. Nature of the 1977–1978 volcanic ash from Usu volcano, Hokkaido, Japan. Kazan, 24: 223–238 (in Japanese).

Kondo, R., Sase, T. and Kato, Y. 1988. Opal phytolith analysis of Andisols with regard to interpretation of paleovegetation. In: D.I. Kinloch, S. Shoji, F.H. Beinroth and H. Eswaran (Editors), Proc. of the 9th Int. Soil Classification Workshop, Japan. 20 July to 1 August, 1987. Publ. by Jap. Committee for 9th Int. Soil Classification Workshop, for the Soil Management Support Services, Washington, D.C., U.S.A., pp. 520–534.

Kuno, H., 1960. High-alumina basalt. J. Petrol., 1: 121–145.

LaManna, J.M. and Ugolini, F.C., 1987. Trioctahedral vermiculite in a 1980 pyroclastic flow, Mt. St. Helens, Washington. Soil Sci., 143: 162–167.

Lipman, P.W., Christiansen, R.L. and Van Alstine, R.E., 1968. Retention of alkalies by calc-alkalic rhyolites during crystallization and hydration. Am. Mineral., 54: 286–291.

Loughnan, F.C., 1969. Chemical Weathering of the Silicate Minerals. Elsevier, New York, pp. 27–66.

Lowe, D.J., 1986. Controls on the rates of weathering and clay mineral genesis in airfall tephras: a review and New Zealand case study. In: S.M. Colman and D.P. Dethier (Editors), Rates of Chemical Weathering of Rocks and Minerals. Academic Press, Orlando, pp. 265–330.

Masui, J., Shoji, S. and Uchiyama, N., 1966. Clay mineral properties of volcanic ash soils in the northeastern part of Japan. Tohoku J. Agr. Res., 17: 17–36.

Meyer, J.D., 1971. Glass crust on intratelluric phenocrysts in volcanic ash as a measure of eruptive violence. Bull. Volcanologique, 34: 358–368.

Miehlich, G., 1984. Chronosequenzen und antropogene Veranderungen andesitischen Vulkanaschboden in drei Klimastufen eines randtropisches Gebirges (Sierra Nevada de Mexico). Habilitationsschrift, Universitat Hamburg, FRD.

Mitchell, W.A., 1975. Heavy minerals. In: J.E. Gieseking (Editor), Soil Components, Volume 2. Inorganic Components. Springer-Verlag, Berlin-Heidelberg-New York, pp. 449–480.

Mizota, C., 1976. Relationships between the primary mineral and the clay mineral composition of some recent Andosols. Soil Sci. Plant Nutr., 22: 257–268.

Mizota, C., 1982. Tropospheric origin of quartz in Ando soils and Red-Yellow soils on basalts, Japan. Soil Sci. Plant Nutr., 28: 517–522.

Mizota, C., 1983. Eolian origin of the micaceous minerals in an Ando soil from Kitakami. Soil Sci. Plant. Nutr., 29: 379–382.

Mizota, C. and Inoue, K., 1988. Eolian dust contribution to soil development on volcanic ashes in Japan. In: D.I. Kinloch, S. Shoji, F.H. Beinroth and H. Eswaran (Editors), Proc. of the 9th Int. Soil Classification Workshop, Japan. 20 July to 1 August, 1987. Publ. by Jap. Committee for 9th Int. Soil Classification Workshop, for the Soil Management Support Services, Washington, D.C., U.S.A., pp. 547–557.

Mizota, C. and Matsuhisa, Y., 1985. Eolian addition to soils and sediments of Japan. Soil Sci. Plant Nutr., 31: 369–382.

Mizota, C. and van Reeuwljk, L.P., 1989. Clay mineralogy and chemistry of soils formed in volcanic material in diverse climatic regions. Soil Monograph 2, ISRIC, Wageningen, The Netherlands, 186 pp.

Mizota, C., Carrasco, J.A. and Wada, K., 1982. Clay mineralogy and some chemical properties of Ap horizons of Ando soils used for paddy rice, Japan. Geoderma, 27: 225–237.

Mokma, D.L. and Buurman, P., 1982. Podzols and podzolization in temperate regions. ISM Monogr. 1. International Soil Museum, Wageningen. 131 pp.

Nagasawa, K., 1978. Weathering of volcanic ash and other pyroclastic materials. In: T. Sudo and S. Shimoda (Editors), Clays and Clay Minerals of Japan. Elsevier, Amsterdam, pp. 105–125.

Nagasawa, K. and Noro, H., 1987. Mineralogical properties of halloysites of weathering origin. Chem. Geol., 60: 145–149.

Nakai, M. and Yoshinaga, N., 1980. Fibrous goethite in some soils from Japan and Scotland. Geoderma, 24: 143–158.

Neall, V.E., 1985. Parent materials of Andisols. In: F.H. Beinroth, W.L. Luzio, F.P. Maldonado, and H. Eswaran (Editors), Proceedings of the Sixth International Soil Classification Workshop, Chile and Ecuador. Part 1: Papers. Sociedad Chilena de la Ciencia del Suelo, Santiago, Chile, pp. 9–19.

Okada, K., Morikawa, S., Iwai, S., Ohira, Y. and Ossaka, J., 1975. A structure model of allophane. Clay Sci., 4: 291–303.

Ossaka, J., 1982. Activity of volcanos and clay minerals. Nendo Kagaku, 22: 127–137 (in Japanese).

Parfitt, R.L., 1990. Allophane in New Zealand — A review. Aust. J. Soil Res., 28: 343–360.

Parfitt, R.L. and Childs, C.W., 1983. Comments on clay mineralogy of two Northland soils New Zealand. Soil Sci. Plant Nutr., 29: 555–559.

Parfitt, R.L. and Childs, C.W., 1985. Estimation of allophane and halloysite in three sequences of volcanic soils. Catena Suppl., 7: 1–8.

Parfitt, R.L. and Childs, C.W., 1988. Estimation of forms of Fe and Al: a review, and analysis of contrasting soils by dissolution and Moessbauer methods. Aust. J. Soil Res. 26: 121–144.

Parfitt, R.L. and Henmi, T., 1980. Structure of some allophanes from New Zealand. Clays Clay Miner., 28: 285–294.

Parfitt, R.L. and Kimble, J.M., 1989. Conditions for formation of allophane in soils. Soils Sci. Soc. Am. J., 53: 971–977.

Parfitt, R.L. and Saigusa, M., 1985. Allophane and humus-aluminum in Spodosols and Andepts formed from the same volcanic ash beds in New Zealand. Soil Sci., 139: 149–155.

Parfitt, R.L. and Wilson, A.D., 1985. Estimation of allophane and halloysite in three sequences of volcanic soils, New Zealand. In: E. Fernandez Caldas and D.H. Yaalon (Editors), Volcanic Soils. Catena Suppl., 7: 1–8.

Parfitt, R.L., Furkert, F.J. and Henmi, T., 1980. Identification and structure of two types of allophane from volcanic ash soils and tephra. Clays Clay Min., 28: 328–334.

Parfitt, R.L., Russell, M. and Kirkman, J.H., 1982. The clay mineralogy of yellow-brown loam soils. In: V.E. Neall (Editor), Soils Groups of New Zealand, Part 6. N.Z. Soc. Soil Sci., Lower Hutt, pp. 48–53.

Parfitt, R.L., Russell, M. and Orbell, G.E., 1983. Weathering sequence of soils from volcanic ash involving allophane and halloysite. Geoderma, 29: 41–57.

Parfitt, R.L., Saigusa, M. and Cowie, J.D., 1984. Allophane and halloysite formation in a volcanic ash bed under different moisture conditions. Soil Sci., 138: 360–364.

Parfitt, R.L., Childs, C.W. and Eden, D.N., 1988. Ferrihydrite and allophane in four Andepts from Hawaii and implications for their classification. Geoderma, 41: 223–241.

Paterson, E., 1977. Specific surface area and pore structure of allophanic soil clays. Clay Miner. 12: 1–9.

Pevear, D.R., Dethier, D.P. and Frank, D., 1982. Clay minerals in the 1980 deposits from Mount St. Helens. Clays Clay Miner., 30: 241–252.

Ping, C.L., Shoji, S. and Ito, T., 1988. Properties and classification of three volcanic ash-derived pedons from Aleutian Islands and Alaska Peninsula, Alaska. Soil Sci. Soc. Am. J., 52: 455–462.

Rich, C.I., 1968. Hydroxy interlayers in expansible phyllosilicates. Clays Clay Miner., 16: 15–30.

Robie, R.A., Hemingway, B.S. and Fisher, J.R., 1978. Thermodynamic properties of minerals and related substances at 298.15 K and 1 bar (10^5 Pascals) pressure and at higher temperatures. U.S. Geol. Surv. Bull. 1452., 456 pp.

Ruxton, B.P., 1988. Towards a weathering model of Mount Lamington Ash, Papua New Guinea. Earth-Sci. Rev., 25: 387–397.

Saigusa, M. and Shoji, S., 1986. Surface weathering in Zao tephra dominated by mafic glass. Soil Sci. Plant Nutr., 32: 617–628.

Saigusa, M., Shoji, S. and Kato, T., 1978. Origin and nature of halloysite in Ando soils from Towada tephra, Japan. Geoderma, 20: 115–129.

Saigusa, J., Shoji, S. and Takahashi, T., 1980. Plant root growth in acid Andosols from northeastern Japan. 2. Exchange acidity Y_1 as a realistic measure of aluminum toxicity potential. Soil Sci., 130: 242–250.

Saito, K. and Shoji, S., 1984. Silica adsorption and dissolution properties of Andosols from northeastern Japan as related to their noncrystalline clay mineralogical composition. Soil Sci., 138: 341–345.

Schnitzer, M., 1977. Recent findings on the characterization of humic substances extracted from soils from widely differing climatic zones. In: Proceedings Symposium on Soil Organic Matter Studies, Braunschweig, International Atomic Energy Agency, Vienna, pp. 117–131.

Schnitzer, M., 1978. Humic substances: Chemistry and reactions. In: M. Schnitzer and S.U. Khan (Editors), Soil Organic Matter. Elsevier Publishing Co., Amsterdam, pp. 1–64.

Schwertmann, U., 1985. The effect of pedogenic environments on iron oxide minerals. Adv. Soil Sci. 1: 171–200.

Schwertmann, U., 1988. Occurrence and formation of iron oxides in various pedoenvironments. In: J.W. Stucki, B.A. Goodman, and U. Schwertmann (Editors), Iron in Soils and Clay Minerals, Reidel, Bordrecht. pp. 267–308.

Schwertmann, U. and Taylor, R.M., 1989. Iron oxides. In: J.B. Dixon and S.B. Weed (Editors), Minerals in Soil Environments, 2nd ed., Soil Science Society of America. Madison, WI, pp. 379–438.

Schwertmann, U., Fitzpatrick, R.W., Taylor, R.M. and Lewis, D.G., 1979. The influence of aluminum on iron oxides. Part II. Preparation and properties of Al-substituted hematites. Clays Clay Min., 27: 105–112.

Shimizu, H., Watanabe, T., Henmi, T., Masuda, A. and Saito, H., 1988. Studies on allophane and imogolite by high-resolution solid-state ^{29}Si- and ^{27}Al-NMR and ESR. Geochem. J., 22: 23–31.

Shoji, S., 1983. Mineralogical properties of volcanic ash soils. In: N. Yoshinaga (Editor), Volcanic ash soil — Genesis, Properties, Classification. Hakuyusha, Tokyo, pp. 31–72 (in Japanese).

Shoji, S., 1986. Mineralogical characteristics. I. Primary minerals. In: K. Wada (editor), Ando Soils in Japan, Kyushu University Press, Fukuoka, Japan, pp. 21–40.

Shoji, S. and Fujiwara, Y., 1984. Active Al and Fe in the humus horizons of Andosols from northeastern Japan: Their forms, properties, and significance in clay weathering. Soil Sci., 137: 216–226.

Shoji, S. and Masui, J., 1969. Amorphous clay minerals of recent volcanic ash soils in Hokkaido: II. Soil Sci. Plant Nutr., 15: 191–201.

Shoji, S. and Masui, J., 1971. Opaline silica of recent volcanic ash soils in Japan. J. Soil Sci., 22: 101–112.

Shoji, S. and Masui, J., 1972. Amorphous clay minerals of recent volcanic ash soils. II. Mineral composition of fine clay fractions. J. Sci. Soil Manure, Japan., 43: 189–230 (in Japanese).

Shoji, S. and Ono, T., 1978. Physical and chemical properties and clay mineralogy of Andosols from Kitakami, Japan. Soil Sci., 126: 297–312.

Shoji, S. and Saigusa, M., 1977. Amorphous clay materials of Towada Ando soils. Soil Sci. Plant Nutr., 23: 437–455.

Shoji, S., Yamada, I. and Masui, J., 1974. Soils formed from the andesitic and basaltic ashes. I. The nature of the parent ashes and soil formation. Tohoku J. Agr. Res., 25: 104–112.

Shoji, S., Kobayashi, S., Yamada, I. and Masui, J., 1975. Chemical and mineralogical studies on volcanic ashes. I. Chemical composition of volcanic ashes and their classification. Soil Sci. Plant Nutr., 21: 311–318.

Shoji, S., Saigusa, M. and Ebihara, M., 1980. Concentration of cobalt in volcanic ash. J. Sci. Soil Manure, Japan, 51: 335–336 (in Japanese).

Shoji, S., Yamada, I. and Kurashima, K., 1981. Mobilities and related factors of chemical elements in the topsoils of Andosols in Japan: 2. Chemical and mineralogical composition of size fractions and factors influencing the mobilities of major chemical elements. Soil Sci., 132: 330–346.

Shoji, S., Fujiwara, Y., Yamada, I. and Saigusa, M., 1982. Chemistry and clay mineralogy of Ando soils, Brown Forest soils, and Podzolic soils formed from recent Towada ashes, N.E. Japan. Soil Sci., 133: 69–86.

Shoji, S., Ito, T., Saigusa, M. and Yamada, I., 1985. Properties of nonallophanic Andosols from Japan. Soil Sci., 140: 264–277.

Shoji, S., Takahashi, T., Ito, T. and Ping, C.L., 1988a. Properties and classification of selected volcanic ash soils from Kenai Peninsula, Alaska. Soil Sci., 145: 395–413.

Shoji, S., Takahashi, T., Saigusa, M., Yamada, I. and Ugolini, F.C., 1988b. Properties of Spodosols and Andisols showing climosequential and biosequential relations in southern Hakkoda, northeastern Japan. Soil Sci., 145: 135–150.

Shoji, S., Kurebayashi, T. and Yamada, I., 1990. Growth and chemical composition of Japanese pampas grass (*Miscanthus sinensis*) with special reference to the formation of dark-colored Andisols in northeastern Japan. Soil Sci. Plant Nutr., 36: 105–120.

Shoji, S., Nanzyo, M., Shirato, Y. and Ito, T., 1993. Chemical kinetics of weathering in young Andisols from northeastern Japan using 10°C-normalized soil age. Soil Sci., 155: 53–60.

Singleton, P.L., McLeod, M. and Percival, H.J., 1989. Allophane and halloysite content and soil solution silicon in soils from rhyolitic volcanic material, New Zealand. Aust. J. Soil Res., 27: 35–42.

Stevens, K.F. and Vucetich, G.C., 1985. Weathering of Upper Quaternary tephras in New Zealand.

Part II. Clay minerals. Chem. Geol., 53: 237–247.

Stumm, W., Furrer, G., Wieland, E. and Zinder, B., 1985. The effects of complex-forming ligands on the dissolution of oxides and aluminosilicates. In, J.I. Drever (Editor), The Chemistry of Weathering. D. Reidel Publishing Co., Boston, pp. 55–74.

Takahashi, T., Dahlgren, R. and van Susteren, P., 1993. Clay mineralogy and chemistry of soils formed in volcanic materials in the xeric moisture regime of northern California. Geoderma (in press).

Tazaki, K., 1986. Observation of primitive clay precursors during microcline weathering. Contrib. Mineral. Petrol., 92: 86–88.

Tazaki, K. and Fyfe, W.S., 1987a. Formation of primitive clay precursors on K-feldspar under extreme leaching conditions. Proc. Int. Clay Conf. Denver., 1985: 53–58.

Tazaki, K. and Fyfe, W.S., 1987b. Primitive clay precursors formed on feldspar. Can. J. Earth Sci., 24: 506–527.

Tazaki, K. and Fyfe, W.S., 1988. Glass-amorphous? In: G.W. Bailey (Editor), Proc. 46th Ann. Meeting Electron Micro. Soc. Amer. San Francisco Press, San Francisco, CA, pp. 472.

Towe, K.M. and Bradley, W.F., 1967. Mineralogical constitution of colloidal "hydrous ferric oxides." J. Colloid Interface Sci., 24: 384–392.

Udagawa, S., Nakada, T. and Nakahira, M., 1969. Molecular structure of allophane as revealed by its thermal transformation. Proc. Int. Clay Conf. 1969, 1: 151–159.

Ugolini, F.C. and Dahlgren, R.A., 1987. The mechanism of podzolization as revealed by soil solution studies. In: D. Righi and A. Chauvel (Editors), Podzols et Podzolization. Assoc. Fr. Estude Sol, Plaisir, France, pp. 195–203.

Ugolini, F.C. and Dahlgren, R.A., 1991. Weathering environments and occurrence of imogolite/allophane in selected Andisols and Spodosols. Soil Sci. Soc. Am. J., 55: 1166–1171.

Ugolini, F.C. and Sletten, R.S., 1991. The role of proton donors in pedogenesis as revealed by soil solution studies. Soil Sci., 151: 59–75.

Ugolini, F.C., Dahlgren, R.A., Shoji, S. and Ito, T., 1988. An example of andosolization and podzolization as revealed by soil solution studies, S. Hakkoda, N.E. Japan. Soil Sci., 145: 111–125.

van Olphen, H., 1971. Amorphous clay materials. Science, 171: 90–91.

Vandickelen, R., De Roy, G. and Vansant, E.F., 1980. New Zealand allophanes: a structural study. J. Chem. Soc. Faraday Trans. I, 76: 2542–2551.

Vempati, R.K. and Loeppert, R.H., 1989. Influence of structural and adsorbed Si on the transformation of synthetic ferrihydrite. Clays Clay Miner., 37: 273–279.

Violante, P. and Wilson, J.J., 1983. Mineralogy of some Italian Andosols with special reference to the origin of the clay fraction. Geoderma, 29: 157–174.

Wada, K., 1978. Allophane and imogolite. In: T. Sudo and S. Shimoda (Editors), Clays and Clay Minerals of Japan. Elsevier, Amsterdam, pp. 147–187.

Wada, K., 1980. Mineralogical characteristics of Andisols. In: B.K.G. Theng (Editor), Soils with Variable Charge, New Zealand Society of Soil Science, Lower Hutt. pp. 87–107.

Wada, K., 1985. The distinctive properties of Andosols. In: B.A. Stewart (Editor), Advances in Soil Science, Springer-Verlag, New York, 2: 173–229.

Wada, K., 1989. Allophane and imogolite. In: J.B. Dixon and S.B. Weed (Editors), Minerals in Soil Environments, 2nd ed., Soil Science Society of America. Madison, WI, pp. 1051–1087.

Wada, K. and Greenland, D.J., 1970. Selective dissolution and differential infrared spectroscopy for characterization of 'amorphous' constituents in soil clays. Clay Min. 8: 241–254.

Wada, K. and Harward, M.E. 1974. Amorphous clay constituents of soils. Adv. Agron., 26: 211–260.

Wada, K. and Henmi, T., 1972. Characterization of micropores of imogolite by measuring retention of quaternary ammonium chlorides and water. Clay Sci., 4: 127–136.

Wada, K. and Higashi, T., 1976. The categories of aluminum- and iron-humus complexes in Ando soils determined by selective dissolution. J. Soil Sci., 27: 357–368.

Wada, K. and Inoue, A., 1974. Adsorption of monomeric silica by volcanic ash soils. Soil Sci. Plant

Nutr., 20: 5–15.

Wada, K. and Kakuto, Y., 1985. Embryonic halloysites in Ecuadorian soils derived from volcanic ash. Soil Sci. Soc. Am. J., 49: 1309–1318.

Wada, K. and Matsubara, I., 1968. Differential formation of allophane, "imogolite" and gibbsite in the Kitakami pumice bed. Trans. Int. Congr. Soil Sci., 9th (Adelaide, S. Aust.), 3: 123–131.

Wada, K., Yamauchi, H., Kakuto, Y. and Wada, S.-I., 1985. Embryonic halloysite in a paddy soil derived from volcanic ash. Clay Sci., 6: 177–186.

Wada, S.-I. and Mizota, C., 1982. Iron-rich halloysite (10A) with crumpled lamellar morphology from Hokkaido, Japan. Clays Clay Miner., 30: 315–317.

Wada, S.-I. and Nagasato, A., 1983. Formation of silica microplates by freezing dilute silicic acid solution. Soil Sci. Plant Nutr., 29: 93–95.

Wada, S.-I. and Wada, K., 1977. Density and structure of allophane. Clay Miner., 12: 289–298.

Wada, S.-I. and Wada, K., 1980. Formation, composition and structure of hydroxy-aluminosilicate ions. J. Soil Sci., 31: 457–467.

Wada, S.-I., Eto, A and Wada, K., 1979. Synthetic allophane and imogolite. J. Soil Sci., 30: 347–355.

White, A.F., 1983. Surface chemistry and dissolution kinetics of glassy rocks at 25°C. Geochim. Cosmochim. Acta., 47: 805–815.

White, A.F. and Claassen, H.C., 1980. Kinetic model for short-term dissolution of a rhyolitic glass. Chem. Geol., 28: 91–109.

Wollast, R. and Chou, L., 1984. Kinetic study of the dissolution of albite with a continuous flow-through fluidized bed reactor. In: J.I. Drever (Editor), The Chemistry of Weathering. Reidel, Dordrecht, pp. 75–96.

Yamada, I., 1988. Tephra as parent material. In: D.I. Kinloch, S. Shoji, F.H. Beinroth and H. Eswaran (Editors), Proc. of the 9th Int. Soil Classification Workshop, Japan. 20 July to 1 August, 1987. Publ. by Jap. Committee for 9th Int. Soil Classification Workshop, for the Soil Management Support Services, Washington, D.C., U.S.A., pp. 509–519.

Yamada, I. and Shoji, S., 1975. Relationships between particle size and mineral composition of volcanic ashes. Tohoku J. Agr. Res., 26: 7–10.

Yamada, I. and Shoji, S., 1982. Retention of potassium by volcanic glasses of the topsoils of Andosol in Tohoku, Japan. Soil Sci., 133: 208–212.

Yamada, I. and Shoji, S., 1983. Properties of volcanic glasses and relationships between the properties of tephra and volcanic zones. Jap. J. Soil Sci. Plant Nutr., 54: 311–318 (in Japanese).

Yamada, I., Shoji, S., Kobayashi, S. and Masui, J., 1975. Chemical and mineralogical studies of volcanic ashes. II. Relationships between rock types and mineralogical properties of volcanic ashes. Soil Sci. Plant Nutr., 21: 319–326.

Yamada, I., Saigusa, M. and Shoji, S., 1978. Clay mineralogy of Hijiori and Numazawa Ando soils. Soil Sci. Plant Nutr., 24: 75–89.

Yonebayashi, K. and Hattori, T., 1988. Chemical and biological studies on environmental humic acids. I. Composition of elemental and functional groups of humic acids. Soil Sci. Plant Nutr., 34: 571–584.

Yoshinaga, N., 1970. Imogolite, a new chain-structure type clay mineral. Nendokagaku, 9: 1–11 (in Japanese).

Yoshinaga, N., 1986. Mineralogical characteristics. II. Clay minerals. In: K. Wada (Editor), Ando Soils in Japan, Kyushu University Press, Fukuoka, Japan, pp. 41–56.

Yoshinaga, N., 1988. Mineralogy of Andisols. In: D.I. Kinloch, S. Shoji, F.H. Beinroth and H. Eswaran (Editors), Proc. of the 9th Int. Soil Classification Workshop, Japan. 20 July to 1 August, 1987. Publ. by Jap. Committee for 9th Int. Soil Classification Workshop, for the Soil Management Support Services, Washington, D.C., U.S.A., pp. 45–59.

Yoshinaga, N. and Aomine, S., 1962. Imogolite in some Ando soils. Soil Sci. Plant Nutr., 8: 22–29.

Chapter 6

CHEMICAL CHARACTERISTICS OF VOLCANIC ASH SOILS

M. NANZYO, R. DAHLGREN and S. SHOJI

6.1. INTRODUCTION

Volcanic ash soils display a wide range of chemical characteristics that strongly reflect the influence of the parent material and degree of weathering as mentioned in Chapters 3 and 5. Of these chemical properties, soil organic matter, active Al and Fe, and variable charge are the most prominent attributes regulating chemical reactions in volcanic ash soils.

Organic matter influences many soil chemical and physical properties and enhances soil biological activity and productivity. The dark color of humus horizons is one of the most important properties involved in the central concept of Japanese Kurobokudo and it is ascribed to the predominance of A-type humic acid (Kumada, 1987). The melanic epipedon has been created to describe these humus horizons that have a thickness of 30 cm or greater and organic carbon contents of 6 percent or more (Soil Survey Staff, 1992). A large amount of organic C accumulates not only in Andisols or Kurobokudo, but also in other soils derived from volcanic ash, such as Spodosols (Ping et al., 1988).

It was common knowledge in the 1960's that the formation of humus-allophane complexes was the major process responsible for the accumulation of organic carbon in Andisols or Kurobokudo (e.g. Kanno, 1961). However, it was recognized in the 1980's that there was an inverse relationship between the accumulation of organic C and formation of allophanic clays in Andisols with udic soil moisture regimes (e.g. Shoji and Fujiwara, 1984). The reason for this inverse relationship is that soil organic matter plays an anti-allophanic role; Al preferentially forms Al-humus complexes and is unavailable to combine with Si to form allophanic clays. Aluminum-humus complexes are very reactive with phosphate and fluoride as compared to allophanic clays.

Phosphorus occurs primarily as acid soluble forms in fresh tephras and its solubility decreases with advance of chemical weathering and formation of active Al and Fe solid-phases. The main forms of active Al and Fe are allophane, imogolite, Al-humus complexes, and ferrihydrite as described in Chapter 5. Thus, availability of phosphorus is not a serious limiting factor to plant growth in ash-derived Entisols as compared to more weathered Andisols. In contrast, in mature Andisols, heavy phosphorus fertilization results in formation of noncrystalline aluminum phosphate materials and phosphorus availability is not as enhanced as would be expected from the amount of phosphorus applied.

Variable charge characteristics are primarily due to noncrystalline materials and soil organic matter. The influence of variable charge surfaces has been recognized in the farming of Andisols. For example, exchangeable bases are easily leached from Andisols under humid climates, but allophanic soils rarely have soil pH(H_2O) values less than 5.0 or toxic levels of exchangeable Al. In contrast to constant charge soils showing repulsion of nitrate, allophanic Andisols with low organic C content can retain nitrate, contributing to a high recovery of applied nitrogen by crops.

Variable charge characteristics in volcanic ash soils were first observed in the variation of CEC as determined by different CEC methods (Birrel and Gradwell, 1956; Aomine, 1958; Egawa et al., 1959). The amount of variable charge is determined by measuring cation and anion retention by allophane and soils at varying pH and electrolyte concentrations of the ambient solutions (Iimura, 1966; Wada and Okamura, 1980). Reflecting such charge characteristics, Andisols show large differences between CEC (0.05 M Ca(AcO)$_2$, pH = 7) and effective CEC (sum of exchangeable bases and KCl-exchangeable Al). For example, the subsurface horizons of Andisols from northeastern Japan have effective CEC values less than 30 percent of CEC. Thus, CEC values are not always useful for predicting cation retention in variable charge soils. It is also noted that allophanic Andisols show a high point of zero net charge (PZNC) as compared to other mineral soils and have low ΔpH values or acric properties when they are low in organic matter.

Review articles related to the content of this chapter have been published by Kosaka (1964), Wada and Harward (1974) Wada (1981, 1985, 1989), Arai et al. (1986), Inoue (1986), Kumada (1987), Parfitt (1988), Arai et al. (1988), Inoue and Higashi (1988), Wada, S-I. (1988), and Van Wambeke (1992).

6.2. INORGANIC CONSTITUENTS

6.2.1. Elemental composition of bulk soils

Tephra, as a parent material, controls soil formation more than any other parent material and the major soil forming process taking place in tephra has been termed andosolization as described in Chapter 3. Andosolization consists of rapid *in situ* weathering and production of a high content of noncrystalline materials. As summarized in Table 6.1, tephras show a wide variation in their chemical composition (The relationship between the content of SiO$_2$ and other elements in fresh tephra was discussed in Chapter 5). The behavior of chemical elements in volcanic ash soils can be inferred by comparing the total elemental analysis of the bulk soil samples with that of the original tephra. Soils show an elemental composition similar to that of unweathered tephras when they are young. As weathering proceeds,

TABLE 6.1

Content of major elements in fresh tephras from Japan (Shoji et al., 1975)

Element	Range (%)
SiO_2	48 –73
Al_2O_3	12 –20
Fe_2O_3 + FeO	2 –12
TiO_2	0.4– 1.2
CaO	2 –11
MgO	0.5– 8
K_2O	0.1– 4
Na_2O	1.5– 5

they become enriched with noncrystalline materials and are highly aluminous or alumino-ferruginous, reflecting the rock type of the parent material.

Changes in elemental composition of tephras under leaching conditions can be demonstrated as a function of the clay content to estimate the degree of weathering and the content of chemical elements in the bulk soil samples. Figure 6.1 shows such relationships obtained between the clay content and elemental concentration (on an ignition basis) using the database of allophanic Andisols from Japan (Wada, 1986). Dacitic tephra, a tephra with intermediate Si concentration (62.7% SiO_2), contains 14.3 percent Al_2O_3 and 7.82 percent Fe_2O_3 as calculated according to the related regression equations. It should be noted that the contents of Al and Fe are nearly compatible with those obtained by the regression equations between silica and other elements in fresh tephras as presented in Table 3.2.

Silica concentrations decrease considerably with the advance of weathering under udic conditions, while aluminum, iron, and titanium steadily accumulate due to their low mobility (Fig. 6.1). Calcium and sodium are the most mobile elements and are intensively leached, resulting in the virtual absence of these elements in highly weathered soils. Both magnesium and potassium tend to decrease with the advance of weathering, but their correlations to the clay content are not as high. The sorting of mafic minerals during tephra deposition may increase the variation of magnesium content in the parent materials. The potassium content in tephras has been observed to be considerably variable depending on the volcanic zones from which the tephras originated; this contributes, in part, to its variability in Andisols (Yamada and Shoji, 1983).

The major chemical elements such as silica, aluminum, and iron characteristically determine many chemical characteristics of Andisols, especially in matured soils. The concentration of these elements can be estimated using the following equations:

$$SiO_2 \% = -0.22 \times clay \% + 62.7 \qquad (r = 0.60^{***}, \quad n = 89)$$

$$Al_2O_3 \% = 0.24 \times clay \% + 14.3 \qquad (r = 0.74^{***}, \quad n = 89) \qquad and$$

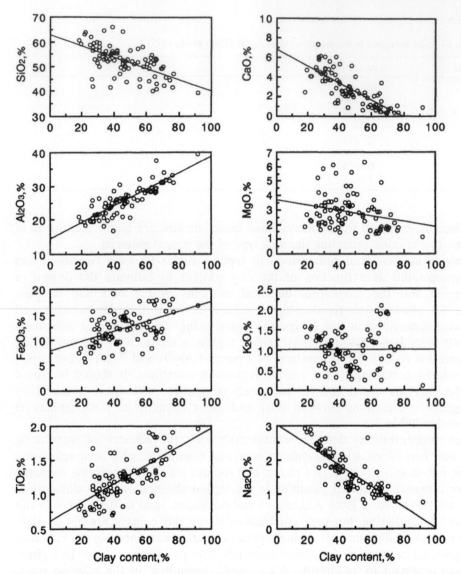

Fig. 6.1. The relationships between clay contents and the contents of major elements in the fine-earth fraction of allophanic Andisols. Rock type of these soils is primarily dacitic (prepared from data of Wada, 1986).

$$Fe_2O_3\ \% = 0.095 \times clay\ \% + 7.82 \qquad (r = 0.50^{***},\ n = 89)$$

Assuming that the most weathered soils show a clay content of 100 percent, their silica, aluminum, and iron contents are calculated as shown in Table 6.2. As compared to the elemental composition of the fresh dacitic tephra (the

TABLE 6.2

The calculated contents of silica, alumina, and iron in the most weathered Andisols and fresh dacitic tephra

Element (%)	The most weathered soils	Fresh dacitic tephra
SiO_2	40.3	62.7
Al_2O_3	38.5	14.3
Fe_2O_3	17.3	7.8

parent material), the most weathered soils are highly aluminous and their Al/Si atomic ratio is calculated to be 1.13. When the parent material is basaltic, the corresponding soils are highly alumino-ferruginous.

The Al/Si atomic ratio of allophane in Andisols under udic soil moisture regimes is approximately 2 as described later (see Fig. 6.6). The difference between the calculated (1.13) and observed (2) ratios strongly suggests that highly weathered Andisols contain some amount of crystalline silicate minerals with lower Al/Si ratios. In fact, such soils usually contain a significant quantity of halloysite which may be transformed from allophane (Wada, 1989).

Under udic soil moisture regimes, the mobility sequence of chemical elements in nonallophanic Andisols is somewhat different from that of allophanic Andisols. As with allophanic Andisols, aluminum and iron show the lowest mobility in nonallophanic Andisols from Tohoku, Japan whose average composition of the parent material is dacitic to rhyolitic (Kurashima et al., 1981). The accumulation of Al results in the enhancement of aluminous properties with the advance of weathering. However, there are significant differences in the behavior of bases between the two groups of Andisols. As shown in Fig. 6.2, the ratio of bases to aluminum are employed to determine the mobilities of bases. Calcium to aluminum and sodium to aluminum ratios indicate that both elements are highly mobile and virtually absent in the most weathered soils, as also observed for allophanic Andisols (Fig. 6.1). In contrast, potassium tends to increase with the advance of weathering though it shows small losses during the initial stages of weathering. Magnesium is somewhat lower in mobility than silica. Such behavior of potassium and magnesium in nonallophanic Andisols is attributable to their clay mineralogy which is characterized by an abundance of 2:1 layer silicates. Potassium can be held in the interlayer positions and magnesium can be incorporated into the octahedral sheets of the 2:1 layer silicates.

Finally, it is interesting to note that both allophanic and nonallophanic Andisols show contrasting mineralogy, as mentioned in Chapter 5, but both show aluminous or alumino-ferruginous properties as described in the following section.

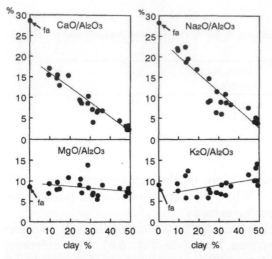

Fig. 6.2. The relationship between clay content and the weight ratio of each base to Al_2O_3 in the fine-earth fraction of soil samples: fa = the mean weight ratio of each base to Al_2O_3 in 11 fresh felsic volcanic ashes (prepared from Kurashima et al., 1981; copyright, Williams & Wilkins, with permission).

6.2.2. Active aluminum and iron

The accumulation of active forms of aluminum and iron is one of the most important properties involved in the central concept of Andisols. Active Al and Fe determine or strongly influence not only the unique chemical and physical properties of Andisols, but also soil productivity. Although active Al and Fe in Andisols are incorporated into and/or are combined with a variety of soil components, they occur primarily as allophane, imogolite, Al/Fe-humus complexes, and ferrihydrite.

The active forms of aluminum and iron are effectively extracted with 0.15–0.2 M acid oxalate solution. The Al/Fe humus complex portion of the active Al and Fe can be preferentially dissolved by 0.1 M sodium pyrophosphate solution (McKeague, 1967; Wada, 1989). Therefore, the content of active Al incorporated into allophane and imogolite can be estimated by subtracting the pyrophosphate-extractable Al (Al_p) from the oxalate-extractable Al (Al_o).

The frequency distribution of Al_p/Al_o ratios in Andisol A horizons from the Andisol TU Database (1992) is shown in Fig. 6.3. It appears that there are two maxima in the Al_p/Al_o ratios occurring at 0.1–0.4 (allophanic Andisols) and 0.8–1.0 (nonallophanic Andisols). This indicates that nonallophanic Andisols as well as allophanic Andisols are widely distributed in various climatic regions as also reported by Parfitt and Kimble (1989).

The distribution of active Al and Fe within profiles of allophanic and nonallophanic Andisols from Japan was summarized by Saigusa et al. (1991), as presented in Fig. 6.4 for active Al. The allophanic group has mean active Al and Fe contents

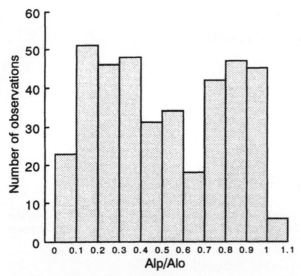

Fig. 6.3. Frequency distribution of the ratio of pyrophosphate-extractable Al to oxalate-extractable Al in A horizons having $Al_o + 1/2\,Fe_o > 0.4\%$ and phosphate retention $\geq 25\%$ (prepared from Andisol TU database, 1992).

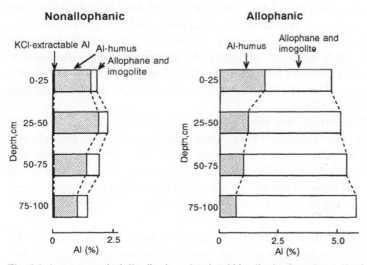

Fig. 6.4. Average vertical distribution of active Al in allophanic and nonallophanic Andisols from Japan. (prepared from data of Saigusa et al., 1991. Copyright Kluwer, with permission).

of 4.7 and 1.8 percent, respectively, in the surface horizons, and shows a gradual increase in these components with increasing depth. In contrast, the nonallophanic group has mean active Al and Fe contents of 1.8 and 1.1 percent, respectively, in

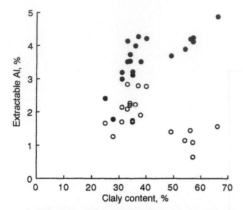

Fig. 6.5. The relationship between clay content and oxalate-extractable Al (open circles) or oxalate-extractable Al plus citrate-extractable Al (closed circles) of nonallophanic Andisols. Citrate extraction was carried out for 30 hours (prepared from data of Ito and Shoji, 1993).

the surface horizon, followed by a slight increase in the second horizon and then decreases with increasing depth. The ratio of Al_p/Al_o showed a decreasing trend with soil depth: from 0.86 to 0.71 in nonallophanic Andisols and from 0.40 to 0.12 in allophanic Andisols (Saigusa et al., 1991).

It is interesting to note in Fig. 6.5 that the active Al of nonallophanic Andisols increases in the early stages of weathering but decreases with the further advance of weathering as indicated by the relationship between the clay content and the amount of active Al. Such behavior of active Al in nonallophanic Andisols is largely attributable to the formation of hydroxy-Al and aluminosilicate interlayers in 2:1 layer silicates and/or the consumption of active Al by layer silicate formation (Ito and Shoji, 1992).

These interlayered materials are preferentially removed by citrate treatment (Tamura, 1958; Wada and Kakuto, 1983). Thus, the sum of Al_o and citrate-extractable Al has a positive linear relationship with clay content or with the advance of weathering (Fig. 6.5), indicating the preferential incorporation of Al into chloritized 2:1 layer silicates during the advanced stages of weathering. Active Fe occurs mainly as ferrihydrite in Andisols, but it is subject to transformation to goethite as weathering proceeds (Ito and Shoji, 1992).

Active Al and Fe occur in various soil components as described earlier. The distribution of the primary soil components can be approximately determined if the elemental content of each component is known. Allophane has two end members showing Al/Si atomic ratios of 2 and 1 (Wada, 1989). As shown in Fig. 6.6, however, all the Andisols with udic soil moisture regimes in the Andisol TU Database show $(Al_o–Al_p)/Si_o$ ratios of approximately 2. This indicates that the allophane occurring in these soils has the same Al/Si atomic ratio as that of imogolite (Al/Si atomic ratio of 2). Thus, the content of allophane and imogolite can be estimated according to the following equation (Parfitt and Henmi, 1982;

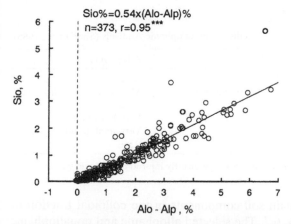

Fig. 6.6. Relationship between acid oxalate extractable Al minus pyrophosphate-extractable Al and acid oxalate extractable Si of Andisols (prepared from Andisol TU database, 1992).

Parfitt and Wilson, 1985):

$$\text{allophane and imogolite } \% = Si_o\% \times 7.1$$

The contents of organic matter and ferrihydrite are approximately calculated as follows (Parfitt, 1988; Childs et al., 1990):

$$\text{organic matter } \% = \text{organic C}\% \times 1.72;$$
$$\text{ferrihydrite } \% = Fe_o\% \times 1.7$$

The content of crystalline clay minerals is obtained as the difference between the clay content and the estimated concentration of noncrystalline materials.

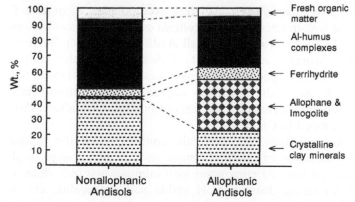

Fig. 6.7. Average composition of colloidal fractions of A horizon soils of nonallophanic ($n = 23$, $Al_p/Al_o \geq 0.8$, $Al_o + Fe_o/2 \geq 2.0\%$) and allophanic ($n = 18$, $Al_p/Al_o \leq 0.3$, $Al_o + Fe_o/2 \geq 2.0\%$) Andisols on an oven-dry basis (prepared from Andisol TU database, 1992).

TABLE 6.3

Comparison of properties between nonallophanic Andisols and allophanic Andisols (Shoji et al., 1985)

Properties	Nonallophanic	Allophanic
Constant negative charge	Significant	None
Soil acidity	Very strongly to strongly acid	Moderately to slightly acid
Critical pH(H$_2$O) [a]	About 5	None
Al saturation	High	Very low to none
KCl-extractable Al	Abundant	Very small to none
Al toxicity	Common	Rare

[a] The highest pH(H$_2$O) at which most common crops are significantly injured by Al toxicity.

The estimated content of the main soil components in the colloidal fraction of Andisol A horizons is shown in Fig. 6.7. The selected allophanic and nonallophanic Andisols show a contrasting distribution of active Al. The main forms of active Al are allophane/imogolite and Al-humus complexes in allophanic Andisols, while they are Al-humus complexes in nonallophanic Andisols. Dissimilarities between the two groups of Andisols are summarized in Table 6.3. Allophanic Andisols are moderate to slightly acid even when the base saturation is very low and rarely contain toxic level of KCl-extractable Al. In contrast, nonallophanic Andisols are very strongly acid when the base saturation is low reflecting significant amounts of constant negative charge. Aluminum toxicity to plant roots is often observed in such Andisols.

6.3. ORGANIC CONSTITUENTS

The abundance of very dark-colored humus dominated by A-type humic acid is one of the important properties defining the central concept of Kurobokudo in Japan. In fact, Andisols in Japan show the highest accumulation of organic C among mineral soils. Some well-drained Andisols contain organic C concentrations greater than 18 percent (Fig. 6.8). However, not all Andisol humus-rich horizons are very dark (value and chroma 2 or less, moist). Fulvic Andisols also contain large amounts of humus, but they appear dark brown because their humus is rich in fulvic acid and P-type humic acids with a low degree of humification. Thus, melanic and fulvic great groups are provided for some suborders of Andisols (Soil Survey Staff, 1992) as mentioned in Chapter 4.

Humus, as well as noncrystalline clay materials, contributes to the unique chemical and physical properties of Andisols such as variable charge, high phosphate retention (reaction with Al complexed with humus), low bulk density, notable friability, weak stickiness, formation of stable soil aggregates, etc. It also greatly influences the productivity of Andisols through its role in supplying nutrient elements, retaining available water for plants, and development of a favorable rooting environment.

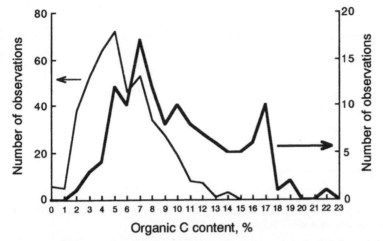

Fig. 6.8. Frequency distribution of organic C contents of A, buried A, and Ap horizons of Andisols (thick line, $Al_o + 1/2\,Fe_o \geq 0.4\%$, prepared from Andisol TU database, 1992) and Ap horizon soils of cultivated Andisols in Japan (thin line, prepared from data of Oda et al., 1987).

6.3.1. Total organic carbon and nitrogen

The striking accumulation of organic C and N in Andisols is indicated by comparing the organic C and N contents between Andisols and other mineral soils. For example, the content of C and N was 5.0 percent and 0.37 percent, respectively, in the A_p horizons of Kurobokudo in Japan, and 1.82 and 0.17 percent, respectively, in the corresponding horizons of other mineral soils (Oda et al., 1987). The mean C/N ratio of cultivated Andisols (14) is higher than that of other mineral soils (11). This suggests that the organic N is more strongly retained in Andisols; however, the total amount of mineralized N is also greater in Andisols as described in Chapter 8.

The frequency distribution of organic C content in Andisols shows a wide range (Fig. 6.8). Both cultivated and uncultivated Andisols have modes at 4–5 percent and 7–8 percent. Many physical properties of Andisols such as bulk density, liquid limit, and plastic limit are strongly affected when organic C contents are greater than 6 percent (Shoji, 1984). Thus, a value of 6% was chosen as the lower limit of organic C for the melanic epipedon and fulvic surface soils (Soil Survey Staff, 1992).

As described in Chapters 3 and 5, soil organic matter plays an anti-allophanic role by forming Al-humus complexes under udic soil moisture regimes, strongly suggesting that nonallophanic Andisols can accumulate more organic C than allophanic Andisols. In fact, this suggestion is supported by the relationship between Al_p/Al_o ratios and the organic C content of humus horizons of Andisols from the Andisol TU database (1992) as shown in Fig. 6.9. The data show that the accumulation of organic C tends to increase with increasing Al_p/Al_o ratios and that

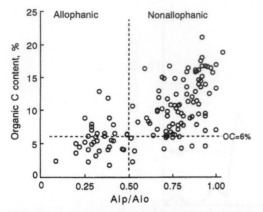

Fig. 6.9. The relationship between Al_p/Al_o values and organic C contents of A_p, A, and buried A horizons of Andisols (prepared from Andisol TU database, 1992).

nonallophanic Andisols occupy a majority of the Andisols with organic C content greater than 6.0%. The Al_p/Al_o ratio can be used to approximately separate allophanic Andisols (ratio < 0.5) and nonallophanic Andisols (ratio ≥ 0.5) as shown by Saigusa et al. (1991).

6.3.2. Humic acids

The very dark color of Andisols has inspired many soil scientists to study the humus composition of these soils. Currently, there is increasing evidence that the very dark color is attributable to the existence of highly humified humic acid, or A-type humic acid. A melanic index of less than 1.70 indicates a dominance of A-type humic acid and has been employed as one of the criteria for the melanic epipedon (Shoji, 1988; Soil Survey Staff, 1992). Thus, this section places emphasis on the properties of A-type humic acid in Andisols.

(1) Humic acid types

Although there are various methods for humus extraction, the methods using dilute NaOH and sodium pyrophosphate solutions are most common for the study of humus in Andisols. The NaOH solution dissolves a great proportion of humus (typically greater than 60 percent using hot NaOH, Adachi, 1973) while pyrophosphate solution preferentially extracts metal-humus complexes. Thus, 0.5 percent NaOH solution has been adopted for extraction of humus in Andisols to determine the melanic index value (Honna et al., 1988). Extractable humus is traditionally fractionated into humic acids and fulvic acids based on differential solubility at low pH values. In Japan, humic acids are further divided into A-, B-, P-, and Rp-types according to relative color intensity (RF) and color coefficient ($\Delta \log K$) (Kumada, 1987). These parameters are defined as follows:

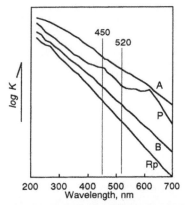

Fig. 6.10. Absorption spectra of humic acids of A-, B-, P-, and Rp-types. Concentrations are arbitrarily chosen (prepared from data of Kumada, 1987; copyright JSSP, with permission).

RF: $K_{600} \times 1000/C$, where C is mL of 0.02 M $KMnO_4$ consumed by 30 mL of humic acid solution used for determining the absorption spectrum.

$\Delta \log K$: $\log K_{400} - \log K_{600}$, where K is absorbance at 400 and 600 nm of the same humic acid solution used for the RF determination.

Thus, RF is regarded as color density of humic acids, and $\Delta \log K$, as the slope of the absorption spectra of humic acids, except for the P-type humic acid. As demonstrated in Fig. 6.10, the absorption spectra of A-, B- and Rp-type humic acids of Andisols in the UV-VIS range (300–700 nm) appear to be almost smooth lines on a logarithmic scale (Kumada, 1987). The magnitude of the slope of A-type humic acid is lower than the other types. In contrast, P-type humic acid has shoulders or peaks at 280, 450, and 570 nm, concave segments in the ranges of 300–400 and 480–580 nm, and a maximum at 615 nm (Kumada, 1987). Thus, A-type humic acid can be distinguished from the other humic acids according to these distinctive features of its absorption spectrum.

Like melanic Andisols, fulvic Andisols often have low ratios of absorbance at 400 nm to absorbance at 600 nm, reflecting the absorption maximum of P-type humic acid near 615 nm. However, a ratio of absorbances at 450 and 520 nm provides high values (>1.70) for fulvic Andisols. Thus, melanic Andisols (melanic indices <1.70) are separated from fulvic Andisols according to the melanic index which is defined as the ratio of absorbance at 450 nm to absorbance at 520 nm (Honna et al, 1988).

Andisols with grass vegetation in Japan show a predominance of A-type humic acid followed by P- and B-type humic acids. However, such a distribution of humus composition is not always the case for Andisols from other countries as shown in Fig. 6.11 (Shoji et al. 1987). For example, Shoji et al. (1987) showed that humus from New Zealand Andisols had a high content of fulvic acid, and the humic acid was dominated by P-type humic acid. Also in New Zealand Andisols,

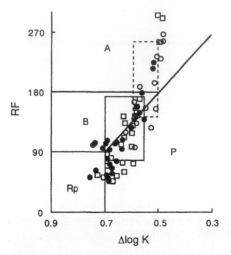

Fig. 6.11. Division of humic acids using color coefficient ($\Delta \log K$) and relative color intensity (RF) into A-, B-, Rp-, and P-types. Humic acids of Andisols from Japan and New Zealand plot mostly inside the squares drawn by dotted line and solid line, respectively. Closed circles, squares, and open circles indicate the humic acids of Andisols from New Zealand, Chili, and Ecuador, respectively (prepared from data of Shoji et al., 1987).

grass vegetation contributes to the formation of A-type humic acid while forest vegetation promotes formation of P-type humic acid (Sase, 1986). The effect of vegetation on the formation of humic acids was described in detail in Chapter 3.

(2) Selected properties of A-type humic acid

The elemental composition, functional group content, aromaticity, etc. provide useful information for the chemical characterization of A-type humic acid as described below. The A-type humic acid is more aromatic and less aliphatic and has a higher carboxyl group content than other types of humic acids. Yonebayashi and Hattori (1988) determined the elemental composition and functional group content of purified A-type humic acid from Andisols in Japan (Table 6.4). The C content ranged between 54.3 and 58.4 percent and the O content between 35.8

TABLE 6.4

Elemental and functional group compositions of A-type humic acid from 10 Andisols in Japan (Yonebayashi and Hattori, 1988)

	C	H	N	O	COOH	Phenol-OH	C=O	Alcohol-OH
	(%)				(mole kg^{-1})			
Maximum	58.4	4.4	4.6	38.5	6.2	1.4	5.8	5.9
Minimum	54.3	2.3	1.9	35.8	3.6	0.1	2.6	0.6
Mean	56.7	3.3	3.0	37.0	4.6	0.7	4.6	3.0

and 38.5 percent. These values are very similar to the intermediate values of C and O contents listed for humic acids of world soils (Schnitzer, 1978). In contrast, the H content ranged between 2.3 and 4.4 percent which was on the low end of H contents listed for humic acids of world soils. These data indicate a relatively high aromaticity and a high content of oxygen-containing functional groups in the A-type humic acid fraction. The N content of 1.9–4.6 percent is relatively high and the C/N ratio of 14–33 is somewhat lower than those of humic acids from world soils (Schnitzer, 1978).

Various functional groups occur in humic acids. Yonebayashi and Hattori (1988) determined that A-type humic acid from Andisols contain considerably high amounts of carboxyl and carbonyl groups as compared with those of B- and P-type humic acids. They also showed that A-type humic acid contains a low content of phenolic hydroxyl groups, and that there is a large variation in alcoholic hydroxyl group content.

The composition of functional groups and aromaticity of humic acids were confirmed by infrared absorption spectroscopy (IR), and nuclear magnetic resonance spectroscopy (^{1}H-NMR and ^{13}C-NMR). The IR spectrum of A-type humic acid shows a weak absorption band at 2920 cm^{-1}, indicating a low content of saturated hydrocarbons (Fig. 6.12a). In contrast, the strong absorption bands at 1720 cm^{-1} and 1600 cm^{-1} are attributed to the high content of carboxyl and carbonyl groups and aromatic rings, respectively (Yonebayashi and Hattori, 1989).

An example of a ^{1}H-NMR spectrum for A-type humic acid from Andisols is shown in Fig. 6.12b. The signals in the spectrum are assigned to six types of protons such as aromatic protons, lactone protons, methoxy or alcoholic protons, and protons attached to alpha, beta, gamma, etc. positions of the side chain of aromatic rings. The A-type humic acid from Andisols has large amounts of protons in the alpha position and aromatic protons, and small amounts of methoxy or alcoholic protons as compared to the other types of humic acids (Yonebayashi and Hattori, 1989).

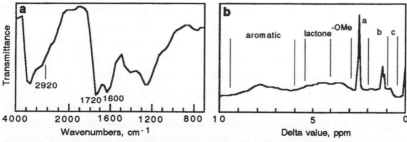

Fig. 6.12. Infrared (a) and ^{1}H-NMR (b) spectra of humic acid extracted from Andisols. The letters a, b, and c denote protons attached to aliphatic carbon alpha to aromatic rings, protons in the region extending from methyl groups beta to aromatic rings or those bound to methylene or methine carbons beyond beta to aromatic rings, and the protons attached to methyl terminal carbons gamma or further from aromatic rings, respectively (prepared from data of Yonebayashi and Hattori, 1989).

The ^{13}C-solid state high resolution NMR spectrum of humic acid extracted from Andisols was reported by Tate et al. (1990). An A-type humic acid from an Andisol in Hokkaido, Japan contained greater amounts of aromatic carbon and a smaller amount of aliphatic carbon than the other humic acids.

The three different spectroscopic analyses described above provide similar information on the chemical structure of A-type humic acid from Andisols: the A-type humic acid from Andisols is more aromatic and less aliphatic than the other types of humic acids.

6.3.3. Nonhumic substances

(1) Saccharides

Carbohydrates in soil organic matter occur primarily as polysaccharides. The polysaccharides after acid hydrolysis produce hexoses (glucose, galactose and mannose), pentoses (arabinose, ribose and xylose), deoxyhexose (fucose and rhamnose), uronic acids (galacturonic and glucuronic acids), etc. (Schnitzer, 1991).

Humus-rich Andisols contain 1 to 3 percent neutral saccharides (hexoses, pentoses and deoxyhexoses) in the bulk soil and display a linear positive relationship between soil organic C content and monosaccharide content. The percentage of neutral saccharide carbon in organic C of Andisols ranges from 4 to 12 percent in modern humus horizons and from 3 to 4 percent in buried humus horizons. These values are considerably lower than those of nonandic soils in Japan (Murayama, 1980, 1984).

A great proportion of soil saccharides is commonly extractable with diluted sodium pyrophosphate or sodium hydroxide solutions, while they are not extracted with diluted sodium sulfate solution. Thus, soil saccharides are considered to consist of plant remains, microbes and/or complexes with soil colloids or metal ions (Murayama, 1984).

Saccharides in humus-rich Andisols consist mostly of hemicellulose and they can be easily extracted with sodium pyrophosphate solution, suggesting that they are largely complexed with metallic ions. In contrast, cellulose is present in only small amounts and is not as easily extracted as soil saccharides. Cellulose is determined after hydrolysis using concentrated sulfuric acid solution.

It is widely accepted that polymerization of degradation products from plant and microbial remains is a plausible mechanism leading to humus formation. The monosaccharide component of plant materials is also altered in this process. Hasegawa and Shoji (1993) examined the effect of vegetation on the monosaccharide composition of Andisols. For this purpose, they selected soils along a biosequential relationship in South Hakkoda, northeastern Japan where Melanudands and Fulvudands formed in *Miscanthus sinensis* and *Fagus crenata* ecosystems, respectively. As mentioned in Chapter 3, both above-ground and below-ground parts of *M. sinensis* produce a large biomass dominated by cellulose and hemicellulose that contribute significantly to the genesis of Melanudands. Hasegawa and Shoji (1993) observed that there was no notable difference in the

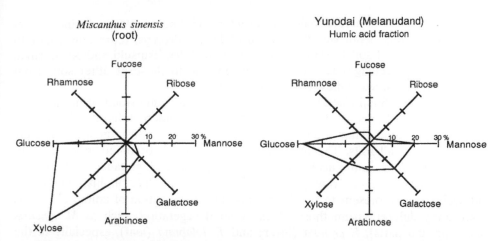

Fig. 6.13. Monosaccharide composition of *Miscanthus sinensis* roots and the humic acid fraction of the A horizon of a Melanudand from South Hakkoda, northeastern Japan (prepared from data of Hasegawa and Shoji, 1993).

monosaccharide composition between the above- and below-ground plant parts. As presented in Fig. 6.13, xylose and glucose comprise approximately 50 and 30 percent of total noncellulosic monosaccharides, respectively. Arabinose, galactose and mannose are found in intermediate amounts while ribose, fucose and rhamnose are present in very small amounts or nearly absent.

The noncellulosic monosaccharide composition of Melanudands was considerably different from that of *M. sinensis*. Organic matter from the Melanudand formed in the *M. sinensis* ecosystem had a saccharide content of 1 percent in the fine-earth fraction and 8.1 and 5.7 percent in the humic acid and fulvic acid fractions, respectively. Since the humic acid fraction is the most stable fraction of soil organic matter, its monosaccharide composition is presented in Fig. 6.13. Glucose, which could be largely from noncellulosic origin, shows the greatest content, occupying about 30 percent of the total content of monosaccharides. It is followed by mannose, galactose, arabinose and xylose. The content of ribose, fucose and rhamnose ranged from 4 to 7 percent, being significantly higher than their concentration in the plant material of *M. sinensis*.

There were also important differences in monosaccharide composition between Fulvudand organic matter and litter under *Fagus crenata* (Hasegawa and Shoji, 1993). The former contains higher percentages of glucose (noncellulosic) and mannose and lower percentages of xylose and arabinose than the latter. It was also noted that the monosaccharide composition of the humic and fulvic acid fractions was similar in the Fulvudands.

(2) Amino acids

The main organic forms of nitrogen occurring in soil organic matter are protein-N, peptide-N, amino acid-N, amino sugar-N, NH_4-N, N in purines and pyrimidines, and N in heterocyclic ring structures (Schnitzer, 1991). Amino acid-N in Japanese Andisols, measured by hot 6 M HCl hydrolysis, represents 20 to 40 percent of the total soil N (Hasegawa and Shoji, 1993; Tsutsuki and Kuwatsuka, 1978). These percentages are similar in magnitude to those of other soils from widely differing climate zones (Schnitzer, 1991).

Hasegawa and Shoji (1993) studied the amino acid composition of humic acids from the surface horizons of Andisols formed under widely different climate and vegetation in Japan. They found that the amino acid composition was relatively constant among all soils. The major amino acids, which occupied more than 10 percent of the total amino acid fraction, were aspartic acid, glutamic acid, glycine, and alanine. Threonine, serine, proline, valine, isoleucine, leucine, histidine, lysine and arginine were present in small amounts. The composition of amino acids was significantly different from those of the related vegetation, such as *M. sinensis* (above-ground parts), *F. crenata* (litter) and *T. dolabrata* (leaf), especially in the percentage of valine, isoleucine, leucine and aspartic acid.

6.4. METAL HUMUS COMPLEXES

Aluminum/iron humus complexes are commonly present in large amounts in the humus horizons of Andisols. Like allophanic clays, metal humus complexes determine many important physical and chemical properties of humus-rich Andisols, such as their unique consistence, high water retention, low bulk density, variable charge characteristics, large phosphate retention, high fluoride reactivity, etc.

Aluminum humus complexes are formed by the anti-allophanic process; the preferential formation of Al humus complexes rather than allophanic clays, as first observed in young Andisols from Hokkaido, Japan by Shoji and Masui (1972). This process occurs intensively in acidic (pH(H$_2$O) < 5) humus horizons of Andisols with udic soil moisture regimes. It contributes to the accumulation of large amounts of organic C which shows a strong positive correlation with the content of pyrophosphate-extractable Al (Al$_p$) or Al complexed with humus as presented in Fig. 6.14. Organic C reaches 6 percent when Al$_p$ is approximately 0.85 percent. In laboratory synthesis experiments of allophane (Inoue and Huang, 1987) and imogolite (Inoue and Huang, 1990), humic substances were shown to inhibit the formation of these minerals. The anti-allophanic process also contributes to the formation of laminar opaline silica or pedogenic opal in the humus horizon of young Andisols because Al preferentially combines with humus and is unavailable to react with Si (Shoji at al., 1982; Shoji and Fujiwara, 1984).

The molar ratio of organic C to Al in metal-humus complexes was examined using various methods. The present authors have obtained a mean ratio of 13 using regression analysis of the relationship between total organic C and Al$_p$ of many Andisols from Japan and USA excluding the uppermost humus layers (Fig. 6.14, Andisol TU database, 1992). According to Yonebayashi and Hattori

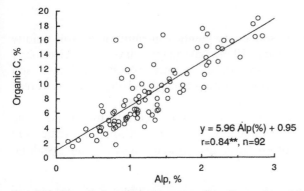

Fig. 6.14. The relationship between pyrophosphate-extractable Al (Al_p) and total organic C of A horizons from Andisols except the uppermost humus layers (prepared from Andisol TU database, 1992). According to the slope of the regression equation, the atomic ratio of C/Al_p is approximately 13.

(1988), A-type humic acids have one mole of carboxyl group per ten moles of organic C on average. Thus, the number of carboxyl groups is greater than that of Al_p. Inoue and Higashi (1988) have shown that the molar ratio of pyrophosphate-extractable C to Al_p is about 6. The ratio of 6 is too low in comparison with the carboxyl group content of humic acid (Yonebayashi and Hattori, 1988) if all the Al_p exists in monomeric forms. In contrast, the ratio of 6 appears possible if Al complexed with humus is partially polymerized and contains a low charge (Higashi, 1983). According to Tsutsuki and Kuwatsuka (1992), humin (insoluble humus) samples prepared from buried volcanic ash soils using HF treatment were also complexed with Al and the average C/Al atomic ratio of five humin samples was calculated to be 16.

Aluminum complexed with humus contains hydroxyl groups which have high reactivity with fluoride similar to allophanic clays. Thus, the pH(NaF) can not adequately separate nonallophanic and allophanic Andisols (Shoji and Ono, 1978). As presented in Fig. 6.21, Al-humus complexes do sorb large amounts of phosphate as indicated by the close relationship between Al_o and the phosphate retention of Andisols which have Al_p/Al_o ratios greater than 0.5 (nonallophanic Andisols). As compared to Al in allophane, Al complexed with humus is highly reactive as discussed later.

6.5. PHYSICO-CHEMICAL PROPERTIES

The important physico-chemical properties which characterize Andisols are acidity, variable charge, specific sorption of phosphate and heavy metals, soil solution, and resistance to reduction, etc. These properties are primarily related to the forms and content of active Al and Fe.

6.5.1. *Soil acidity*

Although there are a variety of soil components contributing to soil acidity (Thomas and Hargrove, 1984; Yoshida, 1979), allophanic clays, chloritized 2:1 minerals, and humus are especially important in most Andisols. These soil components have different mechanisms generating acidity and thus have different acid strengths.

Acidity of allophanic Andisols that have very low contents of humus (organic C content 3 percent or less) and layer silicates (group 1 Andisols) is attributable primarily to the dissociation of protons from the broken edges of noncrystalline aluminosilicates as follows:

$$\equiv SiOH \longrightarrow \equiv SiO^- + H^+$$

A considerable number of allophanic Andisols show $pH(H_2O)$ values of 5.8 to 6.0, although the base saturation of these soils is 10 percent or less (Fig. 6.15). Other Andisols in this group have $pH(H_2O)$ of lower than 5.7, possibly reflecting the existence of small amounts of 2:1 layer silicates. However, allophanic Andisols rarely contain toxic levels of Al which hinders plant root growth as reported by Shoji et al. (1980).

Nonallophanic Andisols (group 2 Andisols) typically have a clay fraction dominated by chloritized 2:1 minerals or hydroxy-Al interlayered 2:1 minerals (Shoji et al., 1985). Some Andisols also have aluminosilicate materials in the interlayer position (Wada and Kakuto, 1983). Chloritized 2:1 layer silicates have various amounts of KCl-extractable Al ions on their constant negative charge sites which contribute to the greater acidity of nonallophanic Andisols (Saigusa et al.,

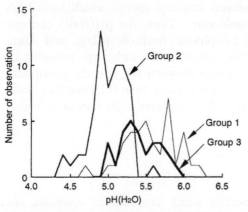

Fig. 6.15. Frequency distribution of $pH(H_2O)$ values of group 1 Andisols (KCl-extractable Al $\leq$ 2 $cmol_c$/kg; organic C $\leq$ 3%), group 2 Andisols (KCl-extractable Al $>$ 2 $cmol_c$/kg), and group 3 Andisols (KCl-extractable Al $<$ 2 $cmol_c$/kg; organic C $\geq$ 6%). All of these samples meet the first criteria of andic soil properties and their percent base saturation is 10% or less (prepared from Andisol TU database, 1992 and Wada, 1986).

1980). KCl-extractable Al ions behave as an acid through the following reaction (Sadiq and Lindsay, 1979):

$$Al^{3+} + H_2O = AlOH^{2+} + H^+ \qquad pK = 5.02$$

Nonallophanic Andisols with low base saturation values show low $pH(H_2O)$ values typically ranging between 4.8 and 5.3, irrespective of the humus content (Fig. 6.15). This indicates that exchangeable Al ions are the primary source of acidity in nonallophanic Andisols as described in the above equation.

Acidity of humus-rich allophanic Andisols (organic C content 6 percent or more, group 3 Andisols) is largely determined by carboxyl groups and is significantly influenced by the formation of Al humus complexes. Although humic acids of Andisols have strongly acidic ($pK_a \leq 3.2$) carboxyl groups, many of them are blocked by complexation with Al (approximately 20 percent of the total carboxyl groups, Yonebayashi and Hattori, 1988). Thus, only a small fraction of the weakly acidic carboxyl groups contribute to the acidity of group 3 soils, which show intermediate $pH(H_2O)$ values between group 1 and 2 Andisols (Fig. 6.15).

The $pH(KCl)$ values also provide a measure of acidity in Andisols and are usually lower than $pH(H_2O)$. The $pH(KCl)$ values as well as $pH(H_2O)$ values of the three groups of Andisols are summarized as follows (Fig. 6.15, Andisol TU database, 1992):

	pH(KCl)	pH(H$_2$O)
Group 1 Andisols (allophanic, humus-poor)	5.0–5.6	5.2–6.0
Group 2 Andisols (nonallophanic)	3.8–4.4	4.8–5.3
Group 3 Andisols (allophanic, humus-rich)	4.3–5.0	5.0–5.7

The magnitude of the differences between $pH(H_2O)$ and $pH(KCl)$ values follows the order: group 2 Andisols > group 3 Andisols > group 1 Andisols. The large difference for group 2 Andisols indicates that the KCl-extractable Al ions on 2:1 layer silicates have the greatest contribution to strengthen the acidity through the hydrolytic reaction of Al ions as described earlier. In contrast, group 1 Andisols show the least difference due to their low KCl-extractable Al. Some Andisols that have large amounts of noncrystalline clays and very low contents of layer silicates and soil organic matter, show $pH(KCl)$ values greater than $pH(H_2O)$ values or the acric property (Uehara and Gilman, 1981). This is attributed to the adsorption of Cl^- ions to positive charge sites of noncrystalline clays, resulting in the net formation of KOH which increases the $pH(KCl)$ of the soil solution. The group 3 Andisols show intermediate differences between $pH(H_2O)$ and $pH(KCl)$ values due to increased dissociation of weakly acidic $-COOH$ and $\equiv SiOH$ upon addition of KCl. The reason for the small pH shift of group 1 Andisols is due to the simultaneous increase in both negative ($\equiv SiO^-$) and positive ($=AlOH_2^+$ and/or $=FeOH_2^+$) charges of noncrystalline minerals in 1 M KCl solution.

6.5.2. Charge characteristics

(1) Variable charge

Andisols display typical variable charge soil characteristics. The main soil components contributing to the variable charge are allophanic clays and humus. These soil colloids have negative charge sites originating from SiO^- of allophanic clays and chloritized 2:1 minerals and $-COO^-$ of humus. The positive charge sites consists of $=AlOH_2^+$ of allophanic clays, hydroxy-Al polymers in the interlayer of chloritized 2:1 minerals, and $=FeOH_2^+$ of ferrihydrite (Wada, 1985).

An example of the variable charge nature of an Alic Melanudand is shown in Fig. 6.16a. The negative charge at low pH values is largely attributable to the constant charge of 2:1 layer silicates. At higher pH values, the 2:1 layer silicates and the variable charge of soil organic matter both contribute to the negative charge. Although the positive charge of the hydroxy-Al interlayers of 2:1 minerals can increase with decreasing pH, it is balanced by the permanent negative charge of the 2:1 mineral so that the net positive charge of the soil is approximately zero. In contrast, the allophanic Bw horizon of a Typic Hapludand shows variable positive charge comparable to the variable negative charge within the common soil pH range resulting in a point of zero net charge (PZNC) at pH values of 5 to 6 as illustrated in Fig. 6.16b.

The amount of variable charge changes greatly with pH and the ionic strength of the ambient solution. These relationships are expressed by the following general equation (Wada and Okamura, 1980; Okamura and Wada, 1983):

$$\log CEC \text{ or } AEC = a\,pH + b\log C + c$$

where CEC ($cmol_c$ kg^{-1}) and AEC ($cmol_c$ kg^{-1}) are the amount of negative and positive charge, respectively, C (M) is the concentration of the electrolyte, and

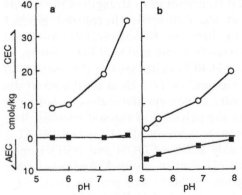

Fig. 6.16. Charge characteristics of an A horizon from an Alic Melanudand (a, nonallophanic) and a Bw horizon from a Typic Hapludand (b, allophanic) (prepared from data of Saigusa et al., 1992).

a, *b*, and *c* are constants. Soils which have constant negative charge in addition to variable charge show greater *c* values.

Andisols commonly contain a mixture of variable charge colloids such as humus-rich allophanic Andisols (allophane and humus), or a mixture of variable and constant charge colloids such as humus-rich nonallophanic Andisols (chloritized 2:1 minerals and humus). These Andisols show variable negative charge characteristics as shown in Fig. 6.16, but their positive charge is usually very small or virtually absent, indicating the depression of positive charge due to humus and layer silicate minerals and the absence of a PZNC within the normal soil pH range.

Although the CEC values of moderately weathered Andisols are not so low compared with other mineral soils, effective CEC which is defined as the sum of exchangeable bases plus KCl-extractable Al is typically low, especially in regions lacking a dry season. The difference between the CEC and the effective CEC values is due to their variable charge characteristics. Most pH(H$_2$O) values of Andisols ranges from 4.8 to 6.5 and these are one to two pH units lower than the pH value for CEC determination (pH = 7). In addition, ionic strength of natural soil solutions (10^{-3}–10^{-4} M) are usually much lower than the 1 M concentrations used for CEC measurements. Thus, most exchange sites on variable charge Andisols are occupied by protons, and nonspecifically sorbing cations and anions are not effectively retained in Andisols.

Charge characteristics of allophanic Andisols are closely related to the acidity properties. An Acrudoxic Hydric Fulvudand from Java, Indonesia was selected to show this relationship. As presented in Fig. 6.17, the A horizon shows high organic C content, a moderate CEC level, very low effective CEC, and low

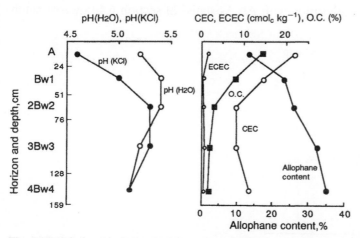

Fig. 6.17. Relationship between charge characteristics and acidic properties of an Acrudoxic Hydric Fluvudand from West Java. While the A horizon has a large amount of organic C, high CEC, and a relatively large ΔpH, the 2Bw2-4Bw4 horizons have a very small amount of organic C, high Si$_o$, very low ECEC, and a very small ΔpH (prepared from data of Nanzyo et al., 1993).

allophane content (low Si_o). The B horizons have a low organic C content, a lower CEC, very low effective CEC, and high allophane content (high Si_o).

The upper horizons have lower pH(KCl) values (4.6–5.0) than the lower horizons, being ascribed to the dissociation of carboxyl groups on soil organic matter. As compared to the upper horizons, the lower horizons show a sum of exchangeable bases and KCl-extractable Al of less than 2 $cmol_c$ kg^{-1} which defines the acric property, and almost the same pH(H_2O) and pH(KCl) values, reflecting the predominance of allophanic clays. It also contains, 1500 kPa water greater than 70 percent in both upper and lower horizons (Nanzyo et al., 1993). Thus, the soil meets the criteria for the acrudoxic hydric subgroup of Andisols (Soil Survey Staff, 1992), indicating that it is highly weathered and highly leached.

The ΔpH, defined as the difference between pH(KCl) and pH(H_2O), can be used for indicating the presence of variable charge minerals and approximate levels of KCl-extractable Al in humus-poor soils (Uehara and Gillman, 1981). Positive to small negative ΔpH values in the lower horizons of the Acrudoxic Hydric Fulvudand indicate the abundance of allophanic clays and a low content of humus and KCl-extractable Al.

Carboxyl groups significantly contribute to the CEC of humus-rich Andisols. According to the relationship between organic C content and the CEC of humus-rich Andisols, which include allophanic and nonallophanic Andisols (Fig. 6.18), one mole of negative charge corresponds to 40 moles of organic C. For comparison, A-type humic acid has one mole of carboxylic acid groups per 10 moles of organic C (Table 6.4). Therefore, 40 moles of organic C should have approximately 4 moles of negative charge, assuming that all the organic matter has one mole of carboxyl group per 10 moles of organic carbon. This indicates that 75 percent of the carboxylic groups do not exist as free negatively charged sites, possibly because they are blocked by complexation with Al. As stated in section 6.4, approximately

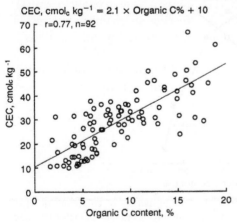

Fig. 6.18. The relationship between organic C content and CEC of A horizons from Andisols (prepared from Andisol TU database, 1992).

13 moles of organic carbon are involved in the complexation of one mole of Al. As a consequence, only 25 percent of the carboxyl groups are free negatively charged sites at a pH of 7.0 in 0.05 M calcium acetate solution.

Dissociation of silanol groups contributes to CEC in allophanic Andisols with low humus content to a large extent. There is a significant correlation between Si_o and CEC with the slope of the regression line indicating that the degree of dissociation of $\equiv SiOH$ groups is about 4 percent at pH = 7.0 in 0.05 M calcium acetate solution. Since the Al/Si atomic ratio of allophanic clays in Udands is approximately 2 (Fig. 6.6), the structure of the Al-rich allophane is similar to imogolite (Shimizu et al., 1988). Therefore, 3 out of 4 hydroxyl groups bonded to Si are coordinated with Al and the remaining hydroxyl group is free. Thus, the number of free silanol groups can be approximated using the content of Si_o.

(2) Ion exchange equilibria

Andisols have cation exchange sites consisting primarily of $-COO^-$, $\equiv SiO^-$ and constant negative charge, and anion exchange sites of $=Al-OH^{2+}$ and $=Fe-OH_2^+$. The exchangeable cations are commonly Ca^{2+}, Mg^{2+}, K^+, Na^+, Al^{3+}, and the exchangeable anions are Cl^-, NO_3^-, SO_4^{2-}. Characteristics of ion exchange reactions in soil systems are: (1) ions are retained to the exchange complex through electrostatic attractive forces forming outer-sphere complexes, (2) the exchange reaction proceeds equivalently and does not disturb charge balance in either the solid or solution phases, and (3) some ions are preferentially retained by the exchange sites relative to others. For example, the cation exchange sites of allophane, imogolite, and humus show a higher affinity for Ca^{2+} than for NH_4^+. In contrast, halloysite shows a higher affinity for NH_4^+ compared with Ca^{2+} (Okamura and Wada, 1984; Delvaux et al., 1989). These ion exchange equilibria are described using the following equation:

$$A^{2+}_{(S)} + 2B^+_{(L)} = 2B^+_{(S)} + A^{2+}_{(L)}$$

The subscripts S and L indicate solid and liquid phases, respectively, and A^{2+} and B^+ are divalent and monovalent cations, respectively. The exchange equilibrium constant for this equation can be defined as:

$$K_X = \frac{(X_{B(S)})^2 \, X_{A(L)}}{X_{A(S)} \, (X_{B(L)})^2} = \left(\frac{X_{B(S)}}{X_{B(L)}}\right)^2 \cdot \frac{1 - X_{B(L)}}{1 - X_{B(S)}}$$

where X is the equivalent fraction. If $X_{B(S)} = X_{B(L)}$, then $K_X = 1$ and the solid shows no preference for ions. If $K_X > 1$, the solid surface has a higher affinity for B^+ than for A^{2+} and vice versa.

Selecting three contrasting volcanic ash soils, the $Ca^{2+}-NH_4^+$ exchange equilibria are described as a function of $X_{NH_4(L)}$ in Fig. 6.19. Andisols commonly consist of mixed exchange complexes having different affinities for cations which results in their exchange equilibria showing unsymmetrical concaved or convexed lines. In

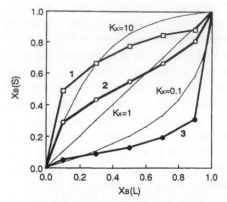

Fig. 6.19. The Ca^{2+}–NH_4^+ exchange equilibria of soils with different exchange materials. Sample names, major exchange materials, and CEC values are as follows: 1 = Weathered volcanic ash, halloysite, 13 $cmol_c$ kg^{-1}; 2 = Bw horizon of an Ando soil, allophane and imogolite, 4 $cmol_c$ kg^{-1}; 3 = A1 horizon of an Ando soil, Al-humus complexes, 12 $cmol_c$ kg^{-1} (prepared from data of Wada, 1981; copyright Springer-Verlag, with permission).

these volcanic ash soils, halloysitic samples show a higher affinity for NH_4^+ while samples dominated by Al-humus complexes show a greater affinity for Ca^{2+}.

(3) Sorption of heavy metal ions

Variable charge soils show a very high affinity for heavy metals as compared with constant charge soils (Wada, 1989). Equilibrium constants comparing heavy metal with Ca^{2+} retention are summarized in Table 6.5 (Abd-Elfattah and Wada, 1981). The selectivity of heavy metal ions in Andisols typically follows the order: Cd^{2+}, $Co^{2+} < Zn^{2+} < Cu^{2+}$, Pb^{2+}. In general, the greater the pK_{so} value of the heavy metal hydroxide and the smaller the p^*K_1 values (negative logarithm of the

TABLE 6.5

Selectivity constants for heavy metal ions in Andisol soil samples. Those of a Red yellow soil and a Gray lowland soil are also shown for comparison (Abd-Elfattah and Wada, 1981)

Soil name: Horizon	Major ion exchange materials	K_{Ca}^M values at [M] soil/CEC (%) = 20 of Ca-saturated soils				
		Pb	Cu	Zn	Co	Cd
And: (B)	A, Im	2600	2200	480	7	12
WVA	Ht	5400	3800	220	19	2
RY: B	Mt	22	9	3	2	1
GrL: Ap	Vt, Hum	780	410	9	4	3
And: A1	Al-hum, A, Im	1700	1700	51	8	7
And: A1	Al-hum	1100	660	17	6	8

Abbreviations: And = Ando Soil; GrL = Gray Lowland Soil; RY = Red Yellow Soil; WVA = Weathered volcanic ash; A = allophane; Im = Imogolite; Ht = Halloysite; Mt = Montmorillonite; Vt = Vermiculite; Hum = Humus.

first hydrolysis constants), the higher the affinity of the heavy metal ion for the variable charge surface. The order of magnitude for pH_{50} values (the pH value at which 50% of the added heavy metal is sorbed) shows a close relationship with the order of pK_{so} values or p^*K_1 values; however, the sorption of heavy metal ions on variable charge surfaces takes place at lower pH values than the hydrolysis of the heavy metal ions (Wada, 1981; Yamamoto, 1982). Both A and Bw horizon soils of Andisols show high affinities for Pb^{2+} and Cu^{2+}. The reason for this high affinity of metal ions (M) is considered to be due to the high stability of the bidentate surface complexes of allophane and humus as described below:

$$\equiv Si - O \diagdown_{\diagdown} M \qquad RCOO \diagdown_{\diagdown} M$$
$$\equiv Si - O \diagup^{\diagup} \qquad RCOO \diagup^{\diagup}$$

The selectivity of heavy metals is affected by pH and the ratio of heavy metal to CEC. The degree of selectivity for heavy metals tends to decrease with decreasing pH values due to proton competition with heavy metals for sorption sites at low pH (Wada and Abd-Elfattah, 1978). The selectivity constant tends to decrease with increasing ratios of heavy metal to CEC values; maximum selectivity occurs at ratios less than approximately 0.2 (Abd-Elfattah and Wada, 1981).

Heavy metal sorption results in the release of protons from Andisol soil materials. The stoichiometry between proton release and heavy metal sorption has been studied to determine the mechanism of heavy metal sorption. The observed value for the ratio of protons released to divalent heavy metal sorbed is approximately 2 in many cases (Yamamoto, 1982). It is interesting to note that in contrast to specific sorption of heavy metals, sorption of anions such as fluoride or phosphate results in the release of hydroxide ions leading to an increase in pH values (see 6.5.3).

6.5.3. Anion sorption

Andisols have a high capacity for phosphate and fluoride sorption due to their high content of active Al and Fe compounds. The phosphate sorption capacity of less weathered or vitric Andisols is smaller than nonvitric Andisols (Ito et al., 1991). Fresh tephras contain 0.004–0.6 g kg^{-1} of acid-extractable phosphorus, possibly occurring as apatite (see Chapter 8). However, the amount of acid-extractable phosphorus in Andisols tends to decrease with the advance of weathering because active Al and Fe form from weathering of the parent material and react to form insoluble Al and Fe phosphate compounds. This reaction may result in phosphorus deficiency for crops grown on these soils.

Phosphate and fluoride sorption differs from ion exchange reactions as follows: (1) the anions react with active Al and Fe compounds to form covalent bonds or inner-sphere complexes; (2) sorption reactions of these anions are not completely reversible and it is difficult for sorbed anions to be desorbed; (3) the sorption capacity for these anions is greater than the amount of positive charge; and (4) the ligands on the surfaces of the active Al or Fe compounds are released by ligand exchange reactions with the sorbing anions. The abundance of active Al and Fe compounds can be estimated by measuring phosphate sorption or the pH rise due to fluoride sorption (pH(NaF)). A discussion of phosphate sorption by Andisols is given below.

(1) Relationship between phosphate sorption and the amount and forms of active aluminum and iron

In general, the amount of phosphate sorption by Andisols is determined by the content and forms of active Al and Fe. Phosphate sorption increases curvilinearly with increasing concentrations of phosphate in solution. The shape of the sorption curve corresponds to the L-curve category of adsorption isotherms given by Sposito (1984). In order to describe this curve mathematically, the Langmuir or Freundlich equation has been applied (Rajan, 1975; Wada and Gunjigake, 1979; Gunjigake and Wada, 1981).

The amount of phosphate sorption by Andisols also depends on the pH of the system (Sekiya, 1972; Imai, 1981; Gunjigake and Wada, 1981). The pH dependency of phosphate sorption is somewhat higher in allophanic Andisols as compared to nonallophanic Andisols. Maximum phosphate sorption occurs between pH values of 3 and 4 and the amount of phosphate sorption decreases with increasing pH as shown in Fig. 6.20. The maximum phosphate sorption becomes more distinct with increasing initial phosphate concentration. Thus, it is necessary to express the

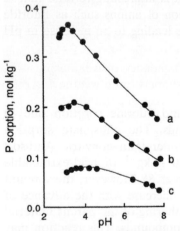

Fig. 6.20. Phosphate sorption in an Andisol as a function of pH and initial phosphate concentration. Twenty mL of phosphate solution ($a = 21.5$, $b = 11.4$, and $c = 3.77$ mM P) was added to a gram of Andisol sample (Sekiya, 1972. Copywright Yokendo, with permission). The amounts of phosphate sorption (Y, mol kg^{-1}) in the pH range of 4–8.5 are expressed using the multiple regression equation reported by Gunjigake and Wada (1981): $\log Y = -0.19\,\mathrm{pH} + 0.691\log C + 1.9, r = 0.96$, where C is the equilibrium concentration in mol L^{-1} (Nanzyo et al., 1984).

amount of phosphate sorption as a function of both the final pH and phosphate concentration. Gunjigake and Wada (1981) employed the extended Freundlich equation to express phosphate sorption as follows:

$$\log Y = a\,\mathrm{pH} + b(\log C) + c$$

where Y is the amount of phosphate sorption in mmol g^{-1}, C is the final phosphate concentration in M and a, b, and c are empirical constants. This equation was successfully applied to describe phosphate sorption for many allophanic and nonallophanic Andisol samples in the pH range of 4–8.5 and phosphate concentrations between 0.001–2 M. Allophanic Andisols have a negative a value, indicating that phosphate sorption decreases with an increase in pH. In contrast, the a value for nonallophanic Andisols increases as the Al_p/Al_o ratio increases and approaches zero or slightly positive values at Al_p/Al_o ratios near unity. This illustrates that the pH dependency of phosphate sorption by Al humus complexes is considerably lower as compared to that of allophanic Andisols.

The lower pH dependency of phosphate sorption by nonallophanic Andisols, whose active Al fraction is dominated by Al humus complexes, can be explained by a hypothesis proposed by Haynes and Swift (1989). Studying the relationship between pH and phosphate sorption using a synthetic Al-humus complex, these researchers proposed that Al is released from humus complexes and is polymerized with increasing pH, resulting in enhanced phosphate reactivity with Al polymers.

Nonallophanic Andisols show a larger phosphate retention than allophanic Andisols, determined by Blakemore's method (Blakemore et al., 1981), when the phosphate retention values are expressed on the basis of $Al_o + 1/2\,Fe_o$ (Fig. 6.21). Almost all nonallophanic Andisols containing $Al_o + 1/2\,Fe_o$ greater than 1.5 percent have phosphate retention values of 85 percent or more, one of the criteria

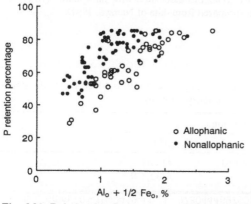

Fig. 6.21. Relationship between $Al_o + 1/2\,Fe_o$ and the phosphate retention percentage of allophanic and nonallophanic Andisols (prepared from Andisol TU database, 1992). Only the phosphate retention percentages ranging from 25 to 85% are shown.

for andic soil properties (Soil Survey Staff, 1992). In contrast, allophanic Andisols require $(Al_o + 1/2 Fe_o)$ values of almost 2.0 percent to meet this criterion.

(2) Ligand exchange reactions associated with phosphate sorption

Phosphate sorption reactions involve ligand exchange and/or precipitation reactions. Ligands such as OH^-, OH_2, $OSi(OH)_3^-$ and $RCOO^-$ are often coordinated with Al and Fe and are exchangeable with phosphate. These ligand exchange reactions alter the amount of positive and negative charge on the solid surface. A schematic representation of ligand exchange reactions associated with phosphate sorption by Na^+-saturated Andisols is shown in Fig. 6.22. The contribution of the individual reactions to phosphate sorption on soil materials depends on pH, the amount of phosphate sorbed, and the forms of active Al and Fe. The forms of active Al involved in ligand exchange reactions with phosphate in allophanic and nonallophanic Andisols at pH values of 4 and 6 are summarized in Table 6.6.

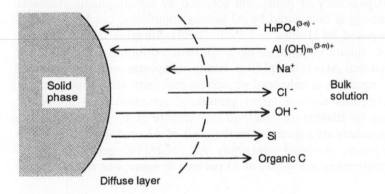

Fig. 6.22. Schematic representation of the chemical reactions associated with phosphate sorption by Na-saturated Andisols in sodium chloride solution (prepared from data of Nanzyo, 1991).

TABLE 6.6

Surface species of active aluminum in allophanic and nonallophanic Andisols which react with phosphate, at the initial pH values of 4 and 6 in 0.1 M NaCl (Nanzyo, 1988b)

Allophanic		Nonallophanic	
pH 4	pH 6	pH 4	pH 6
Dissolved Al	Al-OH	Dissolved Al	Al-OH
Al-OH$_2^+$	Al-OSi	Al-OH	Al-O$_2$CR [a]
Al-OSi	Al-O$_2$CR	Al-O$_2$CR	
Al-OH			
Al-O$_2$CR			

[a] R = organic residue.

When the initial pH is adjusted to 4 by adding HCl, soluble aluminum concentrations are on the order of 10^{-4} M. The ionic species of aluminum is primarily Al^{3+} which forms a solid-phase precipitate with phosphate. At pH = 4.0, allophanic Andisols bear positive charge ($=Al-OH_2^+$) which is balanced with Cl^- in this reaction system. Phosphate sorption results in a decrease of positive charge due to ligand exchange and Cl^- is released. Since the amount of positive charge of nonallophanic Andisols is very small even at pH 4, the contribution of this reaction is negligible. At an initial pH value of 6, this reaction does not contribute to phosphate sorption because the surface charge is primarily negative.

The release of Si increases with increasing amounts of phosphate sorption in allophanic Andisols at initial pH values of both 4 and 6. In contrast, Si release from nonallophanic Andisols is very small at these pH values. This observation indicates that one of the major reactive components in allophanic Andisols is noncrystalline aluminosilicates.

Suspension pH rises as phosphate is sorbed by Andisols. The pH increase is due to release of OH^- by ligand exchange reactions with phosphate. This reaction takes place in both allophanic and nonallophanic Andisols at initial pH values of 4 and 6. The release of silicate ions results in the release of OH^- by the following reactions (Rajan, 1975):

$$=Al-OSi\,(OH)_3 + H_2PO_4^- = =AlH_2PO_4 + H_3SiO_4^-$$

$$H_3SiO_4^- + H_2O = H_4SiO_4 + OH^- \quad (pK = 4.3)$$

The amount of negative charge on the solid-phase increases with phosphate sorption. Negative charge develops by the dissociation of sorbed phosphate, carboxyl, and silanol groups. Sodium ions balance the negatively charged sites in this reaction system.

Organic C is also released from both allophanic and nonallophanic Andisols by phosphate sorption. Nonallophanic Andisol samples release more organic C than allophanic Andisols at initial pH values of both 4 and 6. However, the amount of organic C released is very small compared with the phosphate sorbed, even for nonallophanic Andisols. Accordingly, most of the humus is not solubilized by the reaction with phosphate. Because Al is removed from Al-humus complexes by reacting with phosphate, the humus remains negatively charged due to the presence of noncomplexed carboxyl groups (Nanzyo, 1991).

(3) Phosphate sorption products in Andisols

The important secondary phosphate solid-phase products in Andisols are aluminum phosphate compounds including $Al(OH)_2H_2PO_4$ and $NaAl(OH)_2HPO_4$ (Veith and Sposito, 1977). These minerals were determined to be noncrystalline not only by X-ray diffraction analysis but also by differential diffuse reflectance infrared spectroscopy (Nanzyo, 1987). As presented in Fig. 6.23, they are characterized by a strong broad absorption band at 1130 cm^{-1} due to the P–O stretching vibration; this absorption band is not found in the spectra of metavariscite,

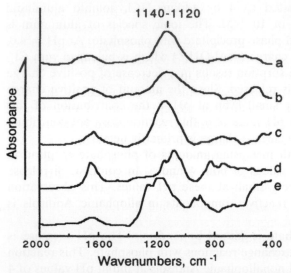

Fig. 6.23. Differential diffuse reflectance infrared spectra of the reaction products of (*a*) superphosphate with a Hapludand, and (*b*) multi-phosphate with a Melanudand. The spectra (*c*), (*d*), and (*e*) are the reference materials, noncrystalline aluminum phosphate, metavariscite, and $Al(H_2PO_4)_3$, respectively, in mixtures with an allophanic clay (adapted from Nanzyo, 1987, p. 752; by courtesy of Marcel Dekker Inc).

variscite, and $Al(H_2PO_4)_3$ (Nanzyo, 1988a, b). A similar noncrystalline aluminum phosphate has also been observed as the main reaction products of superphosphate with allophanic Andisols (Nanzyo, 1987). In this case, Ca^{2+} is preferentially retained on the negatively charged sites.

The phosphate sorption product on poorly crystalline hydrous iron oxides has transitional features in the IR spectrum between noncrystalline iron phosphate and a binuclear surface complex of phosphate on crystalline iron oxides (Nanzyo, 1986). The product is distinctly different from the phosphate sorption product on goethite in the acid pH range (Parfitt, 1978; Nanzyo and Watanabe, 1982).

6.5.4. Solute transport

Solute transport in soils is strongly affected by the interactions between solutes and the soil materials. The primary factors affecting solute-soil interactions in Andisols are variable charge and specific sorption. In placic horizons, precipitation/dissolution and oxidation/reduction reactions also affect solute transport.

Andisols contain large amounts of colloids with variable charge. In A horizons, negative charge predominates reflecting the abundance of soil organic matter, while in B horizons with low organic C content, both negative and positive charge develop due to allophanic clays and ferrihydrite. The effect of these charged surfaces on the transport of nonspecifically sorbed ions is to delay or accelerate their movement as compared with water. Hatano et al. (1988) examined this effect

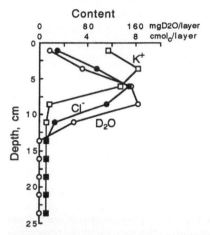

Fig. 6.24. Downward movement of potassium, chloride, and D_2O in a soil column using the Bw horizon of an allophanic Andisol (prepared from data of Hatano et al., 1988).

using the Bw horizon of an Andisol in a soil column experiment. The soil sample had a CEC value of 16.7 $cmol_c$ kg^{-1} and an AEC value of 2.1 $cmol_c$ kg^{-1}, respectively, in 0.2 M KCl. As compared to D_2O, they observed that the downward movement of the potassium ion was delayed the most, followed by the chloride ion as compared to D_2O (Fig. 6.24).

Since the amount of positive charge in the A horizon soils of both allophanic and nonallophanic Andisols is very small, NO_3^- is freely leached in the soil solution. The downward movement of NO_3^- in nitrogen-fertilized fields is about 10 cm per 100 mm of rainfall (Ogawa, 1984; Shoji et al., 1986).

Specific sorption retards the movement of specifically sorbed ions remarkably. Phosphate is an example of a specifically sorbed ion. A large amount of phosphorus fertilizer is commonly applied to Andisols in Japan. The phosphate is specifically sorbed by the soil and accumulates only in the Ap horizons. As a result, there is virtually no leaching of phosphorus from the soil profile as shown in Fig. 6.25. In contrast, the subsurface horizons of heavily fertilized sandy soils formed on nonvolcanic materials contain a considerable amount of water-extractable phosphorus (Nanzyo, 1992). This indicates that the phosphorus is not strongly sorbed by specific sorption reactions and is therefore susceptible to leaching from the soil profile.

6.5.5. Soil solution

Water plays a vital role as a transporting agent and a chemical solvent in soil formation and ecosystem functions. As a result, studies of soil solution chemistry are extremely useful for elucidating current soil forming processes including acid/base chemistry, translocation of solutes, and aqueous activities. Soil solution studies are also useful for examining specific reactions in the soil

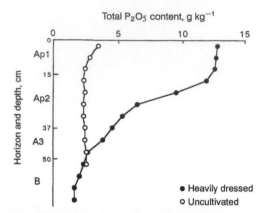

Fig. 6.25. Vertical distribution of total phosphorus content in the solid phase of heavily dressed and uncultivated Andisols in Japan. The difference between the two distribution curves is sorbed phosphorus (prepared from data of Nanzyo, 1992).

environment, such as ion exchange, weathering, precipitation/dissolution, nutrient uptake, oxidation/reduction, etc. Because soil solutions provide an instantaneous record of the response of soil processes to ecosystem perturbations (natural or human-induced), the approach has been termed *dynamic pedology* (Singer et al., 1978). For many studies, this approach of interpreting changes in soil solution composition is considered superior to the traditional approach of collecting and analyzing soil samples, which can only portray an average of past processes rather than the details of the current processes (Jenny, 1980).

The dynamic pedology approach has been successfully used in several studies examining soil forming processes in soils derived from volcanic ash. Ugolini et al. (1988) applied soil solution studies to differentiate the processes of andosolization and podzolization along a climobiotic transect in the southern Hakkoda Mountains, northeastern Japan. Soils along the transect developed on the same volcanic ashes that were deposited approximately 1000 and 5000 yr B.P. Chemical and morphological criteria based on the solid-phase fail, in many cases, to distinguish between Spodosols and Andisols along this transect. In contrast, soil solution studies showed a clear distinction in the soil solution signatures between Spodosols and Andisols.

Podzolization was found to be the dominant soil forming process in the cool, moist sites at higher elevations where the vegetation consists of fir (*Abies mariesii*) and Sasa bamboo (*Sasa kurilensis*). Soil solution composition showed transport of Fe, Al and dissolved organic carbon (DOC) from the O and A horizons to the upper B horizon (Fig. 6.26). Production of organic acids in the O and A horizons resulted in formation of mobile metal-organo complexes and in a lowering of the solution pH preventing dissolution of carbonic acid. Organic acid acidity was removed in the upper B horizons resulting in an increase in the solution pH which allowed carbonic acid to become the major proton donor in

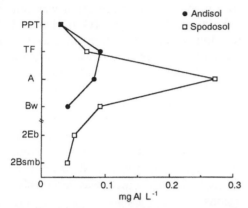

Fig. 6.26. Aluminum concentration in precipitation, throughfall, and soil solutions under various genetic horizons of an Andisol and a Spodosol (prepared from data of Ugolini et al., 1988; copyright Williams & Wilkins, with permission).

the lower soil horizons. In contrast, andosolization was the dominant soil forming process in the warmer, drier sites at low elevation where Japanese pampas grass (*Miscanthus sinensis*) is the dominant vegetation. Andosolization was characterized by an accumulation of Fe, Al and DOC in the A horizon with no significant translocation to the B horizons. Solution pH values were greater than in the Spodosol (0.5–0.7 units higher) and carbonic acid was the dominant proton donor throughout the entire profile. Formation of B horizons in both the Spodosol and Andisol was shown to occur primarily by *in situ* weathering rather than illuviation. Since the HCO_3^- anion does not form soluble complexes with Al or Fe and these metals have a low solubility in the pH range 5–7, Al and Fe are immobile and accumulate in the B horizons. Iron released by carbonic acid weathering forms Fe-oxyhydroxides while the Al can interact with soluble Si to form noncrystalline aluminosilicates, such as allophane and imogolite.

Because soil solutions provide information about current soil processes, their composition is highly sensitive to changes in the vegetation. The effects of Japanese oak (*Quercus serrata*) tree invasion on soils developed under pampas grass were investigated at the Tohoku University Farm, Japan (Dahlgren et al., 1991). Two study sites were chosen that had similar soil-forming factors except for the present vegetation. Soils at both sites developed under a grassland vegetation consisting of Japanese pampas grass; however, one site was invaded by oak vegetation 50 years ago. The soils were classified as medial, mesic Alic Pachic Melanudands and were members of the nonallophanic group of Andisols whose clay-sized mineralogy is dominated by Al/Fe humus complexes and hydroxy-Al interlayered 2:1 minerals rather than noncrystalline aluminosilicates. To examine the soil processes, soil solutions were collected over a 2-year period using tension lysimeters.

The invasion of oak vegetation resulted in a number of changes in the chemistry of the soil solutions. The oak vegetation was responsible for acidifying the soil solutions throughout the upper 50 cm of the profile. This acidification was due primarily to the production of organic acids in the litter layer (O horizons: 5 cm thick), which rapidly formed upon conversion to oak vegetation. In addition, the accumulation of base cations in the oak biomass is partially responsible for acidification within the rooting zone. In contrast, the above-ground biomass of the pampas grass is returned to the soil surface annually and the nutrients are released and returned to the soil by mineralization. Acidification of the surface horizons of the oak site increased the concentration of soluble Al and decreased the dissociation of carbonic acid. Transport of Al was limited to leaching from the surface horizon and arrest in the underlying A horizons. As in the previous study (Ugolini et al., 1988), there was not appreciable translocation of Fe, Al, or DOC to the B horizons and *in situ* weathering, primarily by carbonic acid, was responsible for driving weathering reactions in the B horizons.

The effect of tephra addition on soil processes in tephritic Spodosols was studied in the Cascade Range of Washington following the eruption of Mt. St. Helens (Dahlgren and Ugolini, 1989a). Tephra was applied in 5 or 15 cm depths to soils that had escaped airfall tephra deposition in 1980, but had received periodic additions of tephra during the Holocene. Soil solutions were collected from the major genetic horizons of tephra-treated and nontreated profiles. Soluble salts, composed of basic cations ($Ca^{2+} = Na^+ > Mg^{2+} > K^+$) and SO_4^{2-}, were initially leached from the fresh tephra. Basic cations in the leachate displaced H and Al from the exchange sites in the buried forest floor resulting in a lowering of the solution pH by up to one unit. The soluble salts and Al migrated through the upper portion of the soil profile where their concentrations were greatly attenuated by interaction with the solid-phase. Calcium ion, K^+, Al^{3+} and SO_4^{2-} were strongly retained in the B horizons, whereas Na, Mg and Cl showed little affinity for the solid-phase.

Following the initial removal of soluble salts from the tephra, carbonic acid was the principal proton donor driving weathering reactions in the tephra. Basic cations were liberated by weathering reactions and migrated into the forest floor (O horizons) with the HCO_3^- anion. Upon entering the acidic forest floor, HCO_3^- acquired a proton and was converted to H_2CO_3. Carbonic acid is volatile and diffuses upward into the tephra, where it may participate in another weathering/ transport cycle. The pH of the leachates from the Oa horizon increased by approximately 0.4 unit due to the uptake of protons by HCO_3^- and the retention of basic cations.

No significant enhancement of elemental leaching was measured from the C horizon (1 m depth) during the study. Biological uptake, anion sorption, ion exchange reactions, and neoformation of minerals were responsible for the differential movement and retention of solutes in the soil profile. Podzolization, the prevailing soil forming process in the subalpine zone of the Cascade Range, was relatively unaffected by the addition of tephra.

Chemical equilibrium modeling was used in conjunction with soil solution studies in these same Spodosols in the Cascade Range of Washington to examine the formation and stability of imogolite in these soils. Imogolite was found to be a major component of the clay fraction in the Bs and C horizons, but was absent in the E and Bhs horizons. Following fractionation and speciation of soluble Al into inorganic and organic complexes (Dahlgren and Ugolini, 1989b), saturation indices for various aluminosilicates were calculated (Dahlgren and Ugolini, 1989c). Weathering in the E and Bhs horizons was dominated by organic acids and Al transport occurred primarily as Al-organo complexes. Soil solutions were highly undersaturated with respect to imogolite in the E and Bhs horizons due to the low Al^{3+} activities that resulted from complexation of soluble Al by organic ligands. Therefore, imogolite could not form in these horizons. In contrast, soil solutions from the Bs and C horizons appeared to be in equilibrium with both the hydroxy-Al interlayer of 2:1 layer silicates and imogolite. The interlayer-Al appeared to control Al activities while imogolite regulates H_4SiO_4 activities. This apparent simultaneous equilibrium between hydroxy-Al interlayers and imogolite suggests that the availability of soluble H_4SiO_4 regulates the formation and dissolution of imogolite. While imogolite is a metastable mineral in the Bs and C horizons, its formation is thermodynamically favorable in the lower soil horizons where carbonic acid is the major proton donor. These studies indicate that imogolite formation occurs through *in situ* weathering and synthesis in the lower B horizons rather than through migration of proto-imogolite sols from the E and/or Bhs horizons.

Soil solution studies were also utilized in Andisols under agricultural production to evaluate the fate of phosphorus applied during fertilization. As the application of phosphate was increased (max. = 37,000 kg P_2O_5/ha), the level of available phosphorus increased to a maximum of 7 g P_2O_5 kg^{-1} by the Truog method. At this level of phosphorus fertilization, soil solution activities for Al and phosphate indicate that the solubility product for variscite was exceeded and that solution activities appeared to be regulated by a noncrystalline aluminum phosphate phase (Kato et al., 1987). This result coincides with those from the infrared spectroscopic analysis of phosphate sorbed by Andisols (Nanzyo, 1987).

6.5.6. Reduction of iron under submergence

In Andisols, the reduction of iron under submerged conditions is conspicuously inhibited by surface coatings of active Al (oxalate-extractable Al). Thus, a considerable number of volcanic ash soils continuously used for wetland rice farming lack the iron accumulation layer below the plow layer even if they are well-drained and contain large amounts of active iron and organic matter. In contrast, alluvial paddy soils with moderate to high permeability typically accumulate iron in the subsurface horizons. The iron originates from the plow layer and migrates to the subsurface horizons in the ferrous form where it is subsequently oxidized (Mitsuchi, 1974).

The inhibitory effect of active Al on the reduction of iron has been shown by Noshiroya (1992) in studies of the relationship between the amount of active Al and formation of ferrous iron. As presented in Fig. 6.27, there is an inverse relationship between the amount of active Al and formation of acetate- (pH 2.8) extractable

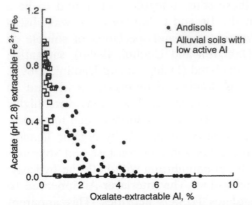

Fig. 6.27. The relationship between oxalate-extractable Al and the ratio of the amount of acetate-(pH 2.8) extractable ferrous ion to oxalate-extractable Fe (prepared from data of Noshiroya, 1992).

ferrous Fe (Kumada and Asami, 1958) as expressed by the ratio of ferrous iron to active Fe (Fe_o). Active Fe is largely reduced to ferrous forms in most alluvial soils which contain only small amounts of active Al (mostly less than 0.5 percent), while there is little reduction in Andisols having active Al concentrations greater than 3 percent.

The inhibition of microbial reduction of Fe by active Al under submerged conditions is further supported by laboratory incubation experiments using additions of glucose and synthetic noncrystalline aluminum hydroxide. Noshiroya (1992) showed that Andisol samples containing active Al concentrations greater than 3 percent had little reduction of ferric Fe, even with the addition of glucose which remarkably contributes to the development of strongly reducing conditions. Moreover, the formation of ferrous Fe in alluvial soils was gradually depressed with increasing additions of noncrystalline aluminum hydroxide and there was virtually no reduction when Al additions exceeded 3 percent. Since the redox potential was observed to be very low (between −200 mV and −100 mV) during the incubation experiment, the attenuation of ferrous Fe formation was primarily attributable to the coating of active Fe by noncrystalline aluminum hydroxide.

REFERENCES

Abd-Elfattah, A. and Wada, K. 1981. Adsorption of lead, copper, zinc, cobalt, and cadmium by soils that differ in cation exchange materials. J. Soil Sci., 32: 271–283.

Adachi, T., 1973. Studies on the humus of volcanic ash soils — regional differences of humus composition in volcanic ash soils in Japan. Bull. Natl. Inst. Agric Sci., Series B, 24: 127–264 (in Japanese, with English abstract).

Andisol TU database, 1992. Database on Andisols from Japan, Alaska, and Northwestern U.S.A. prepared by Soil Science Laboratory, Tohoku University (see Appendix 2).

Aomine, S., 1958. Allophane in soils. Jap. Soil Sci. Plant Nutr., 28: 508–516.

Arai, S., Honna, T. and Oba, Y., 1986. Humus characteristics, In: K. Wada (Editor), Ando Soils in Japan. Kyushu University Press, Fukuoka, Japan, pp. 57–67.

Arai, S., Otsuka, H., Honna, T. and Oba, Y., 1988. Humus characteristics of Andisols. In: D.I. Kinloch, S. Shoji, F.H. Beinroth and H. Eswaran (Editors), Proc. 9th Int. Soil Classification Workshop, 20 July to 1 August, 1987, Japan. Published by Japanese Committee for the 9th Int. Soil Classification Workshop, for the Soil Management Support Services, Washington D.C., U.S.A., pp. 74–80.

Birrell, K.S. and Gradwell, M., 1956. Ion-exchange phenomena in some soils containing amorphous mineral constituents. J. Soil Sci., 7: 130–143.

Blakemore, L.C., Searle, P.L. and Daly, B.K., 1981. Methods for Chemical Analysis of Soils. New Zealand Soil Bureau Scientific Report 10A. Department of Scientific and Industrial Research, New Zealand.

Childs, C.W., Matsue, N. and Yoshinaga, N., 1990. Ferrihydrite in volcanic ash soils of Japan. Soil Sci. Plant Nutr., 37: 299–311.

Dahlgren, R.A. and Ugolini, F.C., 1989a. Effects of tephra addition on soil processes in Spodosols in the Cascade Range, Washington, U.S.A. Geoderma, 45: 331–355.

Dahlgren, R.A. and Ugolini, F.C., 1989b. Aluminum fractionation of soil solutions from unperturbed and tephra-treated Spodosols, Cascade Range, Washington, U.S.A. Soil Sci. Soc. Am. J. 53: 559–566.

Dahlgren, R.A. and Ugolini, F.C., 1989c. Formation and stability of imogolite in a tephritic Spodosol, Cascade Range, Washington, U.S.A. Geochim. Cosmochim. Acta, 53: 1897–1904.

Dahlgren, R.A., Ugolini, F.C., Shoji, S., Ito, T. and Sletten, R.S., 1991. Soil-forming processes in Alic Melanudands under Japanese pampas grass and oak. Soil Sci. Soc. Am. J., 55: 1049–1056.

Delvaux, B., Herbillon, A.J. and Vielvoye, L., 1989. Characterization of a weathering sequence of soils derived from volcanic ash in Cameroon. Taxonomic, mineralogical and agronomic implications. Geoderma, 45: 375–388.

Egawa, T., Watanabe, Y. and Sato, A., 1959. A study on cation exchange capacity of allophane. Proc. Clay Sci., Jap. 1: 260–272 (in Japanese, with English abstract).

Gunjigake, N. and Wada, K., 1981. Effects of phosphorus concentration and pH on phosphate retention by active aluminum and iron of Ando Soils. Soil Sci., 132: 347–352.

Hasegawa, T., and Shoji, S., 1993. In preparation for contribution.

Hatano, R., Jiang, Y.Z., Saito H., Sakuma, T. and Okajima, H., 1988. On the source-sink effect of soil aggregates on the transport of water and solutes — by using D_2O as tracers of water. Jap. J. Soil Sci. Plant Nutr., 59: 253–259 (in Japanese, with Engligh abstract).

Haynes, R.J., and Swift, R.S., 1989. The effects of pH and drying on adsorption of phosphate by aluminum-organic matter associations. J. Soil Sci., 40: 773–781.

Higashi, T., 1983. Characterization of Al/Fe-humus complexes in Dystrandepts through comparison with synthetic forms. Geoderma, 31: 277–288.

Honna, T., Yamamoto, S. and Matsui, K., 1988. A simple procedure to determine melanic index that is useful for differentiating melanic from fulvic Andisols. Pedologist, 32: 69–78.

Iimura, K., 1966. Acidic properties and cation exchange of allophane and volcanic ash soils. Bull. Natl. Inst. Agr. Sci. Series B., 17: 101–157 (in Japanese, with English abstract).

Imai, H., 1981. Phosphate adsorption on volcanic ash soils, The effect of equibrium pH on phosphate adsorption. Jap. Soil Sci. Plant Nutr., 52: 11–19 (in Japanese).

Inoue, K., 1986. Chemical properties. In: K. Wada (Editor), Ando Soils in Japan. Kyushu University Press, Fukuoka, Japan, pp. 69–98.

Inoue, K. and Higashi, T., 1988. Al- and Fe-humus complexes in Andisols. In: D.I. Kinloch, S. Shoji, F.H. Beinroth and H. Eswaran (Editors), Proc. 9th Int. Soil Classification Workshop, 20 July to 1 August, 1987, Japan. Published by Japanese Committee for the 9th Int. Soil Classification Workshop, for the Soil Management Support Services, Washington D.C., U.S.A., pp. 81–96.

Inoue, K. and Huang, P.M., 1987. Effect of humic and fulvic acids on the formation of allophane. In: L.G. Schultz et al., (Ediors) Proc. Int. Clay Conf., Denver, CO., 28 July–2 Aug., 1985. Clay Min.

Soc. Bloomington, IN., pp. 221–226.

Inoue, K. and Huang, P.M., 1990. Perturbation of imogolite formation by humic substances. Soil Sci. Soc. Am. J., 54: 1490–1497.

Ito, T. and Shoji, S., 1993. In preparation for contribution.

Ito, S., Shoji, S. and Saigusa, M., 1991. Classification of volcanic ash soils from Konsen district, Hokkaido, according to the last Keys to Soil Taxonomy (1990). Jap. J. Soil Sci. Plant Nutr., 62: 237–247 (in Japanese, with English abstract).

Jenny, H. 1980. The Soil Resource. Springer, New York, 377 pp.

Kanno, I., 1961. Genesis and classification of main genetic soil types in Japan, 1. Introduction and Humic Allophane Soils. Bull. Kyushu Agr. Expt. Stn., 7: 1–185 (in Japanese, with English abstract).

Kato, H., Oka, N. and Kamewada, K., 1987. Availability and nature of phosphorus in fields. III, Determination of upper limit of soil phosphate concentration in volcanic ash soil in terms of chemical equilibrium. Jap. J. Soil Sci. Plant Nutr., 58: 27–34 (in Japanese).

Kosaka, J., 1964. Chemical properties. In: Ministry of Agriculture and Forestry, Japanese Government (Editor): Volcanic Ash Soils in Japan, Tokyo, pp. 99–126.

Kumada, K., 1987. Chemistry of Soil Organic Matter, Japan Scientific Society Press, Elsevier, pp. 17–56.

Kumada, K. and Asami, T., 1958. A new method for determining ferrous iron in paddy soils. Soil Sci. Plant Food., 3: 187–193.

Kurashima, K., Shoji, S. and Yamada, I., 1981. Mobilities and related factors of chemical elements in the top soils of Andisols in Tohoku, Japan: 1. Mobility sequence of major chemical elements. Soil Sci., 132: 300–307.

McKeague, J.A., 1967. An evaluation of 0.1 M pyrophosphate and pyrophosphate-dithionite in comparison with oxalate as extractants of the accumulation products in Podzols and some other soils. Can. J. Soil Sci., 47: 95–99.

Mitsuchi, M., 1974. Pedogenic characteristics of paddy soils and their significance in soil classification. Bull. Natl. Inst. Agric Sci. Series B, 25: 29–115 (in Japanese, with English abstract).

Murayama, S., 1980. The monosaccharide composition of polysaccharides in Ando soils. J. Soil Sci., 31: 481–490.

Murayama, S., 1984. Origin, degradation, composition of soil saccharides and soil microorganisms. In: H. Wada and S. Tsuru (Editors), Soil Biomass-Population and Metabolism of Soil Microorganisms. Hakuyusha, Tokyo. pp. 65–114 (in Japanese).

Nanzyo, M., 1986. Infrared spectra of phosphate sorbed on iron hydroxide gel and the sorption products. Soil Sci. Plant Nutr., 32: 51–58.

Nanzyo, M., 1987. Formation of noncrystalline aluminum phosphate through phosphate sorption on allophanic Ando soils. Commun. Soil Sci. Plant Anal., 18: 735–742.

Nanzyo, M., 1988a. Phosphate sorption on the clay fraction of Kanuma pumice. Clay Sci., 7: 89–96.

Nanzyo, M., 1988b. Phosphate reactions with Andisols. In: D.I. Kinloch, S. Shoji, F.H. Beinroth and H. Eswaran (Editors), Proc. 9th Int. Soil Classification Workshop, 20 July to 1 August, 1987, Japan. Published by Japanese Committee for the 9th Int. Soil Classification Workshop, for the Soil Management Support Services, Washington D.C., U.S.A., pp. 567–579.

Nanzyo, M., 1991. Chemi-sorption of phosphate on Andisols (volcanic ash soils) — material balance and the sorption product. Phosphorus Res. Bull. 1: 471–476.

Nanzyo, M., 1992. Vertical distribution of phosphorus in the cultivated soils. In: Studies on control of nitrogen and phosphorus loading from arable lands based on the soil management. Publication of the Ministry of Agriculture, Forestry and Fishery, Japan, in press (in Japanese).

Nanzyo, M. and Watanabe, Y., 1982. Diffuse reflectance infrared spectra and ion-adsorption properties of the phosphate surface complex on goethite. Soil Sci. Plant Nutr., 28: 359–368.

Nanzyo, M., Tsuruno, K. and Watanabe, Y., 1984. Method for evaluating phosphate retention capacity of soils at constant final phosphate concentration. Jap. J. Soil Sci. Plant Nutr., 55: 325–331 (in Japanese).

Nanzyo, M., Shoji, S. and Sudjadi, M., 1993. Properties and classification of Andisols from West Java, Indonesia. In preparation for contribution.

Noshiroya, N., 1992. Changes in the soil properties of Andisols due to wetland rice farming, Master thesis. Tohoku Univ. Sendai, Japan (in Japanese).

Oda, K., Miwa, E. and Iwamoto, A., 1987. Compact database for soil analysis data in Japan. Jap. J. Soil Sci. Plant Nutr., 58: 112–132 (in Japanese).

Ogawa, Y., 1984. Effects of leached fertilizers from the volcanic ash upland fields on the fresh water environment and its control. Jap. J. Soil Sci. Plant Nutr., 55: 195–196 (in Japanese).

Okamura, Y. and Wada, K., 1983. Electric charge characteristics of horizons of Ando (B) and Red-Yellow B soils and weathered pumices, J. Soil Sci., 34: 287–295.

Okamura, Y. and Wada, K., 1984. Ammonium-calcium exchange equilibria in soils and weathered pumices that differ in cation-exchange materials. J. Soil Sci., 35: 387–396.

Parfitt, R.L., 1978. Anion adsorption by soils and soil materials. Adv. Agron., 30: 1–50.

Parfitt, R.L., 1988. Variable charge in Andisols. In: D.I. Kinloch, S. Shoji, F.H. Beinroth and H. Eswaran (Editors), Proc. 9th Int. Soil Classification Workshop, 20 July to 1 August, 1987, Japan. Published by Japanese Committee for the 9th Int. Soil Classification Workshop, for the Soil Management Support Services, Washington D.C., U.S.A., pp. 60–73.

Parfitt, R.L. and Henmi, T., 1982. Comparison of an oxalate extraction method and an infrared spectroscopic method for determining allophane in soil clays. Soil Sci. Plant Nutr., 28: 183–190.

Parfitt, R.L. and Kimble J.M., 1989. Conditions for formation of allophane in soils. Soil Sci. Soc. Am. J., 53: 971–977.

Parfitt, R.L. and Wilson, A.D., 1985. Estimation of allophane and halloysite in three sequences of volcanic soils, New Zealand. In: E.F. Caldas and D.H. Yaalon (Editors), Volcanic Soils, Weathering and Landscape Relationships of Soils on Tephra and Basalt, Catena Supplement 7. Catena Verlag, West Germany, pp. 1–8.

Ping, C.L., Shoji, S. and Ito, T., 1988. Properties and classification of three volcanic ash-derived pedons from Aleutian islands and Alaska peninsula, Alaska. Soil Sci. Soc. Am. J., 52: 455–462.

Rajan, S.S.S., 1975. Mechanism of phosphate adsorption by allophane clays, N.Z.J. Sci., 18: 93–101.

Sadiq, M. and Lindsay, W.L., 1979. Selection of standard free energies of formation for use in soil chemistry. Colorado State University Experiment Station Technical Bulletin, 134. pp. 91–92.

Saigusa, M., Shoji, S. and Takahashi, T., 1980. Plant root growth in acid Andosols from northeastern Japan: 2. Exchange acidity Y_1 as a realistic measure of aluminum toxicity potential. Soil Sci., 130: 242–250.

Saigusa, M., Matsuyama, N., Honna, T. and Abe, T., 1991. Chemistry and fertility of acid Andisols with special reference to subsoil acidity. In: R.J. Wright et al. (Editors), Plant-Soil Interactions at Low pH. Kluwer Academic Publishers, Netherlands, pp. 73–80.

Saigusa, M., Matsuyama, N. and Abe, T., 1992. Electric charge characteristics of Andisols and its problem on soil management. Jap. J. Soil Sci. Plant Nutr., 63: 196–201 (in Japanese, with English abstract).

Sase, T., 1986. Plant opal analysis of Andisols in North Island, New Zealand. Pedologist, 30: 2–12 (in Japanese, with English abstract).

Schnitzer, M., 1978. Humic Substances: chemistry and reactions. In: M. Schnitzer and S.U. Khan (Editors), Soil Organic Matter. Developments in Soil Science 8, Elsevier, Amsterdam, pp. 1–64.

Schnitzer, M., 1991. Soil organic matter — the next 75 years. Soil Sci., 151: 41–58.

Sekiya, K., 1972. Phosphate sorption coefficient, In: Committee for Soil Nutrient Determination (Editorial board) Methods of Soil Analysis to Evaluate Soil Fertility. Yokendo, Tokyo, pp. 251–253 (in Japanese).

Shimizu, H., Watanabe, T., Henmi, T., Masuda, A. and Saito, H., 1988. Studies on allophane and imogolite by high-resolution solid-state ^{29}Si- and ^{27}Al-NMR and ESR. Geochem. J., 22: 23–31.

Shoji, S., 1984. Genesis and properties of nonallophanic Andisols. J. Clay Sci. Jap., 24: 152–165 (in

Japanese).

Shoji, S., 1988. Separation of melanic and fulvic Andisols, Soil Sci. Plant Nutr., 34: 303–306.

Shoji, S. and Fujiwara, Y., 1984. Active aluminum and iron in the humus horizons of Andosols from northeastern Japan: their forms, properties, and significance in clay weathering. Soil Sci., 137: 216–226.

Shoji, S. and Masui, J., 1972. Noncrystalline components in young volcanic ash soils, III. Fine clay fraction. Jap. J. Soil Sci. Plant Nutr., 43: 187–193 (in Japanese).

Shoji, S., and Ono, T., 1978. Physical and chemical properties and clay mineralogy of Andosols from Kitakami, Japan. Soil Sci., 126: 297–312.

Shoji, S., Kobayashi, S., Yamada, I. and Masui, J., 1975. Chemical and mineralogical studies on volcanic ashes. I. Chemical composition of volcanic ashes and their classification. Soil Sci. Plant Nutr., 21: 311–318.

Shoji, S., Saigusa, M. and Takahashi, T., 1980. Plant root growth in acid Andosols from northeastern Japan: 1. Soil properties and root growth of burdock, barley, and orchard grass. Soil Sci., 130: 124–131.

Shoji, S., Fujiwara, Y., Yamada, I. and Saigusa, M., 1982. Chemistry and clay mineralogy of Ando soils, Brown forest soils, and Podzolic soils formed from recent Towada ashes, northeastern Japan, Soil Sci., 133: 69–86.

Shoji, S., Ito, T., Saigusa, M. and Yamada, I., 1985. Properties of nonallophanic Andosols from Japan. Soil Sci., 140: 264–277.

Shoji, S., Saigusa, M. and Goto, J., 1986. Acidity of subsoils of Andisols and nitrogen uptake and growth of sorghum, Jap. J Soil Sci. Plant Nutr., 57: 264–271 (in Japanese).

Shoji, S., Ito, T., Nakamura, S. and Saigusa, M., 1987. Properties of humus of Andosols from New Zealand, Chili and Ecuador, 58: 473–479 (in Japanese).

Singer, M., Ugolini, F.C. and Zachara, J., 1978. In situ study of podzolization on tephra and bedrock. Soil Sci. Soc. Am. J., 42: 105–111.

Soil Survey Staff, 1992. Keys to Soil Taxonomy, 5th edition. AID, USDA-SMSS Technical Monograph, No. 19, Blacksburg, Virginia.

Sposito, G., 1984. The Surface Chemistry of Soils. Oxford University Press, New York, pp. 150–154.

Tamura, T., 1958. Identification of clay minerals from acid soils, J. Soil Sci., 9: 141–147.

Tate, K.R., Yamamoto, K., Churchman, G.J., Meinhold, R. and Newman, R.H., 1990. Relationship between the type and carbon chemistry of humic acids from some New Zealand and Japanese soils. Soil Sci. Plant Nutr., 36: 611–621.

Thomas, G.W. and Hargrove, W.L., 1984. The chemistry of soil acidity. In: F. Adams (Editor), Soil Acidity and Liming. 2nd edition, ASA CSSA and SSSA, Wisconsin, U.S.A, pp. 3–56.

Tsutsuki, K. and Kuwatsuka, S., 1978. Chemical studies on soil humic acids, III. Nitrogen distribution in humic acid. Soil Sci. Plant Nutr., 24: 561–570.

Tsutsuki, K. and Kuwatsuka, S., 1992. Characterization of humin-metal complexes in a buried volcanic ash soil profile and a peat soil, Soil Sci. Plant Nutr., 38: 297–306.

Uehara, G. and Gilman, G. 1981. The Mineralogy, Chemistry, and Physics of Tropical Soils with Variable Charge Clays. Westview Tropical Agriculture Series, No. 4, 170 pp.

Ugolini, F.C., Dahlgren, R., Shoji, S. and Ito, T., 1988. An example of andosolization and podzolization as revealed by soil solution studies, southern Hakkoda, northeastern Japan. Soil Sci., 145: 111–125.

Van Wambeke, A., 1992. Andisols. In: Soils of the Tropics, McGraw-Hill, New York, pp. 207–229.

Veith, J.A. and Sposito, G., 1977. Reactions of aluminosilicate, aluminum hydrous oxides, and aluminum oxide with O-phosphate: The formation of X-ray amorphous analogs of variscite and montebrasite, Soil Sci. Soc. Am. J., 41: 870–876.

Wada, K., 1981. Ion exchange and adsorption reactions by soil clays. In: T. Fujisawa and N. Yoshinaga (Editors), Adsorption phenomena by soils. Hakuyusha, Tokyo, pp. 5–57 (in Japanese).

Wada, K., 1985. Distinctive properties of Andosols. Adv. Soil Sci., 2: 173–229.

Wada, K., 1986. Part II. Database, Kurobokudo Co-operative Research Group. In: K. Wada (Editor), Ando soils in Japan. Kyushu University Press, Fukuoka, Japan, pp. 115–276.

Wada, K., 1989. Allophane and imogolite. In: J.B. Dixon and S.B. Weed (Editors), Minerals in Soil Environment. SSSA Madison, WI, U.S.A., pp. 1051–1087.

Wada, K. and Abd-Elfattah, A., 1978. Characterization of zinc adsorption sites in two mineral soils. Soil Sci. Plant Nutr., 24: 417–426.

Wada, K. and Gunjigake, N., 1979. Active aluminum and iron and phosphate adsorption in Ando soils. Soil Sci., 128: 331–336.

Wada, K. and Harward, M.E., 1974. Amorphous clay constituents of soils. Adv. Agron., 26: 211–260.

Wada, K. and Kakuto, Y., 1983. Intergradient vermiculite-kaolin mineral in a Korean Ultisol. Clays Clay Miner., 31: 183–190.

Wada, K. and Okamura, Y., 1980. Electric charge characteristics of Ando A, and buried A horizon soils. J. Soil Sci., 31: 307–314.

Wada, S-I., 1988. Practical significance of the variability of charge, In: D.I. Kinloch, S. Shoji, F.H. Beinroth and H. Eswaran (Editors), Proc. 9th Int. Soil Classification Workshop, 20 July to 1 August, 1987, Japan. Published by Japanese Committee for the 9th Int. Soil Classification Workshop, for the Soil Management Support Services, Washington D.C., U.S.A., pp. 558–566.

Yamada, I. and Shoji, S., 1983. Alteration of volcanic glass of recent Towada ash in different soil environments of northeastern Japan. Soil Sci., 135: 316–321.

Yamamoto, K., 1982. Mechanism of heavy metal sorption by amorphous clay materials. Jap. J. Soil Sci. Plant Nutr., 53: 355–366 (in Japanese).

Yonebayashi, K. and Hattori, T., 1988. Chemical and biological studies on environmental humic acids, I. Composition of elemental and functional groups of humic acid. Soil Sci. Plant Nutr., 34: 571–584.

Yonebayashi, K. and Hattori, T., 1989. Chemical and biological studies on environmental humic acids, II. ^{1}H-NMR and IR spectra of humic acids. Soil Sci. Plant Nutr., 35: 383–392.

Yoshida, M., 1979. Some aspects on soil acidity and its measurement. Jap. J. Sci. Soil Plant Nutr., 50: 171–180 (in Japanese).

Chapter 7

PHYSICAL CHARACTERISTICS OF VOLCANIC ASH SOILS

M. NANZYO, S. SHOJI and R. DAHLGREN

7.1. INTRODUCTION

Volcanic ash soils have many unique physical properties that are attributable directly to the properties of the parent material, the noncrystalline materials formed by weathering, and the soil organic matter accumulated during soil formation. These properties include dark soil color, difficult clay dispersion, unique consistence, low bulk density, and high water holding capacity.

Soil color is the most striking feature observed for volcanic ash soils, especially for their A horizons. As mentioned in Chapters 2 and 6, the color of A horizons is largely determined by the amount of soil organic matter, ratio of humic to fulvic acids, and types of humic acids.

Andisols typically display a large difference in texture when comparing field and laboratory determination methods (Ping et al., 1989). Noncrystalline materials play an important role as cementing agents and react with an excess amount of sodium hexametaphosphate. Furthermore, each of the inorganic colloids shows a different point of zero net charge, so that complete dispersion of mineral particles is virtually impossible. Therefore, the common laboratory determined particle-size analysis shows a very low clay content as compared with field determination (Ping et al., 1989). Ultrasonic pretreatment is effective for clay dispersion (Oba and Kobo, 1965; Ping et al., 1988, 1989), but the optimum conditions for power and duration of ultrasonication have yet to be standardized (Gee and Bauder, 1986).

Andisols generally have low bulk densities that are attributable to the development of highly porous soil structure. Thus, bulk density values less than 0.9 g cm^{-3} are employed as one of the andic soil property requirements (Soil Survey Staff, 1992). Both the low bulk density and the unique consistence greatly contribute to easy tillage of Andisols. Andisols usually show low degrees of stickiness, plasticity, and hardness that result from the abundance of noncrystalline materials and/or soil organic matter. Thus, it is possible to till these soils even when soil water content is lower than the shrinkage limit and somewhat higher than the plastic limit. These physical properties also provide an excellent environment for root growth.

Andisols have well-developed soil structure resulting in high porosity with a range of pore sizes that retain a large amount of water with varying tensions (Furuhata and Hayashi, 1980; Saigusa et al., 1987). Thus, it is not uncommon that

Andisols can hold greater than 30 percent plant available water by volume. The high available water holding capacity and excellent physical properties for root growth contribute to high yields of upland crops (Saigusa et al., 1987).

This chapter examines the physical properties of volcanic ash soils including soil color, soil texture, soil structure and micromorphology, soil consistence, bulk density and porosity, water retention and available water, water permeability, and irreversible changes in physical properties following drying. Reviews concerning the subject matter of this chapter have been published by Yamanaka (1964), Maeda et al. (1977), Warkentin and Maeda (1980), Maeda and Soma (1983), Wada (1985), Maeda and Soma (1986), Warkentin et al. (1988) and Van Wambeke (1992).

7.2. SOIL COLOR

The color of Andisols is largely determined by colloidal coatings such as humified organic matter in the A horizons and free iron oxides and hydroxides in the B horizons. The color of the C horizons is determined by the color of the uncoated mineral grains and any weathering products that may have accumulated.

The surface horizons of Andisols show a variety of soil colors depending on the content of soil organic matter, the ratio of humic to fulvic acids, and the types of humic acids. For example, the A and buried A horizons of Andisols from northeastern Japan typically have hues of 5 YR to 10 YR, values of 3 or less, and chromas of 3 or less when moist, as determined by the Munsell color chart (Otowa et al., 1988). The majority of these horizons meet the melanic or umbric epipedon color requirements and a few, the fulvic color requirement. As described in Chapters 2, 3 and 6, melanic epipedons commonly have hues of 7.5 YR or 10 YR and value and chroma of 2/2 or darker, being attributable to the abundance of A type humic acid (Shoji, 1988). In contrast, the fulvic color is largely determined by the predominance of fulvic acids and existence of P-type humic acid (Shoji, 1988).

Andisol Bw horizons typically have orange to reddish brown colors which are attributable to the type and content of iron minerals such as hematite (red), goethite (yellowish brown), ferrihydrite (brown to red) or lepidocrocite (orange). For example, the Bw horizons of Andisols from northeastern Japan generally show brown colors with hues of 7.5 YR–10 YR and widely varying values and chromas (Otowa et al., 1988). Such colors result primarily from ferrihydrite or ferrihydrite with smaller amounts of goethite (Ito and Shoji, 1992).

Andisol C horizons are generally gray to light yellowish brown in color due to the color of uncoated rhyolite, dacite and andesite tephras mixed with small amounts of iron oxide and hydroxide coatings. Basaltic tephras are reddish brown to dark red in color due to partial oxidation of iron minerals.

Poorly drained Andisols are observed to have low chroma and/or high chroma mottles in the B and C horizons, but it is often difficult to detect these mottles in the A horizons because they are masked by soil organic matter. As described in

Chapter 6, ferric oxides in some Andisols are substantially stabilized with respect to microbial reduction due to surface coatings of active Al (Noshiroya, 1992). As a result, the dominant chromas, moist, of 2 or less are not always observed in Andisols even under strongly reduced conditions.

7.3. SOIL TEXTURE

Soil texture is widely used as an index property for physical properties of soils and is often employed to predict many soil properties including soil productivity. However, soil texture is not an adequate index property for volcanic ash soils because accurate determination of soil texture, by common mechanical analysis, is difficult as described below. Thus, particle-size/mineralogy modifiers are used to replace the names of particle-size classes for Andisols in Soil Taxonomy (Soil Survey Staff, 1992), as mentioned in the families section of Andisols in Chapter 4.

Incomplete dispersion of mineral particles and uncertainty of the unit mineral particles in materials such as weathered pumice are the main factors seriously limiting the accuracy of mechanical analysis for volcanic ash soils (Maeda et al., 1977). Common mechanical analysis requires a number of pretreatments to obtain the maximum dispersion of clay in suspensions. Pretreatments include the removal of cementing agents, vigorous stirring or shaking and dispersion of the clay particles through the introduction of ions that increase the negative charge of particles. Since sodium hexametaphosphate increases negative charge and precipitates calcium and magnesium ions which hinder dispersion, this dispersing agent is widely used (Gee and Bauder, 1986)

In spite of the above-mentioned pretreatments, it is often difficult to obtain complete dispersion of mineral particles in Andisols. Noncrystalline materials that are present in large amounts in volcanic ash soils contribute to the formation of stable aggregates which strongly resist dispersion. Aggregation of allophanic clays results from strong cohesive forces, most probably reflecting the low surface charge density (Kubota, 1976), and cementing of aggregates by iron oxides and hydroxides such as ferrihydrite. It is important to note that sodium hexametaphosphate is not completely effective in obtaining dispersion (Kobo and Oba, 1965; Kobo et al., 1974), possibly because this dispersing agent has a strong affinity for allophanic clays and can form edge to edge flocculated aggregates.

The effect of pH on the dispersion of clay particles is conspicuous in Andisols. It is closely related to the amount of charge on the particle surfaces. For example, imogolite disperses only under acid conditions (Yoshinaga and Aomine, 1962) because its surface shows increased positive charge with an increase in the proton concentration (lowering of pH) of the ambient solution. In contrast, the pH region in which allophane disperses depends on its chemical composition. Allophane with an Al/Si atomic ratio of 2 disperses on both the acid and alkaline side, and allophane with an Al/Si atomic ratio of unity, on the alkaline side. These facts

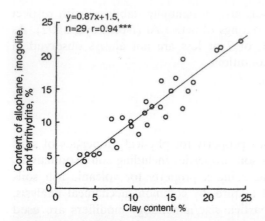

Fig. 7.1. Relationship between the clay content measured by the pipet method with pretreatment of sonication and pH adjustment and the content of allophane, imogolite and ferrihydrite determined by selective dissolution analysis in the B and C horizons of Andisols (prepared from Andisol TU database, 1992).

indicate that maximum dispersion is obtained at pH values far from the point of zero charge (PZC) for the noncrystalline clays. PZC values of allophanes with an Al/Si atomic ratios of 1.2, 1.5, and 1.8 are 5.5, 6.5, and 6.9, respectively (Clark and McBride, 1984). Imogolite showed zero electrophoretic mobility in the pH range 9 to 12 (Horikawa, 1975). Thus, the dispersion of clays in Andisols is conducted at pH 4 or 10 using HCl or NaOH to adjust the pH (Wada, 1986).

Ultrasonic dispersion of clay suspensions with pH adjustment as described above is widely employed in Japan. The intense energy of cavitation produced by sound waves blasts soil aggregates apart, resulting in effective dispersion as first observed by Kobo and Oba (1965). The effectiveness of this method in Andisols is shown by comparing the clay content obtained by mechanical analysis with that estimated by selective dissolution analysis as presented in Fig. 7.1. These data were obtained from Andisol B and C horizon samples, whose clay fraction consisted primarily of allophane with an Al/Si atomic ratio of 2, imogolite, and ferrihydrite (Andisol TU database, 1992). The content of allophane and imogolite, and that of ferrihydrite were estimated using acid oxalate extractable Si and Fe, respectively, as described in Chapter 6. The regression equation shows reasonably good agreement in the clay contents obtained by the two methods, indicating that mechanical analysis with pretreatments of ultrasonication and pH adjustment is reasonably reliable.

There are several advantages of ultrasonic dispersion as reviewed by Edwards and Bremner (1967). However, no standardized procedure for ultrasonication has been established (Gee and Bauder, 1986). For example, ultrasonic power and duration of treatment are soil-dependent, indicating that the optimum conditions for ultrasonication should be determined by trial and error for each soil sample.

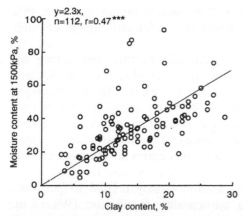

Fig. 7.2. Relationship between the clay content measured by the pipet method with pretreatments of sonication and pH adjustment and moisture content at 1500 kPa (prepared from Andisol TU database, 1992).

Since it is difficult to successfully apply the traditional particle-size determination by mechanical analysis to Andisols, the 1500 kPa water retention has been recommended to estimate clay content in Andisols. The clay content of Andisols, determined by mechanical analysis using ultrasonication and pH adjustment, and their 1500 kPa water retention values show an almost linear relationship up to a clay content of 30 percent as illustrated in Fig. 7.2. Thus, the common particle-size class names are not used for Andisols, but rather substitutes for particle-size class names based on 1500 kPa water retention, such as ashy, medial and hydrous, are used in their place (Soil Survey Staff, 1992). These substitutes were introduced in Chapter 4.

7.4. SOIL STRUCTURE

Soil structure is the arrangement of primary soil particles into secondary particles, units, or peds. Structure changes with horizon, organic matter content, soil texture, soil moisture content, etc. Andisols have various types of soil structure such as granular, angular blocky, subangular blocky, platy, or prismatic.

The surface horizons of uncultivated Andisols commonly have moderate to strong granular structure and sometimes subangular blocky structure. Noncrystalline materials and soil organic matter greatly contribute to the formation and stabilization of the soil structure. For example, Egashira et al. (1983) showed that water-stable aggregates greater than 53 μm in Japanese Andisols contain organic matter concentrations of 2.8–35.3 percent and clay contents of 11.9–71.1 percent. Cultivation of Andisols tends to degrade soil structure, mainly through compaction and reduction of soil organic matter, resulting in subangular blocky or angular blocky structure in the surface horizons. Most Bw horizons of Andisols have moderate subangular blocky structure, followed by angular blocky structure. Prismatic and platy types of soil structure are observed in a limited number of Andisols.

It is well known that soil moisture or alternate wetting and drying has a strong effect on aggregate formation. This was also observed for Andisols showing a hydrosequential relationship in Tokachi district, Hokkaido, Japan (Furuhata and Hayashi, 1980). The A and B horizons of Typic Hapludands, locally called "dry soils", have moderate granular structure while those of Hydric Melanaquands, locally called "wet soils", have moderate blocky structure. A single grain arrangement occurs in all the glassy, ashy, or cindery C horizons of both groups of Andisols.

The size distribution of soil aggregates is more closely related to pore size distribution and water retention than to particle-size distribution (Wu et al., 1990). For example, Furuhata and Hayashi (1980) conducted a correlation study between aggregate size and pore diameters of Andisols from Hokkaido, Japan. They showed that soil aggregates of 250–1000, 100–250, and less than 100 μm make the greatest contribution to the occurrence of macropores greater than 100 μm, mesopores between 100 and 6 μm and mesopores between 6 and 0.4 μm, respectively (Table 7.1). This indicates that the diameters of these macropores and mesopores are primarily determined by the diameters of inscribed spheres in the necks produced by adjacent spheres (Furuhata and Hayashi, 1980). In contrast, micropores smaller than 0.4 μm in diameter have the highest correlation with aggregates greater than 500 μm in diameter. This result suggests that they exist mostly inside the large aggregates (Furuhata and Hayashi, 1980).

Morphology and distribution of pores, primary soil particles, aggregates, and the presence or absence of illuviated clay skins on ped surfaces are examined using soil thin sections. Oriented clay skins are usually absent on the peds of Andisols, indicating little clay migration in these soils. The lack of clay migration reflects the difficulty of clay dispersion as described earlier.

Kawai (1969) classified the fabric of Andisols from Japan into 7 kinds: compacted, fine grained porous, blocky loose, coarse grained porous, foam-like compound, smoothed dense, and cemented porous. Their characteristics are described as follows:

TABLE 7.1

Size of soil aggregates showing the greatest contribution to the occurrence of pores with different diameters in Andisols from Hokkaido, Japan (prepared from data of Furuhata and Hayashi, 1980)

Range of pore diameter (μm)	Size of soil aggregates showing the greatest contribution (μm)
>100	250–1000
100–6	100–250
6–0.4	<100
<0.4	500–2000

Compacted fabric:	The skeleton grains and plasma in the soil are randomly distributed. Though it has a few planes and vugh voids, it does not look porous.
Fine grained porous fabric:	It is highly porous with many small peds of about 0.05 mm in diameter scattered throughout. Organic carbon content of soils with this fabric is as high as 13 to 21 percent.
Blocky loose fabric:	It has ortho, macro-sized vugh voids and numerous channel voids in their vicinity. There are many ortho, micro-sized vugh voids in the plasma and the primary structure is loosely aggregated. Its porosity is low.
Coarse grained porous fabric:	It is dominantly composed of primary peds sized from 0.15 to 0.3 mm in diameter which is 3 to 5 times as large as those for fine grained porous fabric.
Foam-like compound fabric:	It is characterized by the occurrence of foam-like inherited micro-size voids in plasma, originating from volcanic glass, and meta macro-sized vugh voids between peds.
Smoothed dense fabric:	Meso to macro-sized meta voids with a smooth surface are dominant. This fabric is found in halloysitic B horizons and not in A horizons.
Cemented porous fabric:	It has macro-sized voids and skeleton grains whose surfaces have clay coatings. This fabric is generally found in hard pans and also in weathered pumice layers.

Pore size measured by microscopic observation of thin sections is limited to those pores greater than 10 μm in diameter (Kawai, 1969). Since medium to fine textured Andisols are rich in pores less than 6 μm in diameter (Fig. 7.7), more than half the pores in Andisols can not be evaluated using thin sections. Thus, even a soil horizon with compacted fabric, which is observed to have few pores by microscope evaluation has low bulk density.

According to observations of Andisols from Tokachi, Hokkaido, Japan (Furuhata and Hayashi, 1980), soil horizons with granular structure have coarse grained porous fabric. Andisols with blocky structure younger than 10,000 years tend to have blocky loose fabric while those older than 10,000 years have smoothed dense fabric. Andisols with platy structure were shown to have compact fabric.

Sutanto et al. (1988) reported that soils developed on airfall tephra from Mt. Merbabu, central Java, have a very porous structure ranging from granular to spongy. Their micromorphological study suggests that these soils have fine grained porous fabric or coarse-grained porous fabric according to Kawai's classification.

7.5. SOIL CONSISTENCE

Soil consistence is the manifestation of cohesion and adhesion forces acting between soil particles at various water contents. It includes stickiness and plasticity when wet, friability when moist, and hardness when dry. The evaluation of soil consistence is usually conducted by hand in the field. Moderately weathered Andisols have friable consistence and high liquid and plastic limits and show only slight stickiness compared with other mineral soils.

As described in Chapter 2, Andisols rich in organic matter and/or allophanic clays are usually very friable to friable when moist; slightly sticky to nonsticky, and slightly plastic to nonplastic, when wet; and soft to slightly hard when dry. These properties lead to ease of cultivation at moisture contents not only between the shrinkage and plastic limits, but also at moisture contents lower than the shrinkage limit and somewhat higher than the plastic limit. In contrast, nonallophanic Andisols with low organic carbon contents are sticky and plastic when wet so cultivation should be practiced at moisture contents only between the shrinkage and plastic limits.

Smeariness is also observed in the field to determine the thixotropic property of Andisols. Thixotropy is a reversible gel-sol-gel transformation in certain materials brought about by a mechanical disturbance followed by a period of rest (Jumikis, 1967). Andisols containing very large amount of noncrystalline materials often show strong thixotropy as described for the Hilo soil (Acrudoxic Hydrudand) from Hawaii in Chapter 2. Therefore, engineering practices should be carefully evaluated for these soils.

Atterberg limits express the plastic properties of soils quantitatively as determined by laboratory methods. The properties used to characterize Andisols are liquid limit, plastic limit, and plasticity index. Atterberg limits are affected by various factors such as particle size distribution, clay mineral composition, soil organic matter content, initial water content, exchangeable cations, surface area, etc. In this section, emphasis is placed on the effect of the clay mineral composition on the Atterberg limits.

Inahara (1989) studied the relationship between clay contents and liquid limit using three groups of Japanese Andisols with different clay minerals. Soil groups were dominated by allophanic clays, chloritized 2:1 minerals or halloysite, and all soils contained small amount (less than 6 percent) of organic C. Though the soil samples were divided into three groups according to their clay mineral

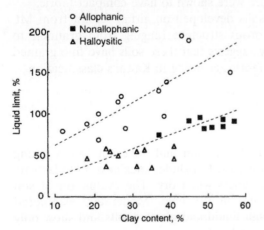

Fig. 7.3. Relationship between the clay content and liquid limit of Andisols. The soil samples are divided into allophanic Andisols, nonallophanic Andisols (chloritized 2:1 minerals), and halloysitic soils (prepared from data of Inahara, 1989).

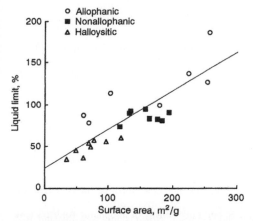

Fig. 7.4. Relationship between the surface area and liquid limit of Andisols (prepared from data of Inahara, 1989).

composition, the final analysis showed that they could be grouped into two categories, allophanic and other soils. This grouping was based on the relationship between clay content and liquid limit as demonstrated in Fig. 7.3. The allophanic soils have very high liquid limit values compared to the other soils. Inahara (1989) also found similar relationships between the clay content and the plastic limit and plasticity index for the same samples.

Inahara (1989) further studied the relationship between the surface area and liquid limit of Andisols. As presented in Fig. 7.4, the relationship can be expressed by a linear regression equation indicating that the liquid limit of Andisols is primarily determined by their surface area. Similar results were obtained between the plastic limit and surface area. These results support the water film theory described by Baver et al. (1972) that plasticity of soils is caused by forces associated with water films around and between soil particles. In addition to soil mineralogy, soil organic matter also strongly affects the liquid limit and plastic limit of Andisols (Aragaki et al., 1987; Inahara, 1989) because organic matter also increases specific surface area. This effect of organic matter becomes conspicuous in Andisols with organic C contents greater than 6 percent (Inahara, 1989).

The liquid and plastic limits of Andisols usually show strong drying hysteresis. These values are substantially decreased with drying and the decrease for Andisols is much greater than for most other soils. The degree of drying hysteresis is largely determined by the content of noncrystalline materials and soil organic matter, and the initial water content or drying experienced by the soil.

7.6. BULK DENSITY AND POROSITY

Low bulk density is a characteristic feature of Andisols. Bulk density typically ranges between 0.4 and 0.8 g cm^{-3} in moderately weathered Andisols. It should

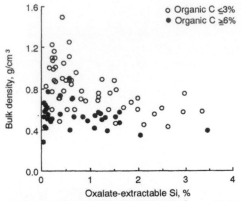

Fig. 7.5. Relationship between the oxalate-extractable Si and bulk density of allophanic Andisols (prepared from Andisol TU database, 1992).

be noted that allophanes in Andisols have particle densities of 2.5 to 2.7 g cm^{-3} (Maeda et al., 1977; Bielders et al., 1990), which is similar to the particle density of other mineral soils. Thus, the development of porous soil structure is the primary factor responsible for the low bulk density of Andisols.

Bulk densities of fresh rhyolitic and dacitic ashes are approximately 1.5 g cm^{-3} or more and decrease with the advance of weathering. Weathering proceeds with rapid development of porous soil structure to which noncrystalline materials and soil organic matter contribute to a large extent.

Allophane is one of the most important noncrystalline materials contributing to the low bulk density of Andisols through the development of porous soil structure. As illustrated in Fig. 7.5 (Andisol TU database, 1992), the bulk density values of allophanic Andisols with low organic C tend to decrease with an increase in their oxalate-extractable silica (Si$_o$) or allophane content (%Si$_o$ × 7.1). The bulk density becomes less than 0.9 g cm^{-3} (andic soil property requirement) when the allophane content is greater than approximately 5 percent. Bulk densities of humus rich allophanic Andisols tend to be lower than those of humus poor allophanic Andisols. Bulk density values of nonallophanic Andisols are largely determined by soil organic matter as presented in Fig. 7.6. They show a tendency to decrease with increasing organic C content and become smaller than 0.9 g cm^{-3} when the organic C content is greater than approximately 3 percent.

The porosity of single-grained sand is about 40 percent while a clayey nonandic soil with an abundance of crystalline clay minerals shows almost the same porosity based on calculations of equi-size spheres. The porosity and pore size distribution of Andisols are dependent on the development of soil structure. For example, the C horizons of young Andisols (Tarumae-b soils, 600 years old) have porosities of about 60 percent while both A and B horizons of moderately weathered Andisols (Tarumae-c soils, 2000 years old) have porosities of about 80 percent as demonstrated in Fig. 7.7 (Furuhata and Hayashi, 1980). The two Andisols show

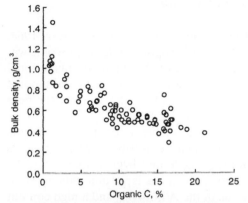

Fig. 7.6. Relationship between organic C content and bulk densities of Alic Andisols (prepared from Andisol TU database, 1992).

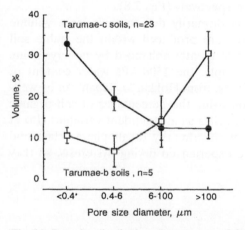

Fig. 7.7. Pore size distribution of two groups of Andisols (Tarumae-b soils, 600 years old and Tarumae-c soils, 2000 years old) from Hokkaido, Japan (prepared from data of Furuhata and Hayashi, 1980).

contrasting pore size distributions. The young Andisols have a greater amount of macropores larger than 100 μm in diameter and a lower amount of micropores (<0.4 μm) and mesopores (0.4–6.0 μm). In contrast, the moderately weathered Andisols have a large amount of micropores (<0.4 μm) and mesopores (0.4–6.0 μm), contributing to the large plant available water holding capacity.

The unusually high amount of micropores in allophanic Andisols is attributable to the intra- and inter-particle pores of allophane. This mineral has a hollow spherical structure and the inside diameter of the sphere is 3 to 4 nm. Moreover, Parfitt and Henmi (1980) suggested that the hollow spherical aluminosilicate sheet of allophane has small holes with diameters of 0.3 to 0.5 nm which allow water molecules to pass through (see Chapter 5).

7.7. WATER RETENTION AND PLANT AVAILABLE WATER

Andisols retain a large amount of water in soil pores of varying size (Iwata, 1968; Furuhata and Hayashi, 1980; Saigusa et al., 1987). Micropores hold hygroscopic water, mesopores retain capillary water, and macropores accommodate gravitational water or air. Plant available water within the soil occurs primarily in the mesopores and is generally defined as the difference between 33 kPa water content (or field capacity) and 1500 kPa water content (or permanent wilting point).

Young Andisols retain a considerable amount of available water and low amounts of hygroscopic water (>1500 kPa). As Andisols mature and weathering products accumulate, they hold large amounts of both available and hygroscopic water. An example of a mature soil is an Alic Melanudand which has large amounts of Al/Fe humus complexes and chloritized 2:1 minerals in the A horizons and a high content of noncrystalline materials in the Bw horizons. This soil holds available water of 21–40 and 21–27 percent (volume) and hygroscopic water of 28–38 and 35–36 percent (volume) in the A and B horizons, respectively (Fig. 7.8).

The high water retention of Andisols is primarily due to their large volume of mesopores and micropores. These pores are produced within the stable soil aggregates. Formation of these aggregates is greatly enhanced by noncrystalline materials and soil organic matter. For example, the 1500 kPa water content of allophanic Andisols, excluding the A1 horizons, from Hokkaido, Japan can be well-expressed by a multiple-regression equation using the percentages of soil organic matter and percentage of noncrystalline materials as independent variables (Ito et al., 1991). The noncrystalline materials consist primarily of allophanic clays and ferrihydrite. The A1 horizon samples have experienced drying hysteresis, so they were excluded in this regression analysis.

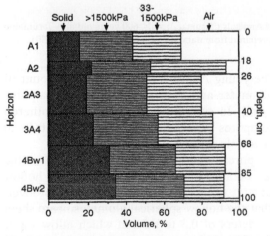

Fig. 7.8. Vertical distribution of plant available water and hygroscopic water of a Melanudand (prepared from data of Shoji, 1984).

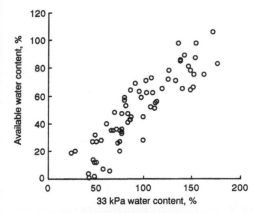

Fig. 7.9. Relationship between 33 kPa water content and plant available water content of Andisols from Japan (prepared from data of Saigusa et al., 1987).

The definition of available water for Andisols was confirmed in greenhouse experiments by Saigusa et al. (1987). They compared the water uptake of test plants and the available water content and observed that 1500 kPa water content determined by the pressure membrane method was an accurate measure of the wilting point for plants grown on Andisols. Furthermore, Saigusa et al. (1987) obtained a strong linear relationship between 33 kPa water content and the amount of available water as presented in Fig. 7.9. It appears that about 60 percent of the 33 kPa water is available to plants.

Allophane greatly contributes to the retention of high tension water because its fine particle-size and hollow spherical structure can accommodate water molecules in both intra- and inter-spherical pores. Henmi (1991) determined that water retained in the inter-spherical pores may be removed by heating at 85°C, while heating to temperatures greater than 140°C is required to remove water inside the allophane spherules. This explanation is at least partially supported by the observation that heating allophane at temperatures of 110 and 150°C does not affect the intensity of the infrared absorption band at 1630 cm^{-1} which is due to OH bending vibrations of adsorbed water molecules. However, the intensity of the 1630 cm^{-1} band is greatly attenuated by heating at temperatures greater than 200°C (Kitagawa, 1975). Weight loss of allophane between 200 and 105°C when heated for 24 hours is 8 percent (Kitagawa, 1977).

7.8. WATER PERMEABILITY

High water permeability is a distinctive physical property of volcanic ash soils under both saturated and unsaturated conditions. It is favorable for upland farming because it provides good drainage necessary for optimum growth of upland crops, but it is often unfavorable for wetland rice farming because it contributes to

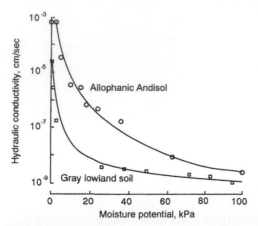

Fig. 7.10. Comparison of unsaturated hydraulic conductivity of an allophanic Andisol and a Gray low-land soil as a function of moisture potential (prepared from data of Hasegawa, 1986).

leaching of irrigation water and plant nutrients.

Saturated hydraulic conductivity and unsaturated hydraulic conductivity are used to describe water permeability characteristics of soils. The unsaturated hydraulic conductivity depends strongly on water content and rapidly decreases with decreasing water content. The saturated hydraulic conductivity of Andisols used for wetland rice ranges from 10^{-3} to 10^{-4} cm sec^{-1} and is 10–100 times greater than Gray lowland soils or other paddy soils (Motomura, 1979). In order to use Andisols for wetland rice farming, percolation is reduced by subsoil compaction or bentonite application to lower the water percolation rate to 20–30 mm day^{-1}, including evaporation of about 5 mm day^{-1}.

Under unsaturated conditions, Andisols have greater hydraulic conductivity than other mineral soils such as clayey alluvial soils (Hasegawa, 1986) and red/yellow soils (Iwata, 1966; Iwata et al., 1988). For example, Fig. 7.10 shows the relationship between moisture tension and unsaturated hydraulic conductivity of an Andisol and a clayey alluvial soil from Ibaraki, Japan. The Andisol has hydraulic conductivities greater than the alluvial soil in the moisture tension range of zero to 100 kPa. According to the soil moisture/potential curve, the Andisol holds approximately twice as much water as the clayey alluvial soil between tensions of 10 and 100 kPa (Hasegawa, 1986). This fact indicates that the ratio of cross sectional area for unsaturated water flow to total soil cross sectional area is high in Andisols as compared with the alluvial soil. This is the main reason for the higher unsaturated hydraulic conductivity of Andisols.

7.9. IRREVERSIBLE CHANGES IN SOIL PHYSICAL PROPERTIES WITH DRYING

Irreversible changes in soil physical properties occur upon drying in Andisols. Physical properties affected by drying include water retention, clay dispersibility,

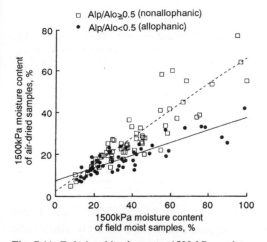

Fig. 7.11. Relationships between 1500 kPa moisture content of field moist Andisols (having less than 100% 1500 kPa water retention) and those of air-dried Andisols (prepared from Andisol TU database, 1992).

liquid limit, plastic limit, etc. In addition, there are some irreversible changes in chemical properties such as an increase in ammonium acetate-extractable (1 M, pH 4.8) Al (Kubota, 1976).

It is widely recognized that water retention in volcanic ash soils shows strong hysteresis with drying. Field moist Andisols show a considerable decrease in 1500 kPa water content after drying. For field moist samples having 1500 kPa water contents between 30 and 100 percent, the degree of reduction in 1500 kPa water content with drying is larger in allophanic than in nonallophanic Andisols (Fig. 7.11). Regression equations between 1500 kPa water content of field-moist samples (x) (having less than 100 percent water content) and that of air-dried samples (y) are as follows:

$$y = 0.63x + 2.0, \qquad r = 0.90, \; n = 65, \quad \text{for nonallophanic Andisols;}$$

$$y = 0.30x + 7.4, \qquad r = 0.85, \; n = 56, \quad \text{for allophanic Andisols}$$

Field moist samples having 1500 kPa water contents greater than 100 percent, or hydric Andisols, show a large decrease in 1500 kPa water content following air drying. Although these hydric samples are grouped as nonallophanic, the magnitude of the reduction in 1500 kPa water content with air-drying is similar to those for allophanic Andisols (Andisol TU database, 1992).

There are various soil constituents affecting the irreversible reduction of 1500 kPa water content. The effect of individual soil components can be evaluated by determining the ratio of 1500 kPa water content of air-dried soil samples to that of field moist soil samples. Correlation analysis between the degree of hysteresis and selected soil constituents shows that oxalate-extractable Al, Fe, and

TABLE 7.2

Correlation analyses between hysteresis in 1500 kPa water content and selected soil constituents. The degree of hysteresis in 1500 kPa water content with air-drying was evaluated by determining the ratio of 1500 kPa water content of air-dried sample to that of field moist samples (prepared from Andisol TU database, 1992)

Soil constituents	n	Correlation coefficients
Total organic carbon	121	0.25 **
Oxalate-extractable Al	121	−0.58 ***
Oxalate-extractable Fe	120	−0.40 ***
Oxalate-extractable Si	112	−0.54 ***
Pyrophosphate-extractable Al	121	−0.15

Si have significant correlation coefficients (0.1 percent level) (Table 7.2). These results strongly suggest that noncrystalline inorganic materials, namely allophanic clays, make the greatest contribution to this irreversible change. In contrast, organic matter does not appear to play an important role in this hysteresis. These results coincide with the difference in the degree of reduction in 1500 kPa water content with air-drying between allophanic and nonallophanic Andisols as shown in Fig. 7.11.

Tsutsumi et al. (1977) compared the water retention at 33–1500 kPa of field-moist, air-dried and oven-dried Andisols and found that the irreversible reduction in the water retention occurred following the air-drying process. There was no further significant decrease after oven-drying the samples.

Irreversible flocculation is a major concern to the reproducibility of particle-size determination by mechanical analysis in Andisols. Kubota (1976) showed that there were irreversible changes in particle-size distribution of allophanic Andisols depending on the intensity of the drying pretreatment (Fig. 7.12). Irreversible

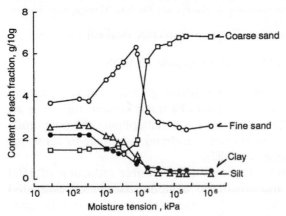

Fig. 7.12. The relationship between particle-size distribution and intensity of dehydration treatment of a Bw horizon soil of an allophanic Andisol (prepared from data of Kubota, 1976).

flocculation starts at 10^2–10^3 kPa moisture tension as indicated by the increase in fine sand content and decrease in the clay and silt contents. The effect continues to 10^4–10^5 kPa moisture tension as indicated by the rapid decrease of fine sand content and the rapid increase of coarse sand content. It was also observed that a reduction of water holding capacity proceeds along with the irreversible flocculation of the soil materials (Kubota, 1976).

The liquid and plastic limits of volcanic ash soils greatly decrease with drying as compared to other mineral soils. Though liquid and plastic limits of volcanic ash soils are not affected by the decrease in the initial water content to a certain threshold, the liquid and plastic limits decrease with further drying beyond this value. The plasticity index also decreases with decreasing initial water content (Maeda et al., 1983; Maeda and Soma, 1986; Tsutsumi et al., 1977). However, the effect of drying on the liquid and plastic limits of soils is small when the soils are air-dried.

The moisture tension at which the liquid limit begins to decrease is about 1200 kPa (Soma, 1978). This moisture tension is almost identical to the moisture tension at which the irreversible flocculation and the irreversible decrease in water holding capacity occur.

REFERENCES

Andisol TU database, 1992. Database on Andisols from Japan, Alaska, and Northwestern U.S.A. prepared by Soil Science Laboratory, Tohoku University (see Appendix 2).

Aragaki, M., Adachi, T., Miyauchi, S. and Nagata, N., 1987. Effects of the organic matter on water retention and structure of the Kuroboku Soil: Especially on nonallophanic soil. Trans. Jpn. Soc. Irrig. Drain. Reclam. Eng., 131: 59–67 (in Japanese).

Baver, L.D., Gardner, W.H. and Gardner, W.R., 1972. Soil Physics, 4th edition. John Wiley and Sons, New York, pp. 89–90.

Bielders, C.L., De Backer, L.W. and Delvaux, B., 1990. Particle density of volcanic soils as measured with a gas pycnometer. Soil Sci. Soc. Am. J., 54: 822–826.

Clark, C.J. and McBride, M.B., 1984. Cation and anion retention by natural and synthetic allophane and imogolite. Clays Clay Miner., 32: 291–299.

Edwards, A.P., and Bremner, J.M., 1967. Dispersion of soil particles by sonic vibration. J. Soil Sci., 18: 47–63.

Egashira, K., Kaetsu, Y. and Takuma, K., 1983. Aggregate stability as an index of erodibility of Ando soils. Soil Sci. Plant Nutr., 29: 473–482.

Furuhata, A. and Hayashi, S., 1980. Relation between soil structure and soil pore composition: case of volcanogenous soils in Tokachi district. Res. Bull. Hokkaido Natl. Agric. Exp. Stn., 126: 53–58 (in Japanese, with English abstract).

Gee, G.W., and Bauder, J.W., 1986. Particle-size analysis. In: A. Klute (Editor), Methods of Soil Analysis, Part 1, Physical and Mineralogical Methods, 2nd edition. ASA and SSSA, Madison, Wisconsin, pp. 383–411.

Hasegawa, S., 1986. Soil water movement in upland fields converted from paddy fields. Soil Physical Conditions and Plant Growth, Japan, 53: 13–19 (in Japanese).

Henmi, T., 1991. Idea and methodology on the study of amorphous clays. J. Clay Sci. Soc. Jap., 31: 75–81 (in Japanese).

Horikawa, Y., 1975. Electrokinetic phenomena of aqueous suspensions of allophane and imogolite. Clay Sci., 4: 255–263.

Inahara, M., 1989. Studies on the consistence and compaction of Andisols with different colloidal composition. Master thesis. Faculty of Agriculture, Tohoku University, Sendai, Japan (in Japanese).

Ito, T. and Shoji, S., 1993. In preparation for contribution.

Ito, T., Shoji, S. and Saigusa, M., 1991. Classification of volcanic ash soils from Konsen district, Hokkaido, according to the last Keys to Soil Taxonomy. Jap. J. Soil Sci. Plant Nutr., 62: 237–247 (in Japanese, with English abstract).

Iwata, S., 1966. The characteristics of water movement in soils during drainage with special reference to field moisture capacity. Bull. Natl. Inst. Agric. Sci., B17: 149–176 (in Japanese, with English abstract).

Iwata, S., 1968. Soil moisture in volcanic ash soils. Soil Physical Conditions and Plant Growth, Japan, 18: 18–26 (in Japanese).

Iwata, S., Tabuchi, T. and Warkentin B.P., 1988. Soil-Water Interactions: Mechanisms and Applications. Marcel Dekker, New York, pp. 337–346.

Jumikis, A.R., 1967. Introduction to Soil Mechanics, D. Van Nostrand Company, Inc. Princeton, New Jersey, pp. 49–50.

Kawai, K., 1969. Micromorphological studies of Andosols in Japan. Bull. Natl. Agr. Sci. Series B., 20: 77–154 (in Japanese, with English abstract).

Kitagawa, Y., 1975. Dehydration, micromorphology and chemical composition of allophane. Bull. Natl. Inst. Agr. Sci. Series B., 26: 95–131 (in Japanese, with English abstract).

Kitagawa, T., 1977. Determination of allophane and amorphous inorganic matter in soils. Bull. Natl. Inst. Agr. Sci. Series B., 29: 1–48 (in Japanese, with English abstract).

Kobo, K. and Oba, Y., 1965. Dispersion of volcanic ash soils by ultrasonic treatment. Jap. J. Soil Sci. Plant Nutr., 36: 207–210 (in Japanese).

Kobo, K., Oba, Y. and Oishi, K., 1974. Genesis and characteristics of volcanic ash soil in Japan, V. The relationship between the degree of weathering and parent material composition in clay fraction and dispersion of clay. Jap. J. Soil Sci. Plant Nutr., 45: 8–11 (in Japanese).

Kubota, T., 1976. Surface chemical properties of volcanic ash soil — especially on phenomenon and mechanism of irreversible aggregation of the soil by drying. Bull. Natl. Inst. Agric. Sci., B-28: 1–74 (in Japanese, with English abstract).

Maeda, T., and Soma, K., 1983. Physical and engineering characteristics of volcanic ash soils. In: N. Yoshinaga (Editor), Volcanic Ash Soils — Formation, Characteristics, and Classification. Hakuyusha, Tokyo, pp. 99–139 (in Japanese).

Maeda, T. and Soma, K., 1986. Physical properties. In: K. Wada (Editor), Ando Soils in Japan. Kyushu Univ. Press, Fukuoka, Japan, pp. 99–111.

Maeda, T., Takenaka, H. and Warkentin, B.P., 1977. Physical properties of allophane soils. Adv. Agron., 29: 229–264.

Maeda, T., Soma, K., Adachi, T., Takenaka, H. and Tsutsumi, S., 1983. Problems on the measurements relating to physical characteristics of organo-volcanic ash soils and organic clayey soils. Trans. Jap. Soc. Irrig. Drain. Eng., 103: 1–12 (in Japanese).

Motomura, S., 1979. Physical properties of soil groups distributed in the lowland. In: Soil Physical Conditions and Plant Growth. Youkendo, Tokyo, pp. 60–71 (in Japanese).

Noshiroya, N., 1992. Changes in the soil properties of Andisols due to wetland rice farming. Master thesis, Faculty of Agriculture, Tohoku University, Sendai, Japan (in Japanese).

Oba, Y. and Kobo, K., 1965. Clay aggregates in fine sand fraction of volcanic ash soils. Jap J. Soil Sci. Plant Nutr., 36: 203–206 (in Japanese).

Otowa, M., Shoji, S. and Saigusa, M., 1988. Allic, melanic and fulvic attributes of Andisols. In: D.I. Kinloch, S. Shoji, F.H. Beinroth and H. Eswaren (Editors), Proc. 9th Int. Soil Classification Workshop, Japan, 20 July to 1 August, 1987. Publ. by Japanese Committee for the 9th International Soil

Classification Workshop, for the Soil Management Support Services, Washington D.C., USA., pp. 192–202.

Parfitt, R.L. and Henmi, T., 1980. Structure of some allophanes from New Zealand, Clay Clay Miner., 28: 285–294.

Ping, C.L., Shoji, S. and Ito, T., 1988. Properties and classification of three volcanic ash-derived pedons from Aleutian islands and Alaska peninsula, Alaska. Soil Sci. Soc. Am. J., 52: 455–462.

Ping, C.L., Shoji, S., Ito, T., Takahashi, T. and Moore, J.P., 1989. Characteristics and classification of volcanic-ash-derived soils in Alaska. Soil Sci., 148: 8–28.

Saigusa, M., Shoji, S. and Nakaminami, H., 1987. Measurement of water retention at 15 bar tension by pressure membrane method and available moisture of Andosols, Jap. J. Soil Sci. Plant Nutr., 58: 374–377 (in Japanese).

Shoji, S., 1984. Genesis and properties of nonallophanic Andisols. J. Clay Sci. Jap., 24: 152–165 (in Japanese).

Shoji, S., 1988. Separation of melanic and fulvic Andisols, Soil Sci. Plant Nutr., 34: 303–306.

Soil Survey Staff, 1992. Keys to Soil Taxonomy, 5th edition. AID, USDA-SMSS Technical Monograph, No. 19, Blacksburg, Virginia.

Soma, K., 1978. Studies on the relationship between Atterberg limits and initial water content of soil. Soil Physical Conditions and Plant Growth, Jpn., 38: 16–22. (in Japanese)

Sutanto, R., DeConinck, F., and Doube, M., 1988. Mineralogy, Charge Properties and Classification of Soils on Volcanic Materials and Limestone in Central Java (Indonesia). Geological Institute, Belgium, 233 pp.

Tsutsumi, S., Adachi, T. and Takenaka, H., 1977. Change of physical and engineering properties of organic soils from the view point of drying process. Trans. Jap. Soc. Irrig. Drain. Reclam. Eng., 71: 7–15 (in Japanese, with English abstract).

Van Wambeke, A., 1992. Andisols. In: Soils of the Tropics. McGraw-Hill. New York et al., pp. 207–232.

Wada, K., 1985. The distinctive properties of Andosols. Adv. Soil Sci., 2: 173–229.

Wada, K., 1986. Part II. Data Base, Kurobokudo Co-operative Research Group. In: K. Wada (Editor), Ando Soils in Japan. Kyushu University Press, Fukuoka, Japan, pp. 115–276.

Warkentin, B.P. and Maeda, T., 1980. Physical and mechanical characteristics of Andisols. In: B.K.G. Theng (Editor), Soils with Variable Charge. New Zealand Soc. of Soil Sci. Lower Hutt, New Zealand, pp. 281–301.

Warkentin, B.P., Maeda, T., and Soma, K., 1988. Physical characteristics for classification of Andisols. In: D.I. Kinloch, S. Shoji, F.H. Beinroth and H. Eswaren (Editors), Proc. 9th Int. Soil Classification Workshop, Japan, 20 July to 1 August, 1987. Publ. by Japanese Committee for the 9th International Soil Classification Workshop, for the Soil Management Support Services, Washington D.C., USA., pp. 97–107.

Wu, L., Vomocil, J.A. and Child, S.W., 1990. Pore size, particle size, aggregate size, and water retention. Soil Sci. Soc. Am. J., 54: 952–956.

Yamanaka, K., 1964. Physical properties. In: Ministry of Agriculture and Forestry, Japanese Government (Editor), Volcanic Ash Soils in Japan. Tokyo, pp. 69–91.

Yoshinaga, N. and Aomine, S., 1962. Allophane in some Ando soils. Soil Sci. Plant Nutr., 8: 6–13.

Chapter 8

PRODUCTIVITY AND UTILIZATION OF VOLCANIC ASH SOILS

S. SHOJI, M. NANZYO, and R. DAHLGREN

8.1. INTRODUCTION

Volcanic ash soils or Andisols are among the most productive soils in the world which explains why these soils have a high human carrying capacity (Leamy, 1984). This is clearly illustrated by the fact that the most densely populated area of the world, having more than 400 persons per km^2, is located in the volcanic ash soil zone of central Java, Indonesia. For example, Mohr (1938) showed that the most fertile soils in Indonesia are juvenile volcanic ash soils based on the relationship between soils and population density obtained from the Census Returns for 1930. He also described soils and population densities in the area surrounding Mt. Merapi in central Java as a remarkable case. The areas with the youngest ash deposits had population densities exceeding 400 per km^2, while an area without such ash deposits showed only 245 per km^2. In contrast, the foot slope areas surrounding Mt. Merapi had population densities from 800 to more than 1000 per km^2. He noted that the parent material of these fertile ash soils had a basaltic composition with high concentrations of calcium, magnesium, iron, phosphorus, etc.

Furthermore, it is interesting to note that well-known vestiges of early civilization such as the old capital Yogyakarta, Candi Prambanan, Borobudor cathedral and the Java man are situated in the vicinity of the active volcanoes, Mts. Merapi and Merbabu. These volcanoes intermittently ejected volcanic ash as shown by the multiple strata of ash layers in the area surrounding these volcanoes (Miyake, 1983; Miyake et al., 1984; Sutanto et al., 1988).

In contrast, volcanic ash soils or Kurobokudo in Japan were originally regarded as impoverished soils because of high amounts of active aluminum, very low contents of available phosphate, low concentrations of exchangeable bases and strong acidity (Nomoto, 1958). Thus, the extensive utilization of volcanic ash soils for common upland crops was significantly delayed as compared to alluvial soils used for lowland rice farming that was started more than 2000 years ago.

The chemical and physical properties of volcanic ash soils influence their productivity in various ways. Accurate and extensive information on these properties is necessary for developing programs for improving productivity, sustainable utilization, and conservation of these soils. This chapter describes the productivity of volcanic ash soils, the chemical and physical properties relating to soil productivity, and finally, the utilization and management of these soils in Japan.

8.2. SOIL PRODUCTIVITY

Soil productivity is the capacity of a soil to produce a certain yield of a specific crop or sequence of crops under optimum management practices. Its evaluation varies with changes in the target yield which is often determined by advances in agrotechnology. Its evaluation is also influenced by the changing social and economic environment.

In the tropical regions where low input farming is common and weathering of soil materials is rapid, volcanic ash soils are generally evaluated as the most productive among arable soils, especially when their parent material is basaltic. Although other unique properties are associated with volcanic ash soils, as described below, it appears that the high productivity of these soils is largely due to their rapid release of nutrients as compared to associated Oxisols and Ultisols which show a wide distribution in tropical regions.

As already described, volcanic ash soils (Andisols and Entisols) on the foot-slopes of Mts. Merapi and Merbabu, central Java are the most productive in the region (Mohr, 1938; Miyake, 1983; Miyake et al., 1984; Sutanto et al., 1988). These soils have several important properties relating to their high productivity as follows:

(1) The parent material consists of cumulative deposits of fine-grained ashes having basalt or andesite composition, resulting in the formation of a composite morphology,

(2) The soils have a deep, unrestricted rooting zone,

(3) Humus horizons are thick and contain large amounts of organic N,

(4) Apatite (calcium phosphate) is relatively abundant in the parent material, and

(5) Plant available water is abundant.

These properties of volcanic ash soils provide a soil environment conducive for deep rooting and can supply the large amounts of nutrients and water necessary for vigorous plant growth.

Volcanic ash soils in the Philippines, which include not only Andisols, but also Mollisols, Inceptisols, etc., are generally highly productive (Otsuka et al., 1988). Although phosphorus is often the yield limiting element for common crops grown on volcanic ash soils, Otsuka et al. (1988) showed that the parent materials are generally rich in this nutrient. Furthermore, all the soils accumulate a relatively large amount of phosphorus in their humus horizons. Some horizons contain more than 0.5 percent phosphorus as P_2O_5. It is also interesting to note that about half of the accumulated phosphorus occurs in organic forms. This strongly suggests that organic phosphorus is subjected to rapid mineralization, resulting in the maintenance of high phosphorus fertility in these tropical volcanic ash soils.

The most important Andisols in Hawaii, U.S.A. are Typic Hydrudands and Mollic Hapludands. These soils have high amounts of plant available water and a high nutrient supplying ability, especially in the case of bases and minor elements. The productivity of Typic Hydrudands and Mollic Hapludands is evaluated to be moderate and high, respectively (H. Ikawa, 1991, personal communication).

Volcanic ash soils in Ethiopia, Uganda, Rwanda, Brundi, Kenya, Tanzania and Zaire in Africa are also very productive. These areas have a high human carrying capacity of approximately 200 persons per km^2 (Wakatsuki, 1990). It is also noted that these soils are rich in phosphorus (Kosaki, 1989; Mizota, 1987).

Productivity of volcanic ash soils in the temperate regions is evaluated considerably different according to variations in climatic conditions and types of farming. In regions having dry or relatively dry summers, volcanic ash soils except vitric or coarse-textured soils are considered to be highly productive. These soils are subject to weak leaching so that few have pH(H$_2$O) values less than 5.2 and thus have only trace amounts of KCl-exchangeable Al or toxic Al. They have considerable amounts of available nutrients and excellent physical properties such as high friability, free drainage, and high plant available moisture content. For example, on the North Island of New Zealand where grassland farming is common, volcanic ash soils are generally allophanic and are evaluated to be fertile. These soils have few problems related to the supply of major elements for most agricultural crops, except for the maintenance requirements of phosphorus and potassium. They also have no significant minor element deficiencies (Hewitt, 1989).

In contrast, in humid temperate regions having rainy summers as in Japan, volcanic ash soils (Kurobokudo) used for upland cropping are generally ranked in capability class III (cultivated soils in Japan are grouped into four capability classes, from class I to class IV with class I being the best). These soils have many limitations or hazards for crop production and they require fairly intensive management practices. In contrast, the use of these soils for wetland rice results in a capability class II, indicating some limitations or hazards for crop production and the need for some management practices to obtain normal crop production (Minist. Agr. Forest. and Fish., Japan, 1983).

Some Andisols in Japan are also very productive for growing upland crops. The properties providing the most productive soils were summarized by Shoji (1979) as follows:

(1) Parent material: basaltic ash,

(2) Morphology: pachic (thick humus-rich horizon) indicating intermittently repeated ash deposition,

(3) Physical properties: medium-textured, friable, free drainage, high plant available moisture content, and unrestricted rooting zone, and

(4) Chemical properties: high base saturation, pH(H$_2$O) > 6.0, absence of toxic Al, accumulation of high amounts of organic matter, and high natural supply of nutrients.

In cold regions as in Alaska where podzolization commonly takes place in well-drained soils, Andisols are more productive than Spodosols. For example, the productivities of two important Andisols, Fulvicryands and Haplocryands, in south central Alaska are ranked as high and medium for pastures and grasslands, respectively (C.L. Ping, 1991, personal communication).

Volcanic ash soils are also highly productive forest soils. In Oregon and Washington, U.S.A., the coastal western hemlock and Sitka spruce forests are located on Andisols (Meurisse, 1988). These forests are among the most productive forests in the world. Pumice soils on the North Island of New Zealand are ideally suited to the growth of exotic conifers such as *Pinus radiata*. These pine trees show extremely rapid growth rates under the climatic conditions prevailing in the area. The unrestricted deep rooting zone and high moisture retention capacity of volcanic ash soils are the most favorable factors for the growth of tree species (Leamy, 1982).

8.3. CHEMICAL PROPERTIES RELATING TO SOIL PRODUCTIVITY

8.3.1. Nitrogen

Nitrogen is often the yield limiting nutrient for agricultural plants grown on most soils including volcanic ash soils. Nitrogen is required in large quantities by most agricultural plants throughout their growth period. The use of N fertilizer has dramatically increased yields of agricultural plants, but the recovery of N from ordinary N fertilizer sources such as urea and ammonium sulfate by plants is generally not very high. Consequently, losses of N from the soil are excessively high, especially in intensively cropped areas. Thus, N fertilizer application is considered to be a significant source of environmental pollution. Reducing N losses and hence, controlling environmental degradation is now a major concern of scientists working in many disciplines. Development of an efficient N fertilization program requires an integration of information concerning N uptake by each crop to obtain the optimum yield, N dynamics in the soil, and recovery of fertilizer and soil N by the plant.

(1) Accumulation and stability of organic N in Andisols

Andisols often accumulate high concentrations of organic matter which contains appreciable quantities of organic nitrogen. According to the database of cultivated Andisols (Kurobokudo) in Japan (Oda et al., 1987), the mean organic N content of the plow layer is 0.44 percent for paddy fields and 0.37 percent for upland fields. If an Andisol has pachic morphology, its pool of organic N is estimated to be greater than 15 ton N/ha, by assuming a 50 cm humus-rich horizon, organic N contents greater than 0.6 percent and a bulk density of 0.5 g cm^{-3}. Thus, the mineralization rates associated with these large accumulations of organic N determine the nitrogen fertility status of Andisols.

The percentage of readily mineralizable organic N in the pool of total organic N is small in Andisols as compared to that of nonandic soils in Japan. Saito (1990)

TABLE 8.1

Nitrogen mineralization parameters of Andisols and nonandic soils from northeastern Tohoku, Japan (incubated under aerobic conditions) (From Saito, 1990)

Soils	Number of samples	N_0 [a] (mg N/100 g dry soil)		N_0/N_t [b] (%)		k [c] (per day, 25°C)	
		Mean	Range	Mean	Range	Mean	Range
Andisols	16	17.6	10.9–30.6	3.5	1.7– 7.2	0.0043	0.0015–0.0063
Nonandic soils	8	10.0	14.7–15.0	8.2	4.7–16.0	0.0059	0.0038–0.0081

[a] Nitrogen mineralization potential.
[b] Total organic nitrogen.
[c] Rate constant.

compared the N mineralization potential (N_0) to the pool of total organic N (N_t) between Andisols and nonandic soils from northeastern Japan which have experienced various soil management and cropping practices. As presented in Table 8.1, the percentage of mineralizable N (N_0/N_t values) is considerably smaller in Andisols (mean: 3.5 percent) compared to nonandic soils (mean: 8.2 percent). This fact strongly suggests that organic N in Andisols is fairly resistant to microbial decomposition. Although various mechanisms have been proposed to explain the stabilization of organic N (Stevenson, 1982), it is most likely that the reaction of proteinaceous constituents with humified organic matter and the formation of Al-organic matter complexes are especially important mechanisms for the stabilization of organic N in Andisols. In addition, well-developed stable microstructures are commonly observed in both the surface and subsurface horizons of Andisols. The micropores in these microstructures appear to be sterically unfavorable for enzymatic reactions with organic N compounds.

(2) Assessment of soil N availability

The uptake and recovery of fertilizer N by plants is also dependent on the mineralization rate of soil organic N. Thus, development of accurate methods for assessing the N supplying capacities of soils (or the natural N supply) is of great importance for N fertilization management programs.

There are a variety of methods used to estimate soil N availability such as vegetative, microbial, total soil analysis, chemical extraction, and indirect methods (Meisinger, 1984; Stanford, 1982). Of these procedures, microbial or incubation methods are most commonly employed in Andisols. Aerobic incubation is usually adopted for upland soils and anaerobic incubation for wetland rice soils.

Incubation methods are employed to determine the amount of N which is mineralized from soil organic N under a given set of conditions. A kinetic approach has been introduced by Stanford and Smith (1972) and Sugihara et al. (1986) to mathematically depict the mineralization process. Saito (1990) compared the mineralization process of organic N in Andisols and nonandic soils, using a first-order reaction as follows:

$$N = N_o \left[1 - \exp\left(-kt\right)\right]$$

where N is the amount of organic N mineralized by aerobic incubation, N_o is the N mineralization potential introduced by Stanford and Smith (1972), k is the rate constant and t is incubation time in days.

Using the above equation Saito (1990) showed that andic and nonandic soils have distinctly different values for the N mineralization parameters (Table 8.1). The N_o values were considerably greater for Andisols (mean: 17.6) compared to nonandic soils (mean: 10.0). On the other hand, the k values are smaller in Andisols (mean: 0.0043) than in nonandic soils (mean: 0.0059). These data indicate that Andisols have a greater N mineralization potential (due to a greater pool of total N); however, the rate constant is slower than for nonandic soils.

The N mineralization process in paddy soils assessed by anaerobic incubation indicates release of N from two pools of organic N as expressed below:

$$N = N_{o,1} \left[1 - \exp\left(-k_1 t\right)\right] + N_{o,2}\left[1 - \exp\left(-k_2 t\right)\right] + C$$

where N is the amount of organic N mineralized by anaerobic incubation, $N_{o,1}$ is the N mineralization potential for an easily decomposable component (EDC), with rate constant k_1, $N_{o,2}$ is the N mineralization potential for a slowly decomposable component (SDC) with rate constant k_2, C is an empirical constant, and t is incubation time in days. The EDC is present in significant amounts in paddy soils from northeastern Japan which have been cultivated for wetland rice for many years. However, the EDC amount in Andisols is smaller than the SDC (Ando and Shoji, 1986). The N mineralization potential of the EDC ($N_{o,1}$) is significantly influenced by soil drying and is closely related to the N supply for plants during their early growth stage. In contrast, SDC ($N_{o,2}$) is basically unaffected by soil drying and this fraction is the primary source of N during the middle and late growth stages.

(3) Inorganic N in Andisols

Inorganic forms of N commonly occur in the subsurface horizons of Andisols used for upland farming in northeastern Japan (Kitamura et al., 1986; Kitamura, 1988). They are dominated by nitrate-N which is the most important form of N available to plants under upland conditions. Inorganic N in the subsurface horizons is commonly taken up by summer crops in the middle and late growth stages.

The inorganic pool of N can significantly contribute, not only to the yield of upland crops, but also to pollution of surface water and ground water by leaching. Excess uptake of N especially in the later growth stages causes deterioration in the quality of harvest. This detrimental effect has been observed in wheat, water melon, sweet potato, tobacco, and other crops grown on Andisols in Japan. To avoid this problem, the level of inorganic N in the soil needs to be determined using conventional laboratory techniques. The data obtained will serve as the basis for deciding the amount of fertilizer N required for optimum yield and harvest quality along with the timing of N topdressing.

(4) Uptake of N by agricultural plants in Andisols

The uptake of fertilizer and soil N by plants can be determined using tagged N fertilizers. The data obtained for several agricultural crops such as corn, sorghum, winter wheat, sugar beets and tobacco grown on Andisols in northeastern Japan have shown that the amount of soil-derived N absorbed by these crops range from 62 to 101 kg N/ha (Saigusa et al., 1983; Nishimune, 1984; Kitamura et al., 1986; Shoji et al., 1986, 1991). A comparison of soil-derived N with the amount of fertilizer N in the plants indicates that the supply of soil-derived N or the natural supply of N is very important even for plants grown under heavy N fertilization.

The uptake process of soil N and readily soluble basal fertilizer N by agricultural plants shows contrasting patterns (Saigusa et al., 1983; Nishimune, 1984; Kitamura et al, 1986; Shoji et al., 1986, 1991). For example, soil N was taken up mainly at the middle and late growth stages while ammonium sulfate N was absorbed at the early and middle growth stages as shown in Fig. 8.1. This was observed in corn (*Zea mays* L. Pioneer brand 3747) grown on an Alic Melanudand at Tohoku University Farm which was basally fertilized with ammonium sulfate (AS) and polyolefin coated urea-70 (POCU-70) (Shoji et al., 1991). The absorption of large amounts of soil-N at the middle and late growth stages when temperature is normally high indicates that mineralization of soil organic N is enhanced by the higher soil temperature.

POCU-70 is a controlled release fertilizer whose release is temperature-dependent and can release 80 percent of the total N in 70 days at 25°C (Fujita et al., 1983; Gandeza et al., 1991). The release of N from POCU-70 is very

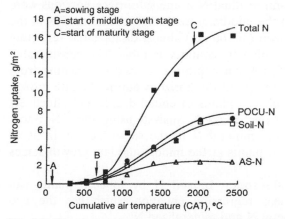

Fig. 8.1. Nitrogen uptake by corn from POCU-Urea (40-0-0)-70, AS (ammonium sulfate), and soil sources in an Alic Melanudand from Tohoku University Farm, northeastern Japan (prepared from Shoji et al., 1991). CAT: sum of mean daily air temperature after sowing.

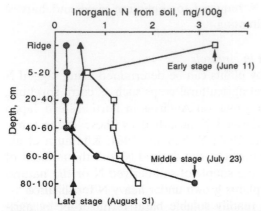

Fig. 8.2. Occurrence of mineral N(NH_4 + NO_3) in the profile of a medial Thaptic Hapludand at Morioka Tobacco Farm, northeastern Japan (prepared from Kitamura et al., 1986).

sensitive to temperature as determined by soil temperature studies in field trials. The uptake of POCU-N by corn had almost the same pattern as that of soil-N. Thus, controlled release fertilizers such as POCU are very useful for farming in Andisols under humid conditions. In contrast, readily soluble N fertilizer is subject to leaching in soils with high permeability making it a less effective fertilizer under these conditions.

Residual mineral N in the subsurface horizons of Andisols was also an important N source available at the middle and late growth stages of upland plants. This is illustrated in Fig. 8.2, showing the distribution of mineral N in the profile of a medial Thaptic Hapludand at Morioka Tobacco Farm, northeastern Japan (Kitamura et al., 1986). At the early growth stage, both mineral N derived from organic N in the Ap horizon and residual N in the subsurface horizons were present in considerable amounts. At the middle growth stage, the mineral N in the Ap horizon and the subsurface horizon (10–60 cm) almost disappeared reflecting N uptake by tobacco plants and leaching by continuous rains. The considerable increase in the residual N in the 80–100 cm depth was caused by leaching from the rain. Mineral N almost disappeared in all the horizons during the late growth stage when the root system of the tobacco plants extended deeply and effectively absorbed mineral N in all the horizons. Plant N analysis using the ^{15}N tracer method also showed that the residual mineral N in the subsurface horizons of Andisols is an important N source for plants at the middle and late growth stages (Kitamura et al., 1986).

Kitamura (1988) showed that absorption of soil N by tobacco plants can be predicted using a multiple linear regression equation which has the two independent variables, residual mineral N and mineralized N:

$$Y = 11.13\,X_1 + 4.86\,X_2 - 13.29 \qquad r^2 = 0.95$$

where Y is the amount of soil N absorbed by tobacco plants (kg ha^{-1}), X_1 is the amount of residual mineral N in the subsurface soil of 20–100 cm depth (kg ha^{-1}), and X_2 is the amount of mineralized N in the surface soil which is obtained by aerobic incubation at 30°C for 4 weeks (kg N ha^{-1}). The quantity X_1 is more than twice that from X_2, indicating a greater contribution of residual N from the subsurface soil.

8.3.2. Phosphorus

Phosphorus is commonly the growth-limiting nutrient element in Andisols with natural ecosystems. Similarily, P is often the growth-determining nutrient element for agricultural crops grown on Andisols because its supply is often very low. Phosphate is strongly sorbed by noncrystalline aluminum and iron materials making it sparingly available for plant uptake.

This section focuses on the chemistry of soil phosphorus and the assessment of available phosphorus in Andisols receiving heavy applications of phosphorus fertilizers. Heavy application of phosphorus has been practiced for a long time in Japan because of the desire for short-term maximization of crop production, especially in soils used for intensive farming. This practice has resulted in the accumulation of a significant amount of phosphorus in the soil (Yoshiike, 1983) because of its low recovery by crops and negligible mobility in the soil profile (cf. Fig. 7.13).

The chemistry and assessment of phosphorus availability have been extensively studied in Japan. The primary objectives of these studies were to maximize crop production with the optimum use of natural phosphorus resources, prevent soil degradation due to excessive application of chemical fertilizers, and to minimize the detrimental effects of chemical fertilizers on the environment. The voluminous knowledge on soil phosphorus in Andisols is highly applicable toward solving many of the P deficiency problems and minimizing the environmental impacts of P fertilization. Innovative fertilization such as *co-situs* application using controlled release fertilizers is a very promising alternative for maximizing yields while at the same time minimizing the deterioration to the environment and the amount of phosphorus fertilizer applied (Shoji and Gandeza, 1992).

(1) Occurrence and solubility of phosphorus in fresh tephras

Fresh tephras contain a certain amount of phosphorus which is readily soluble in acid solutions. For example, fresh tephra samples from Tohoku and Hokkaido, Japan contain total phosphorus ranging from 0.074 to 0.275 percent as P_2O_5, with an average of 0.146 percent. The total phosphorus is higher in felsic tephras compared to mafic tephras (Nanzyo et al., 1992, unpublished). Apatite appears to be the primary phosphorus-bearing mineral in tephras. The moderately rapid dissolution of apatite in acid solutions contributes to rapid revegetation on tephra deposits, even when these deposits are very thick.

Fig. 8.3 shows the relationship between total phosphorus and Truog-soluble phosphorus (available P). The Truog-soluble phosphorus of tephra samples, except those

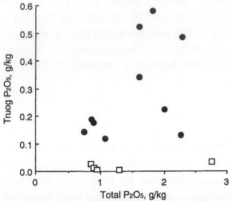

Fig. 8.3. Relationship between total phosphorus and Truog soluble phosphorus in fresh tephras from Tohoku and Hokkaido, Japan. (Nanzyo et al., 1993, unpublished). Circles and squares indicate felsic and andesitic tephras and basaltic tephras, respectively.

with basaltic composition, tends to increase with increasing total phosphorus. About half of the tephra samples show very high dissolution of Truog-phosphorus. In many cases, Truog-phosphorus constitutes greater than 20 percent of the total phosphorus and some tephra samples contain Truog-phosphorus greater than 0.3 g per kg as P_2O_5. Thus, phosphorus is not the limiting nutrient element in revegetation of tephra-covered lands, contrasting with the fact that many crops suffer from severe phosphorus deficiency when they are grown on newly reclaimed Andisols.

Some tephras contain unusually high concentrations of total and available phosphorus. For example, Kawasaki et al. (1991) reported that the Handa pyroclastic flow deposit in Oita Prefecture, southern Japan contains total phosphorus of 1.6 g P_2O_5 kg^{-1} and Truog phosphorus of 0.36 g P_2O_5 kg^{-1} and that the primary phosphorus bearing mineral is apatite.

Walker and Syers (1976) proposed a model to describe the weathering sequence of phosphorus in volcanic ash soils. According to their model, Ca-phosphate in tephras is completely converted to occluded and organic forms in approximately 22,000 years in New Zealand. It is most likely that plant availability of phosphorus rapidly decreases as weathering proceeds.

Plants can absorb soluble phosphorus and weakly bound forms of P. However, most phosphorus reacts rapidly with weathering products such as noncrystalline aluminum and iron materials, resulting in the formation of insoluble metal-phosphorus compounds. According to a kinetic study of chemical weathering in young Andisols with udic soil moisture regimes from northeastern Japan (Shoji et al., 1993), it can be predicted that only a few hundred years are required to form Andisols with acid oxalate extractable Al + 1/2 Fe ≥ 0.4 percent and P retention > 25 percent. Thus, even young Andisols may show phosphorus deficiencies for most agricultural plants. In this case, farming of these Andisols requires heavy applications of phosphate fertilizers to obtain optimum crop yield.

(2) Status of phosphorus in Andisols

Among the most important factors influencing the status of phosphorus in Andisols are biocycling, fertilization, and chemical weathering. The role of these factors is demonstrated by examining total P contents of Japanese Andisols. The characterization data on 25 Andisol pedons from various districts of Japan were employed to study the frequency distribution of total phosphorus (Wada, 1986). The soil samples were divided into three groups (group 1: B, BC, and C horizons; group 2: A, AB, and AC horizons; group 3: Ap horizons) and the distribution and average value of total phosphorus for each group are shown in Fig. 8.4.

Group 1 soils show total phosphorus concentrations ranging from 0.1 to 0.4 percent with an average of 0.19 percent. These values show an increase in both the range and average as compared with those of the parent materials (fresh tephras) described earlier. It appears that these differences are attributable primarily to chemical weathering which results in the loss of silica and bases, and consequently in a relative enrichment of phosphorus.

The relative increase of total phosphorus and the wide range of the frequency distribution are more obvious in group 2 soils as compared with group 1 soils. In addition to chemical weathering, biocycling of phosphorus contributes to the relative increase of phosphorus in this group due to the inclusion of A horizons.

Group 3 soils show the widest range and highest average of total phosphorus (0.39 percent). It appears that fertilization along with chemical weathering and biocycling of phosphorus substantially increase the concentration of phosphorus in the Ap horizons. This is substantiated by the fact that heavy application of phosphorus fertilizer has been practiced in Japan to increase the phosphorus fertility of Andisols. The amount of fertilizer applied was determined according to the phosphorus adsorption coefficient of the soils. As a result, most cultivated Andisols

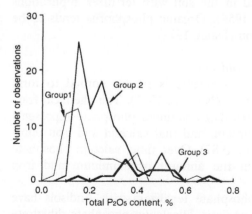

Fig. 8.4. Frequency distribution of total phosphorus content in Andisol samples of 25 pedons from Japan. *Group 1*: B, BC, and C horizon soils (mean = 0.19 percent, n = 48); *Group 2*: A, AB, AC, and buried A horizon soils (mean = 0.22 percent, n = 86); *Group 3*: Ap horizon soils (mean = 0.39 percent, n = 15) (prepared from data of Wada, 1986).

show high available phosphorus levels in their Ap horizons (Yoshiike, 1983).

Continuous heavy application of phosphorus fertilizer increases soil phosphorus to levels that may exceed the plant requirement. For example, Kamewada (1987) showed a quadratic relationship between Truog P and the yield of spinach grown on a humus-rich allophanic Andisol. The yield increased with increasing Truog phosphorus levels and reached a maximum yield at P levels between 1.30 and 2.20 g kg^{-1} as P_2O_5. Yields decreased at higher P levels, indicating the detrimental effect of excess phosphorus on plant yields.

(3) Heavy phosphorus application to newly reclaimed Andisols

Most uncultivated Andisols show very low phosphorus fertility and very low recovery of applied phosphorus fertilizer by agricultural plants. In order to increase the phosphorus supply to crops, heavy phosphorus fertilization was practiced for newly reclaimed Andisols in Japan. The criterion to determine the amount of phosphorus fertilizers to broadcast was a rate equal to 10 percent of the phosphorus adsorption coefficient of the Ap horizon soils (Yamamoto and Miyasato, 1976). This large amount of phosphorus fertilizer was enough to obtain high crop yields for five successive croppings.

Exchangeable Ca and Mg may become partially insoluble by reacting with phosphate in Andisols. In this case, Ca and Mg contained in phosphate fertilizers are useful for supplying these nutrient elements to agricultural plants. Not only low phosphorus supply, but also aluminum toxicity and low base saturation are common limitations for crop production in alic Andisols and tephra-derived Spodosols. Thus, liming prior to phosphate application is recommended for these soils. In ash-derived Haplocryods in Alaska, U.S.A., barley and forage yields were greatly increased by the application of triple superphosphate. The yield was further improved by liming prior to phosphorus fertilization (Michaelson and Ping, 1987).

The forms of phosphorus accumulated in the soil with fertilizer applications are mostly inorganic (Kosaka and Abe, 1958). Organic phosphorus tends to be mineralized and depleted during cultivation (Dalal, 1977).

(4) Reactions of phosphorus fertilizers in Andisols

The phosphorus added to Andisols as fertilizers is readily sorbed to form noncrystalline aluminum phosphate materials (Nanzyo, 1987). The chemical fractionation of inorganic phosphorus indicates that aluminum phosphate represents 42–66 percent of the total phosphorus content, and that calcium and iron phosphates comprise 14–27 percent (Takesako, 1987). Over time, calcium phosphate decreases with a subsequent increase in the amount of aluminum and iron phosphates (Kamewada, 1987).

No crystalline reaction products of phosphate fertilizers with Andisols have been identified in field samples by X-ray analysis. Dicalcium phosphate dihydrate which is easily formed by the incongruent dissolution of superphosphate, is not found in Andisols even following heavy applications of phosphorus fertilizers (Nanzyo, 1987). In contrast, some crystalline phosphate compounds are precipitated under special conditions in the laboratory. For example, the addition of 1 M KH_2PO_4 or $NH_4H_2PO_4$ to allophane followed by incubation at 30°C for a few weeks resulted in formation of K-taranakite and NH_4-taranakite (Wada, 1959). Such phosphate minerals are easily identified by X-ray analysis when present in sufficient concentrations. These minerals were shown to form in soils when phosphate solutions were added to Andisols and the pH was kept at approximately 4 (Nanzyo, 1992, unpublished).

A notable effect of phosphate fertilizers on the chemical properties of Andisols is a change in the variable charge characteristics. For example, phosphate fertilizer applications increased negative charge of Andisol samples by 0.84 mol per mole of P added by fertilization. After a year however, the soil samples lost approximately 40 percent of the increased negative charge (Takesako, 1987).

(5) Assessment of available soil phosphorus

A variety of soil tests used to estimate available soil phosphorus have been extensively studied (Kamprath and Watson, 1980). Among these the Truog method appears to be most commonly employed for Andisols, followed by the Bray method (Ayres and Hagihara, 1952; Adachi, 1985). Mehlich method No. 3 is commonly used for Andisols in Alaska (Michaelson and Ping, 1986) and a method using 2.5% acetic acid solution, for some African Andisols (Mizota, 1987; Kosaki, 1989). The Truog method employs an extracting solution of 10^{-3} M H_2SO_4 containing 3 g of ammonium sulfate per liter (buffered at pH 3), a soil: solution ratio of 1:200, and extraction for 30 min at room temperature (Truog, 1930). Truog-P levels higher than 0.10 g kg^{-1} as P_2O_5 are recommended to farmers in Japan to meet the phosphorus requirements of most crops.

According to this criterion, the amount of superphosphate fertilizer that must be applied to Andisols can be determined according to the relationship between the amount of superphosphate applied and the corresponding Truog-P concentration. For example, the available P level of an uncultivated Hapludand increases with the amount of superphosphate applied. From this relationship it is determined that 1.90 g kg^{-1} of P_2O_5, or 9.5 g kg^{-1} of superphosphate must be applied in order to increase the available phosphorus level to 0.10 g kg^{-1} as P_2O_5 in the soil (Yasuda and Fujii, 1983). Broadcasting of such large amounts of phosphorus fertilizer is resource consuming and not economical. As a result, band placement of phosphorus fertilizers is commonly practiced in order to avoid the rapid chemical reaction of phosphate with active Al and Fe compounds that results in a decrease of phosphorus availability. Shoji and Gandeza (1992) proposed co-situs application of controlled release fertilizers to enhance the recovery of P fertilizer. These new fertilizers do not burn plant buds or roots and show a high recovery by the plants, thus reducing the amount of fertilizer that must be applied.

Recovery of fertilizer phosphorus by agricultural plants in Andisols is commonly less than 20 percent. For example, a fertilizer experiment using seven types of crops, carried out in Hokkaido, northern Japan for 19 years, indicated that the average phosphorus recovery from fertilizer across all crops was 12 percent. The total amount of phosphorus fertilizer applied over the 19 years amounted to 4.2 ton ha^{-1} as P_2O_5 and all the residual phosphorus was retained in the surface horizon (the upper 26 cm) (Yasuda and Saito, 1982).

8.3.3. Potassium

Potassium, along with nitrogen and phosphorus, is one of the major nutrient elements required in large quantities by plants. The mineralogical properties of tephras, chemical weathering regime, vegetation, and cropping practices influence the availability of K in volcanic ash soils. Although K is present in considerable amounts in fresh tephras, the amount in available forms is often insufficient for continuous cropping, especially under humid climates. In contrast with N, the formation of the available forms of K is primarily governed by chemical processes.

(1) Occurrence of potassium in Andisols

Potassium in tephras. Like common volcanic rocks, tephras contain total potassium ranging from 0.5 to 4.0 percent as K_2O (Shoji et al., 1975). Thus, a 10 cm depth of volcanic ash contains a large quantity of potassium ranging from 7.5 to 60 ton per ha as K_2O, assuming a bulk density of 1.5 g cm^{-3}. Total K_2O is greatest in rhyolitic tephras and is least in basaltic tephras. However, it should be noted that the correlation between total K_2O and total SiO_2 is the poorest among all the major and minor elements studied (Shoji et al., 1975; Yamada, 1988). The poor correlation between SiO_2 and K_2O is attributed to differences in the volcanic zones from which the magma source originated (Shoji, 1986; Yamada and Shoji, 1983a) as described in Chapter 5.

The mineral composition of tephras varies widely according to the rock type. However, in all rock types, volcanic glass is the most abundant component, commonly followed by plagioclase and pyroxenes (Yamada and Shoji, 1975; Shoji, 1986). Felsic tephras from the alkali basalt zone generally contain only small amounts of biotite (Yamada and Shoji, 1983a) and alkali feldspar (Machida et al., 1981). Thus, volcanic glass is an overwhelmingly important K-bearing component in all tephras. In general, the total K_2O content of volcanic glasses is considerably greater than that of the original magma from which the glass formed. Potassium concentrations are greatest in volcanic glass from rhyolitic tephras (>4.0 percent) and least in glass from basaltic tephras (<1.0 percent).

Potassium in Andisols. The fate of K in Andisols under humid climates is primarily determined by whether the weathering reactions lead to formation of nonallophanic or allophanic Andisols. Accumulation of K proceeds with the advance of weathering in nonallophanic Andisols (Kurashima et al., 1981). This weathering occurs in strongly acid surface soils derived from noncolored glass-rich tephras and leads to formation of considerable amounts of 2:1 layer silicates (as described in Chapters 3, 5 and 6). Potassium is largely retained by these 2:1 layer silicates and tends to accumulate with the advance of weathering. Weathering to allophanic materials occurs in tephras of all compositions under a variety of soil conditions, except in strongly acid surface soils as mentioned above. Since allophanic clays do not show preferential retention of K, the amount of K tends to decrease with the advance of weathering in humid climates.

Volcanic glass is the most important K-bearing component in tephras as stated earlier. Since the occurrence of biotite and alkali feldspars is limited to a few tephras and the weathering of these minerals is described in detail elsewhere (Bertsch and Thomas, 1985; Sparks and Huang, 1985), only the weathering of volcanic glass is discussed here.

Yamada and Shoji (1983b) found two types of chemical alteration in noncolored glass during the early stages of weathering in tephras. The a type glass alteration shows a large gain in potassium and a large loss of sodium in strongly acid surface soils (pH < 5). Such alteration is explained by ion exchange reaction between Na^+ and H^+ plus K^+ without significant change in the glass structure. It is believed that this process contributes to the accumulation of potassium in the early weathering stages of nonallophanic Andisols.

The b type glass alteration is characterized by large gains in sodium and/or calcium and a significant loss of potassium in the early stages of weathering. Such alteration takes place under a variety of soil conditions, except those specified for the a type alteration. Large losses of both sodium and calcium occur in the middle and late stages of b type glass alteration. Thus, all the alkali and alkaline-earth elements are lost during glass alteration to allophanic clays under humid climates.

The potassium status of Andisols also reflects the soil climate and colloidal components of the soils. The NH_4OAc extractable potassium (exchangeable) of uncultivated Andisols is greater under dry climates than humid climates reflecting the leaching intensity. For example, a Xerand in Idaho contains considerable amounts of exchangeable K ranging between 0.9 and 1.0 $cmol_c$ kg^{-1} (Kimble, 1986). On the other hand, Udands in Japan show mean exchangeable K levels of approximately 0.2 $cmol_c$ kg^{-1} (Wada, 1986). Nonallophanic Andisols contain more exchangeable potassium than allophanic Andisols indicating potassium selectivity by 2 : 1 layer silicates (Andisol TU Database, 1992).

Application of potassium fertilizers influences the content of exchangeable potassium in Andisols. Continuous heavy applications of potassium have remarkably increased the exchangeable potassium levels in cultivated Udands of Japan. The content differs widely and only a small number of Andisols show values less than 0.3 $cmol_c$ kg^{-1} which is the critical level for potassium deficiency in many agricultural plants (Kamata, 1978). Excess accumulation of exchangeable K may induce Mg deficiencies in certain plants.

(2) Assessing plant available potassium in Andisols

According to their mineralogical composition, Andisols are divided into two groups, namely allophanic and nonallophanic Andisols as described earlier. The major colloids contributing to cation retention in these soils are allophanic clays and soil organic matter in most allophanic soils, and chloritized 2 : 1 minerals (hydroxy-Al

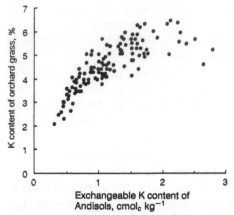

Fig. 8.5. Relationship between exchangeable K in the surface soils of grasslands and K uptake by orchard grass (*Dactylis glomerata*) in northeastern Japan (prepared from Kurashima, 1980).

interlayered 2 : 1 minerals) and soil organic matter in most nonallophanic soils. Since the two groups of Andisols contain few micaceous minerals, the plant available K can be assessed with reasonable accuracy using exchangeable K levels.

The forms of soil K available or partially available to plants consist of soluble, exchangeable, and nonexchangeable K. Although soluble K is readily available to plants, its concentrations are highly variable with changes in soil moisture conditions, especially under humid climates. Therefore, it is not a reliable measure of plant available K throughout the growing season of agricultural plants. In contrast, NH_4OAc-extractable or exchangeable K is considered a more reliable index to assess K availability to plants since it is not susceptible to variations in soil conditions. Numerous studies have shown that exchangeable K levels correlate well with plant uptake of K.

As presented in Fig. 8.5, Kurashima (1980) indicated that there is a close relationship between exchangeable K levels in the surface soils of grasslands and K uptake by orchard grass (*Dactylis glomerata*) in northeastern Japan. In this case, potassium uptake by the plant increased up to exchangeable K of approximately 2 $cmol_c$ kg^{-1}. In some instances, K concentrations in the plants reached 6 percent; a level considered luxury consumption. Potassium deficiency occurs in orchard grass when exchangeable K levels in the surface soil fall below approximately 0.3 $cmol_c$ kg^{-1}, assuming that other elements especially Ca and Mg are not limiting.

It was also reported by Kamata (1978) that common upland crops show potassium deficiency at almost this same level of exchangeable K values as observed for orchard grass. It is well known that the percent K saturation influences not only Mg and Ca uptake by plants, but also the quality of harvested crop. Thus, it is recommended that the proportions of exchangeable Ca, Mg and K in the soil should be 50, 20 and 10 percent (molar basis of charge), respectively, in order to have a balanced supply of these elements for most crops (Kamata, 1978).

8.3.4. Micronutrients

Although deficiencies of various micronutrients in Andisols have been reported, the most prevalent among them are Cu, Zn, and Co. The abundance and availability of these micronutrients in Andisols are dependent on the abundance of the elements in the parent tephras and on their release rates by chemical weathering. Since the capacity of the soil to provide these micronutrients is a chemical property, chemical soil tests are most suitable for estimating their availability. The major categories of micronutrient extractants presently in use are dilute acids and/or solutions containing chelating agents (Sims and Johnson, 1991).

(1) Copper

Copper is an essential element for both plants and animals. It commonly occurs at levels of 5 to 30 ppm in plants. Copper deficiency in plants grown on Andisols was first observed in wheat grown on an Alic Melanudand in Isawa, Iwate prefecture, northeastern Japan in 1957 (Kurosawa et al., 1965). The existence of healthy ripened wheat only under a copper telephone cable in an unripened wheat field led to this finding.

Occurrences of Cu in tephras and Andisols. The concentration of Cu in tephras is closely related to the rock type of the tephras as shown in Fig. 8.6. Kobayashi and Shoji (1976) showed that Cu concentrations vary from 3 to 141 ppm and decrease with increasing total silica contents in tephras from Japan. Concentrations are less than 5 ppm in almost all rhyolitic and dacitic tephras, suggesting that Cu deficiency can readily occur in some plants grown on Andisols derived from these tephras (Lucas and Knezik, 1972). In contrast, basaltic tephras have the highest Cu concentrations and Andisols formed from such tephras typically have sufficient Cu levels for the growth of most plants (Kobayashi and Shoji, 1976; Saigusa et al., 1976).

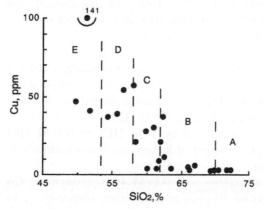

Fig. 8.6. Relationship between rock types and Cu concentration of tephras (prepared from Kobayashi and Shoji, 1976). A = rhyolite; B = dacite; C = andesite; D = basaltic andesite; E = basalt.

Copper concentrations vary widely between the different primary minerals in rhyolitic, dacitic, and andesitic tephras (Kobayashi and Shoji, 1976). The average Cu concentration is 9 ppm in volcanic glass, 6 ppm in plagioclase, 17 ppm in pyroxenes and 25 ppm in titanomagnetite. Copper concentrations in glass of rhyolitic and dacitic tephras are very low ranging from 2 to 4 ppm. Submicroscopic grains of Cu sulfide may occur in some tephras, but their abundance is very low. Although the Cu content of volcanic glass is less than those of heavy minerals, the high proportion of volcanic glass in all tephras results in volcanic glass being the most important mineral source of Cu in Andisols.

Total Cu in Andisols, especially young ones, is substantially determined by the abundance and weathering rate of Cu in the parent tephras. However, biocycling of Cu results in the accumulation of Cu in the humus horizons of Andisols. This reflects the formation of Cu-organic matter complexes which are very stable (Stevenson and Ardakani, 1972) and result in low mobility of Cu in the soil profile. For Towada Andisols formed from rhyolitic and dacitic tephras (Kobayashi, 1979), the Cu content of humus horizons shows a marked increase compared to that of nonhumus horizons and the Cu accumulation is obvious even for buried A horizons (Fig. 8.7).

The total Cu content of humus horizons shows a close correlation with the organic carbon content of these horizons. As reported by Masui et al. (1972), the relation is as follows:

$$\text{total Cu (ppm)} = 8.5 \,(\text{organic C}\%) - 9.0 \qquad r = 0.84^{**}$$

This equation shows that the total Cu content increases with increasing organic matter and the slope is determined by the formation constant for Cu-humus complexes.

The amount of plant available Cu in Andisols may be very limited due to the chelating effect of organic matter. Humified organic matter in humus rich Andisols causes not only Cu deficiency in plants but also attenuates the effect of Cu fertilization as reported by Tsutsumi et al. (1967, 1968). In this context, the critical level of total Cu in the soil at which Cu deficiency in plants can occur is considerably higher in humus rich Andisols than in other soils.

Plant available Cu in Andisols. Although various soil tests for estimating available Cu levels have been proposed (Sims and Johnson, 1991), the methods using dilute mineral acids and chelating agents have been recommended for Udands (Kurosawa, 1970; Kurosawa et al., 1965; Minami, 1970; Chiba et al., 1984). Kurosawa et al. (1965) showed that Cu extracted by 1 M HCl or 0.1 M HCl extraction is closely correlated with the Cu content of leaves and stems from wheat and barley in Iwate, northeastern Japan.

The critical levels of 1 M HCl extractable and 0.1 M HCl extractable Cu for plant deficiencies were observed to be 10 ppm and 0.2 ppm, respectively. In contrast, the critical concentration of Cu in wheat and barley tissue was observed to be 5 ppm in the leaves and stems at harvest time (Chiba et al., 1984). The high

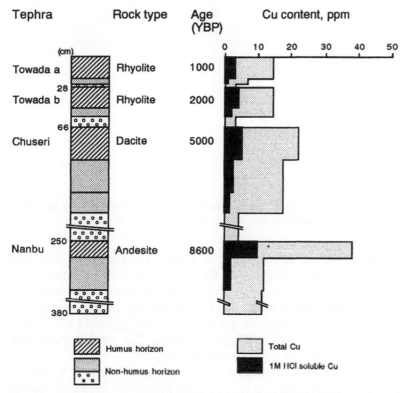

Fig. 8.7. Distribution of total and HCl soluble Cu in a Towada Andisol (prepared from Kobayashi, 1979).

critical values of both total and 1 M HCl extractable Cu suggest that adsorption of Cu by clay minerals and formation of Cu-organic matter complexes contribute to the low concentrations of this element in soil solutions. The role of Cu-organic matter complexes is illustrated by comparing concentrations of HCl extractable Cu with total Cu levels in the humus horizons of the Towada Andisol (Fig. 8.7). Copper deficiencies in plants are commonly corrected by broadcast application or foliar application of $CuSO_4 \cdot 5H_2O$ or Cu chelates.

(2) Zinc

Zinc is an essential element at low concentrations for both plants and animals. It normally occurs at levels of 20 to 100 ppm in plants and its deficiency has been recognized in a variety of agricultural plants grown on Andisols. Although Zn is a heavy metal, its occurrence in tephras and Andisols is significantly different from that of Cu.

Occurrences of Zn in tephras and Andisols. Zinc concentrations in tephras from Japan range from 60 to 100 ppm. Kobayashi and Shoji (1976) observed that

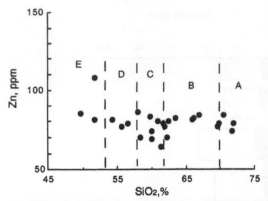

Fig. 8.8. Relationship between rock types and Zn concentration of tephras (prepared from Kobayashi, 1979). A = rhyolite; B = dacite; C = andesite; D = basaltic andesite; E = basalt.

the total Zn concentrations of tephras do not closely correlate with the total silica content of these tephras (Fig. 8.8). However, Zn does demonstrate a close relationship with Fe^{2+} and total Fe contents of the primary minerals. The concentration of Zn in the primary minerals of these tephras varies considerably, ranging from 36 to 78 ppm (average, 56 ppm) in volcanic glass, 14 to 31 ppm (average, 20 ppm) in plagioclase, 147 to 371 ppm (average, 221 ppm) in pyroxenes, and 356 to 1089 ppm (average, 668 ppm) in titanomagnetite.

According to the proportion of primary minerals in andesitic tephras, Kobayashi and Shoji (1976) found the distribution of Zn in the primary minerals as follows: pyroxenes > volcanic glass > titanomagnetite > plagioclase. However, volcanic glass comprises greater than 90 percent of the bulk weight of rhyolitic and basaltic tephras, so that the glass component contains the greatest proportion of the total Zn in these tephras. Since glass is very susceptible to weathering, it is the most important source of Zn in Andisols.

The accumulation of Zn in modern humus horizons of Andisols in northeastern Japan is remarkable and a considerable fraction of this element occurs in an available form as shown in Fig. 8.9 (Kobayashi, 1979). In contrast, such accumulations are not observed in the buried humus horizons of the Towada Andisol. This fact indicates that Zn-organic matter complexes in Andisols are not as stable as the Cu-organic matter complexes described earlier. This also shows that Zn in the surface horizons is enriched by biocycling, but is largely leached after burial in humid climates. Thus, there is no relationship between the total Zn content of humus horizons including buried ones and their organic carbon concentrations as was observed for Cu (Masui et al., 1972).

Zinc deficiencies in plants in northeastern Japan occur in truncated Andisols, coarse textured Andisols, and humus rich Andisols which are low in acid extractable Zn (Chiba et al., 1984; Kobayashi et al., 1964).

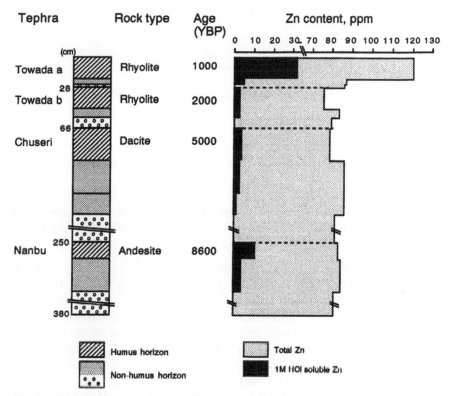

Fig. 8.9. Distribution of total and HCl soluble Zn in a Towada Andisol (prepared from Kobayashi, 1979).

Plant available Zn in Andisols. Although a variety of soil tests for Zn have been evaluated (Sims and Johnson, 1991), soil tests using dilute mineral acids and chelating agents have been recommended for Udands (Minami, 1970; Sekizawa et al., 1973; Chiba et al., 1984). Zinc deficiencies have been observed in a variety of upland crops and vegetables grown on Andisols in Japan. The critical level of 0.1 M HCl extractable Zn in soils is 0.1 ppm. Deficiencies were observed in upland rice when total Zn content of leaf and stem tissues were less than 20 ppm (Chiba et al., 1984). Zinc deficiency is commonly corrected by basal application of zinc sulfate.

(3) Cobalt

Cobalt is an essential micronutrient for animals and is especially important for ruminants. Deficiencies of Co in plants growing on Andisols were first found in New Zealand in about 1900 where it was responsible for "bush sickness" of sheep (Lee, 1974). Cobalt is also essential for N_2 fixation by free-living bacteria, blue green algae and symbiotic relationships.

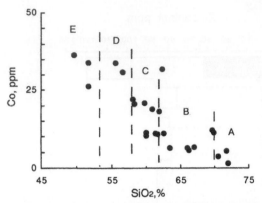

Fig. 8.10. Relationship between rock types and Co concentration of tephras from Japan (prepared from data of Shoji et al., 1980a). A = rhyolite; B = dacite; C = andesite; D = basaltic andesite; E = basalt.

Occurrences of Co in tephras and Andisols. Similar to Cu, Co in tephras shows a close correlation with rock types or total silica content as presented in Fig. 8.10 (Shoji et al., 1980a). Cobalt concentrations are less than 10 ppm in silica rich rhyolitic tephras and increase to more than 25 ppm in silica-poor basaltic tephras. Cobalt concentrations in tephras are closely related with those of Fe and Mg, indicating that Co largely occurs in heavy minerals (Shoji et al., 1980a). Sand mineralogy and chemistry of Andisols also demonstrate the accumulation of Co in the mafic mineral fraction (Kobayashi, 1981).

Kobayashi (1981) reported that total Co concentrations in surface horizons of Andisols used for pastures in Japan range from 3.0 to 42.3 ppm with an average of 16.2 ppm (n = 114). Kobayashi (1982a) also showed that Co released from primary minerals is coprecipitated with Mn oxides as Taylor and McKenzie (1966) reported. The coprecipitation of Co in other minerals reduces the plant availability of Co.

Plant available Co in Andisols. According to the occurrence of Co in Andisols described above, Kobayashi (1982a) proposed a new method using a 4% tannic acid extraction to determine plant available Co. This method provided better estimates of plant response than the 2.5% acetic acid method used by earlier workers. He further showed that the critical levels of tannic acid soluble Co in Andisols and total foliar Co in pasture species are 1 ppm and 0.05 ppm, respectively (Kobayashi, 1982b).

(4) Other micronutrient deficiencies

In addition to that of Cu, Zn, and Co, deficiencies of B, Mn, and S have been observed in Andisols. Boron concentrations in volcanic rocks, especially in basaltic rocks are very low as compared to sedimentary rocks. Thus, B deficiency in a variety of upland crops and vegetables occurs in strongly acid Andisols (pH < 5) subjected to intense leaching. Boron deficiencies were observed in Andisols of Japan when hot-water soluble B concentrations were less than 0.2 to 0.5 ppm (Chiba et al., 1984; Yoshida et al., 1966). One ppm of hot-water soluble B is essential for normal growth of a high B requiring tropical forage legume in Hawaii, U.S.A. (Hue et al., 1988). The concentrations of Mn are high in basaltic tephras and Mn deficiency occurs only in rhyolitic to andesitic tephras with coarse textures. Manganese deficiency in Andisols was first observed for elephant foot (*Amorphophallus Konjae* K. Koch) grown on pumice-derived Andisols in Gunma Prefecture, Japan (Kobayashi et al., 1964). The deficiency level of this element in Andisols is 5 ppm as MnO in exchangeable form and 30 ppm as MnO in easily reducible form (Chiba et al., 1984). Available S is very low in coarse textured Andisols which have low organic matter contents (Tsuji et al., 1980). Sulphur deficiency in pastures of young pumice soils in New Zealand was reported by Toxopeus (1970).

8.3.5. Soil acidity

It was recognized early that there are few Andisols which have more than traces of KCl-extractable Al and pH values less than 5.2 even though their base saturation is well below 10 percent (Smith, 1978). This concept was altered by the discovery of nonallophanic Andisols in northeastern Japan which have a clay fraction dominated by chloritized 2:1 minerals, very strong to strong acidity, and the presence of large amounts of KCl-extractable Al (Shoji and Ono, 1978). In contrast with allophanic Andisols, nonallophanic Andisols may cause serious acid injury and/or aluminum toxicity to plant roots (Shoji et al., 1980b; Saigusa et al., 1980). The wide distribution of nonallophanic Andisols and the necessity for soil management practices to avoid aluminum toxicity have led to the creation of the "alic" subgroup in Andisol classification (Leamy, 1988). The alic subgroup emphasizes the potential for aluminum toxicity and is characterized by having a soil layer with greater than 2 $cmol_c$ kg^{-1} of KCl-extractable Al within the rooting zone.

(1) Acidity of uncultivated Andisols

Acidic Andisols occur extensively in humid areas where leaching of bases from the soil is intense. Their acidity is basically determined by the mineralogy of the colloidal fraction and the degree of base saturation. According to the clay mineralogy, Andisols are divided into allophanic and nonallophanic groups whose acidity chemistry was described in Chapter 6. Allophanic Andisols in northeastern Japan show slight to medium acidity (pH 5.2–6.0), even though they have very low base saturation (<10%). Allophanic Andisols with low organic C content have somewhat higher pH values than those with high organic C content. Thus, it appears that soil organic matter contributes to acidity of allophanic Andisols to some extent. Furthermore, few allophanic Andisols have toxic levels of soluble Al and KCl-exchangeable Al.

Nonallophanic Andisols have base saturation less than 20 percent and pH values ranging from 4.8 to 5.3. This indicates that chloritized 2:1 minerals have a strong buffering capacity due to the formation of nonexchangeable polymeric hydroxy-Al in the interlayers of expansible 2:1 layer silicates. Organic matter is also present in various amounts in nonallophanic soils but its acid sites are largely blocked by the formation of Al organic matter complexes. Thus, the contribution of soil organic matter to the acidity of nonallophanic Andisols is unclear.

(2) Soil acidity and plant growth

Although various factors affecting plant growth in acid soils have been described, Al toxicity as indicated by high levels of neutral salt extractable Al is considered to be extremely important in acidic Andisols (Thomas and Hargrove, 1985). Shoji et al. (1980b) and Saigusa et al. (1980) showed that Al toxicity does not occur in allophanic Andisols even though the base saturation of these soils is extremely low. In contrast, Al toxicity was apparent in nonallophanic Andisols if the pH was lower than a critical value of about 5.0, or even higher in some cases. These workers observed that KCl-extractable Al was the principal source of Al contributing to acid injury of plant roots and that its contribution was closely related to root growth as shown in Fig. 8.11.

A KCl-extractable Al value greater than 2 $cmol_c$ kg^{-1} was shown to be critical to the growth of most agricultural plants susceptible to aluminum toxicity (Saigusa et al., 1980; Saigusa, 1991). For instance, edible burdock (*Arctium lappa*) shows sensitivity to aluminum toxicity at a critical level near 2 $cmol_c$ kg^{-1} (Fig. 8.11). In most crops the symptoms of aluminum toxicity are severe and the root length is remarkably reduced with increasing amounts of KCl-extractable Al. Thus, a value of 2 $cmol_c$ kg^{-1} has been employed as the chemical criterion for classifying alic subgroups of Aquands, Cryands and Udands (Soil Survey Staff, 1992).

Aluminum toxicity was further studied by determining the relationship between acid injury of plant roots and the soluble and KCl-extractable concentrations of Al. As presented in Fig. 8.12, both allophanic and nonallophanic Andisols contain only trace amounts of soluble Al at pH values greater 5.0, but serious acid injury was

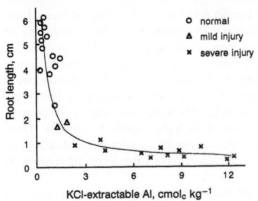

Fig. 8.11. Relationship between KCl-extractable Al contents of Andisols and root growth of Burdock (prepared from data of Saigusa et al., 1980; copyright Williams & Wilkins, with permission).

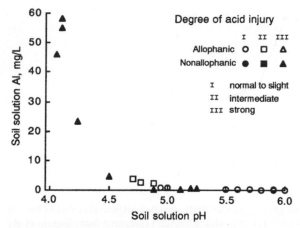

Fig. 8.12. Relationship between soil solution pH and soil solution aluminum in allophanic and nonallophanic Andisols (prepared from data of Nishiya, 1987).

clearly observed only in nonallophanic Andisols. Since KCl-extractable Al is present in considerable amounts in nonallophanic soils, it is suspected to be the main source of aluminum causing aluminum toxicity. The concentration of Al in the soil solution increases rapidly with decreasing pH values below 5.0, so that both KCl-extractable Al and soluble Al are contributing factors to acid injury of plant roots in the pH range 5.0–4.5. The concentration of soluble Al at pH values below 4.5 is high enough to be the single factor causing aluminum toxicity for most plants.

Allophanic Andisols rarely have pH values less than 5.0 and thus, few of these soils experience aluminum toxicity problems. However, some plants on these soils have been observed to have retarded growth and development, suggesting that soil factors other than toxic levels of aluminum are limiting crop performance. Saigusa et al. (1990a) recognized Ca deficiency in corn (*Zea mays*, cv. Choko 1) grown on Andisols which contain NH_4OAc-extractable Ca^{2+} less than 0.2 $cmol_c$ kg^{-1} dry soil. A considerable number of subsoil horizons in Andisols under humid climates have very small amounts of NH_4OAc extractable Ca^{2+}. Since Ca translocation from the roots in Ap horizons to those in the subsoil horizons is very limited (Marschner and Richter, 1974), plant roots in subsoil horizons are subject to Ca deficiency. This results in the prevention of root proliferation and the development of shallow rooting systems. From the foregoing, it is concluded that not only the high level of KCl-extractable Al in nonallophanic Andisols but also the very low levels of NH_4OAc-extractable Ca^{2+} in allophanic Andisols may severely affect the growth and yield of agricultural plants in these soils.

(3) Agronomy of acid Andisols

One of the most effective ways of overcoming the factors limiting plant growth in acidic Andisols, such as high levels of toxic aluminum and low amounts of bases, is application of lime. This method is rather easy to perform if the objective

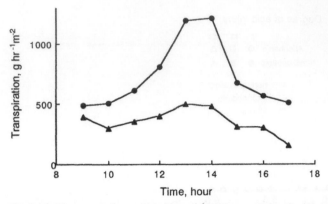

Fig. 8.13. Changes in the transpiration of sorghum grown on Andisols with weakly acidic subsoil (Towada soil, closed circle) and strongly acidic subsoil (Yakeishi, closed triangle) (prepared from Saigusa et al., 1989).

is only to apply lime to the surface soil. In the case of acidic subsoils, which substantially influence plant growth and yield because of their storage of large amounts of water and various amounts of plant nutrients, liming is very difficult to implement. An unlimed subsoil poses a limitation to crop performance because it restricts root elongation resulting in the development of a shallow rooting system. This is one of the most important factors contributing to the infertility of acidic Andisols.

The effect of subsoil acidity is clearly observed in the uptake of water and nitrogen by upland plants (Saigusa et al., 1983; Shoji et al., 1986). Figure 8.13 shows the transpiration of water by sorghum (*Sorghum vulgare L.*) grown on two Andisols during a dry summer. The plants on the Andisol having an allophanic subsoil showed normal growth and absorbed more than twice as much water at midday. In contrast, reduced transpiration of water and mid-day wilting were observed for the plants grown on the Andisol with the acidic nonallophanic subsoil. Since a large quantity of available water was also present in the nonallophanic subsoil, the prevention of root elongation into the subsoil is the major factor contributing to the plant water deficit.

Fertilizer and soil-mineralized N are leached out of the shallow rooting zone in nonallophanic soils by rain water and irrigation. This N could be utilized by plants if they were to develop a deeper rooting system. Therefore, prevention of a deep rooting system by aluminum toxicity contributes to nitrate pollution of surface water and ground water.

Table 8.2 shows the N uptake and yield of barley (*Hordeum vulgare* CV. Norin-24) grown on two Andisols during a rainy summer (Saigusa et al., 1983). Since barley is very sensitive to aluminum toxicity, the rooting zone of this crop is completely restricted to the Ap horizon in the Andisol having the acidic nonallophanic subsoil. In contrast, the rooting system extends deeply (>100 cm) into the subsoil

TABLE 8.2

Nitrogen uptake by barley (*Hordeum vulgare* CV. Norin-24) grown on Andisols having either nonallophanic or allophanic subsoils (from Saigusa et al., 1983)

Plot (Subsoil)	N uptake from different N sources (g/m^2)				N recovery (%)	
	Total	Basal	Topdress	Soil	Basal	Topdress
1. Nonallophanic	6.39	0.65	3.02	2.72	6.5	75.4
2. Allophanic	11.93	4.82	3.25	3.86	48.2	81.3

horizons of the Andisol with allophanic clay mineralogy. Differences in N uptake, especially basal fertilizer and soil N, by plants between the two experimental plots clearly reflect the differences in root elongation into the subsoil. Uptake of basal fertilizer and soil N is much larger in the allophanic subsoil plot than in the nonallophanic subsoil plot. Thus, the yield of barley is significantly greater in the former than in the latter plot. Topdressing of N by surface application was carried out during the middle growth stage after the plants had developed an extensive rooting system in the Ap horizons of both experimental plots. As a result, recovery of topdressed N is high and shows no significant difference between the two experimental plots. In contrast, notable differences were observed in the recovery of basal N (7 versus 48%). Thus, the limited uptake of basal fertilizer and soil N due to restricted root development in the subsoil can be largely avoided by topdressing of N.

As already described, acidic Andisols have two factors affecting plant growth, namely, aluminum toxicity and Ca deficiency. Neutralizing soil acidity with lime to pH values greater than 5.5 can entirely alleviate toxic Al in nonallophanic Andisols. In contrast, soil pH measurements are not a reliable criterion for Ca-deficiencies in allophanic soils because some soils may have pH values greater than 6.0 when the base saturation is very low (<10%). Thus measuring the exchangeable Ca^{2+} content is necessary for determining the application rates of amendments such as calcium sulfate and calcium carbonate which can increase available Ca not only in the surface soil but also in the subsoil (Saigusa et al., 1990b).

8.4. PHYSICAL PROPERTIES RELATING TO SOIL PRODUCTIVITY

Most Andisols have excellent physical properties such as high water holding capacity, favorable tilth, and strong resistance to water erosion. These properties significantly contribute to maintaining the high productivity of Andisols. Development of aggregates in Andisols is closely related to the retention of large amounts of plant available water. Low bulk density and friable consistence contribute to favorable soil tilth leading to easy tillage, seedling emergence and root development. Stable aggregates and high water permeability of Andisols greatly contribute to minimizing water erosion.

8.4.1. Plant available water

(1) Status of water in Andisols

Most Andisols have large amounts of water held at varying matric potentials, indicating that water is retained in pores with varying sizes. As presented in Chapter 7, moderately weathered Andisols have a total porosity of about 78 percent; macropores (13 percent), mesopores (33 percent) and micropores (32 percent). Water retained in the mesopores and micropores corresponds to plant available and adsorbed water, respectively. The macropores serve a major role in soil aeration and water infiltration.

Andisols containing a large amount of noncrystalline materials have considerable amounts of water held at low matric potentials and show hysteresis of water retention when they are dried to matric potentials lower than 1500 kPa (Kubota, 1976). Decreasing the adsorbed water by drying the soil may actually increase plant available water.

(2) Occurrence and assessment of plant available water

Plant available water is generally defined as the amount of water held between field capacity and permanent wilting point. It is commonly estimated as the difference in water content between soil matric potentials of 33 and 1500 kPa. Saigusa et al. (1987) confirmed that a matric potential of 1500 kPa is an acceptable value for the permanent wilting point of many plants grown on Andisols. In Japan, a matric potential of about 10 kPa has been employed as field capacity because the natural moisture content of most Andisols in Japan is equivalent to this matric potential. The field capacity and permanent wilting point are usually determined by the pressure plate and pressure membrane methods, respectively. The latter can be replaced by a centrifugation method to obtain approximate data in a short time period (Saigusa et al., 1990b).

In general, Andisols have a large capacity to provide plant available water, ranging from 83 to 490 kg m^{-3} (270 kg m^{-3} on average) (Ito et al., 1991; Saigusa et al., 1991). Saigusa et al. (1987) showed that there is a close relationship between field capacity (33 kPa) and plant available water as presented in Fig. 7.9. According to the regression equation, the ratio of available water to field capacity is about 0.60 in the surface horizons and 0.5 in the subsurface horizons (Saigusa et al., 1987).

The large water supplying capacity of Andisols has been verified by field experiments. Shiga and Nakajima (1985) compared plant available water between a rain sheltered plot and a control plot of a medial mesic Thaptic Hapludand planted to tobacco in Morioka, northeastern Japan. The automatic shelter was installed to eliminate additions of rainfall to the tobacco plants. Figure 8.14 shows the changes in the vertical distribution of plant available water content as the growing season progressed. Since the rainfall amounted to 560 mm during the growing season (May to August), the control plot maintained a high plant available water content. In contrast, the rain-sheltered plot showed a gradual decrease in plant available water content throughout the growing season and only a small amount of plant available

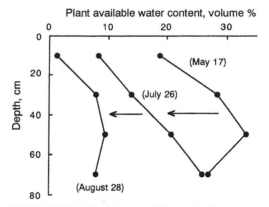

Fig. 8.14. Changes in the vertical distribution of available water content in a tobacco-planted medial mesic Thaptic Hapludand under rain shelter. This soil (to the depth of 80 cm) supplied 157 mm of water to tobacco plants during the growing period from transplanting to harvest stage without any apparent water deficiency (prepared from data of Shiga and Nakajima, 1985).

water remained by late August at harvest time (about 25% of the original available water). However, the plants grown on the rain-sheltered plot showed normal growth without any apparent water deficiency. They developed an extensive rooting system into the deeper soil horizons (below 1 m) and absorbed not only more water, but also more nitrogen as compared to the plants of the control plot (Shiga and Nakajima, 1985). It was noted that both the yield and quality of tobacco leaves were higher in the rain-sheltered plot than in the control plot (yield: 3.1 vs. 2.4 ton per ha). These results clearly show that Andisols have a large water supplying capacity and that the subsurface horizons provide a great portion of that capacity.

8.4.2. Soil tilth

Tilth is the physical condition of a soil related to its ease of tillage, fitness as a seed-bed, and its suitability for seedling emergence and root elongation. Most Andisols manifest excellent tilth. Although the ease of tillage is determined by a number of factors, bulk density and moist consistence are especially important factors affecting the ease of tillage operations in Andisols. Cultivation of Andisols with low bulk density and friable consistence requires less energy and can easily produce a favorable seed- and root-bed.

Compaction or an increase in bulk density may take place as a result of repeated traffic or pressure by heavy agricultural machinery. Andisols show not only a resistance to compaction, but also the ability to recover from compaction following repeated cycles of wetting and drying. Low bulk density is also observed for cultivated Andisols in Japan which have been intensively used for upland farming for a long period time (Akiyama, 1979). This indicates that the aggregates of Andisols are highly stable, being cemented by noncrystalline materials and soil organic matter. It also appears that the high organic matter content of humus

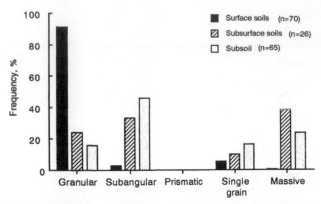

Fig. 8.15. Frequency distribution of soil structure type in cultivated Andisols from Japan (prepared from data of Akiyama, 1979).

horizons is significantly responsible for the low bulk density. Thus, Andisols can maintain good tilth despite continuous tillage operations, even with the use of heavy equipment.

As shown in Fig. 8.15, Andisols in Japan have granular structure in almost all Ap horizons even after continuous intensive upland farming and granular or blocky structure in the subsurface horizons. Such soil structure in the Ap and subsurface horizons contributes to the rapid infiltration of rainfall, high porosity, high available water retention, good drainage, and good aeration.

8.4.3. Water erosion

Water erosion often drastically diminishes the productivity of a soil. In general, Andisols show a strong resistance to water erosion. The related factors include rapid rainfall infiltration which reduces runoff and strong resistance to dispersion of soil aggregates. The rapid infiltration rate is determined by the content of macropores, water permeability of the soil profile, and the water content of the soil. As described earlier, most Andisols have large amounts of macropores and are highly permeable so that they have a high ability to accept large amounts of rainfall before runoff occurs. As often experienced in mechanical analysis, microaggregates in Andisols show strong resistance to dispersion because the aggregates are strongly cemented by noncrystalline materials and soil organic matter. Thus, sonication is required to break aggregates and disperse mineral particles for particle-size analysis as described in Chapter 7.

8.5. UTILIZATION AND MANAGEMENT OF VOLCANIC ASH SOILS IN JAPAN

The islands of Japan are located along the most active part of the circum-Pacific volcanic zone, and vulcanian and plinian explosions are common in Japanese volcanic eruptions. Westerly winds at high altitude commonly prevail over the archipelago. Therefore, felsic and andesitic tephras which were deposited from these volcanoes show a wide distribution primarily to the east of their source volcanoes.

Volcanic ash soils in Japan consist mainly of Andisols (Kurobokudo) with lesser amounts of Entisols. Volcanic ash soils occur mostly on flat uplands and on gentle slopes of volcanic piedmonts while alluvial soils dominate the lowland areas. With regard to land use, Andisols are used for 12 percent of the total paddy fields, 52 percent of total upland fields and 22 percent of total orchard acreage (Minist. Agr. Forest. and Fish., Japan, 1979).

8.5.1. Andisols for common field crops

Most Andisols in Japan are classified as capability class III soils indicating that these soils have many limitations or hazards for crop production and require fairly intensive management practices. However, advancement of agrotechnology enabled Andisols to become highly productive in the 1960's. Today, the important chemical and physical properties unfavorable for the growth of upland crops can be easily ameliorated.

Upland farming in Japan changed significantly with the enactment of the Basal Law of Agriculture in 1961. This law drastically changed the Japanese economic structure leading to the rapid increase in importation of agricultural products, except for rice since the 1970's. This economic environment decreased production of the major crops, and vegetables and cash crops were extensively introduced to farmed Andisol areas, mainly by making skillful use of their excellent physical properties as described later.

Intensive continuous cropping of selected vegetables has become a common practice since the 1970's. While this practice has significantly increased vegetable production, it has created various hazards due to deterioration of both chemical properties and biological factors, such as the occurrence of soil-borne diseases, nematodes, etc.

(1) Soil chemical limitations and management
Since section 8.3 of this chapter describes in detail the chemical constraints for plant growth and their amelioration, this section will briefly discuss the chemical limitations and management of Andisols for common crops.

Low nitrogen fertility: Some Andisols show low nitrogen fertility because they have a low content of organic matter or organic matter dominated by highly humified organics. Intense continuous cropping also lowers nitrogen fertility. Low organic matter coupled with slow mineralization rates often lead to nitrogen deficiencies. In addition to commercial sources of nitrogen fertilizers, applications of compost, stable manure and plant residues are employed to increase nitrogen fertility. Controlled release fertilizers such as polyolefin resin coated fertilizers began recently for use with common crops in Japan. The fertilizers allow single basal *co-situs* application, high recovery by the plant, and minimize environmental pollution (Shoji and Gandeza, 1992).

Low phosphorus fertility: Most Andisols show an inherent low phosphorus fertility. Heavy application of phosphorus fertilizers coupled with efficient placement has been recommended to enhance phosphorus fertility.

Low content of exchangeable bases: Exchangeable bases in Andisols are easily leached from the soil profile under humid climates. Toxic aluminum levels may also occur in some Andisols whose clay fraction is dominated by 2:1 minerals when the base saturation is low. Liming and application of mixed fertilizers are necessary to ameliorate the low base status. Amendments are applied to obtain the recommended exchangeable Ca:Mg:K ratio of 50:20:10 (Kamata, 1977).

Low content of available Cu, Zn and Co: Some Andisols show deficiencies of plant available micronutrients; Cu, Zn and Co being most commonly deficient. Application of inorganic metal salts or metal-chelates has been recommended.

(2) Soil physical problems and management

Most Andisols have excellent physical properties for plant root growth. However, some Andisols have an extremely coarse texture, hard pans, solidified tephra layers, etc. which seriously restrict root development. Deep plowing is often effective for improving coarse-textured Andisols and tephra-derived Entisols by mixing with finer textured horizons. Coarse-textured Andisols occupy about 15 percent of the total area of volcanic ash soils in Japan (Otowa, 1986). Most of these soils show cumulative morphology, reflecting intermittent tephra deposition and repeated soil formation. Their coarse-textured horizons have formed in pumice or scoria.

Cases where deep plowing is necessary can be described as follows (Kikuchi, 1981):

(1) A coarse-textured surface horizon underlaid by a weathered medium- or fine-textured horizon,

(2) A thick coarse-textured subsurface horizon overlaid by a weathered medium- or fine-textured surface horizon,

(3) A coarse-textured subsurface horizon is sandwiched between medium- or fine-textured surface and subsoil horizons.

Mixing of coarse-textured and fine- or medium-textured materials by deep plowing and rotavation can successfully improve physical properties of Andisols resulting in a favorable rooting environment and improved water movement through the profile.

Hard pans and solidified tephra layers seriously limit root development and drainage. Mechanical breaking of these pans and layers removes the physical limitations and significantly enhances soil productivity.

(3) Biological problems and soil management

Hazards posed by soil-borne diseases and nematodes are serious problems for agricultural plants on Andisols, especially when a monoculture is practiced with heavy application of chemical fertilizers. Such cropping systems typically lack cereal crops, such as winter barley or winter wheat which can produce a large amount of dry matter rich in cellulose and lignin. High microbial activity can be maintained by the gradual decomposition of these organic components, reducing the occurrence of soil-borne diseases. Thus, cropping systems which include cereal crops have been recommended to farmers (Agriculture, Forestry and Fisheries Research Council of Japan, 1981). Some plants antagonistic to nematodes are introduced to help control nematodes by natural mechanisms. Maligold (*Tagetes* spp.) is known to be an antagonistic plant for nematodes such as *Pratylenchus penetrans* and *Pratylenchus pratensis*. Since this plant produces nematocides in its roots, it is introduced into cropping systems with Japanese radish as the main crop (Okubo, 1981).

8.5.2. Recent farming of some vegetables suitable for volcanic ash soils in Japan

Vegetables suitable for growth in volcanic ash soils have been extensively introduced and localities especially suited for their production have been developed since 1970 as described earlier. Examples of the special production localities and the plant-soil relationships are briefly described in Table 8.3. The establishment of

TABLE 8.3

Vegetables suitable for growth on Andisols and important soil properties contributing to high quality production at special production localities

Plants	Special production localities	Important soil properties
Chinese yam	Towada, Aomori Pref.	1. Low bulk density and friable consistence 2. Uniform soil physical and chemical properties
Water melon	Obanazawa, Yamagata Pref. Ueki, Kumamoto Pref.	1. Abundance of low and high tension water to cause gradual water stress to the plant
Edible burdock	Mito, Ibaraki Pref.	1. Good drainage 2. Uniform soil physical and chemical properties 3. Absence of toxic Al
Japanese radish, cabbage, and water melon	Miura, Kanagawa Pref.	1. Stable soil structure 2. Friable consistence 3. Fertile soils derived from basaltic tephras
Japanese radish, edible burdock, taro, sweet potato and ginger	Miyazaki Pref.	1. Friable consistence 2. Stable soil structure 3. Abundance of plant available water

such localities substantially relies upon the excellent physical properties of volcanic ash soils (Andisols). Of the various soil properties, excellent tilth and high water retention are the most important. Only a few examples of vegetables will be described below.

(1) Chinese yam in Aomori Prefecture, northeastern Japan

Chinese yam (*Dioscorea batatas*) is rich in starch, polysaccharides and gluco-proteins. The Japanese are fond of this glutinous vegetable and eat its mashed raw product. High quality yam shows a baseball butt shape that is easily broken. Therefore, excellent tilth of Andisols is essential not only for smooth elongation of the roots but also for ease of digging at harvest.

The Towada area in Aomori prefecture is the largest production center for Chinese yams which were extensively introduced into this area in the 1970's (Shoji and Otowa, 1988). The climate is characterized by cool summers for which root crops are well adapted. The major soils of this area are allophanic Melanudands (Thick High-humic Andosols) and Hapludands (High-humic Andosols), (Aomori Pref. Agricultural Experiment Station, 1972). They show cumulative morphology, reflecting intermittent tephra deposition and repeated soil formation as described earlier. These soils have many properties suitable for growing Chinese yam as follows:

(1) Low bulk density and friable consistence,

(2) Uniform physical and chemical properties from the surface to a depth of 100 cm can be attained by deep cultivation,

(3) Medium to coarse texture, and

(4) Good drainage.

Deep narrow strip cultivation (15–20 cm wide and 100 cm deep) is commonly practiced to obtain uniform physical and chemical properties throughout the soil. The soil material is mixed by a special kind of deep-reaching heavy rotavator that mixes the soil from the surface to a depth of 100 cm. In order to minimize hazards from soil-borne diseases and nematodes, heavy application of well-decomposed compost (20 ton per ha) and crop rotation is recommended.

(2) Intensive farming of Japanese radish-cabbage-water melon in Kanagawa Prefecture, Kanto

Miura Peninsula, Kanagawa Prefecture is well known for its production of high quality winter Japanese radish and spring cabbage. Advantages for intensive farming of these plants include warm winters (mean winter air temperature is 9.0°C), a short distance to large markets such as Tokyo, succession of advanced agrotechnology, and productive cumulative Andisols formed from basaltic tephras. The culture of Japanese radish dates back to the Tokugawa period (>100 years ago), and marketing of "Miura" Japanese radish in Tokyo started in the mid-1920's. Miura Peninsula is also one of the leading production centers for spring cabbage marketed in Tokyo.

The dominant soils in the Miura region are Melanudands and Hapludands (Thick High-humic Andosols and High-humic Andosols), (Kamata et al., 1986).

They have excellent physical properties such as excellent tilth, high water retention, and stable soil structure from the surface to the subsurface horizons which are suitable for growing root crops. They have high fertility and productivity reflecting the basaltic nature of the parent material.

Hazards of soil-borne diseases and nematodes are common for intense upland farming in this region. Thus, soil management practices have been carefully prepared according to soil testing in the Miura region as follows (Council of Agriculture Improvement in Miura Peninsula, 1988):

(1) Deep cultivation (deeper than 40 cm),

(2) Subsoil cultivation and pan breaking (the subsoils are often compacted by heavy equipment),

(3) Application of compost and soil-ameliorating materials, and

(4) Growth of Maligold (*Tagetes* spp.) and green-manure crops which suppress nematode activity.

(3) Edible burdock in Ibaraki Prefecture, Kanto

Edible burdock (*Arctium lappa*) is an example of a typical root crop which is very sensitive to aluminum toxicity. It is a popular fiber food in Japan which is rich in cellulose, and enhances the activity of the human intestine. Edible burdock grows straight, reaching depths of about 1 m or more if soil conditions are favorable. Burdock farming started on deep alluvial soils in the flood plains of the Naka River running through Mito City, Ibaraki Prefecture in the 1890's. Today, this prefecture produces the greatest amount of edible burdock, primarily on Andisols. Mito City located in this prefecture is known as the center for burdock production. The success of burdock farming appears to rely upon the excellent properties of allophanic Andisols.

Burdock grows well on allophanic Andisols which do not have toxic Al even when the base saturation is low. Friable consistence is favorable for narrow deep cultivation (width of 12–15 cm; depth of 70 cm or more) to obtain uniform soil properties and for smooth elongation of roots. The soil is also suitable for easy digging at harvest.

The soil management recommendations for growth of high quality burdock include the necessity of good drainage, uniform physical properties from the surface to a depth of 70–100 cm, liming, and prevention of hazards from soil-borne diseases and nematodes (Naka Extension Office, Mito City, 1990).

(4) Intensive farming of root crops in Miyazaki Prefecture, southern Kyushu

Miyazaki Prefecture is the largest production center for various root crops such as Japanese radish, edible burdock, taro, sweet potato, potato, and ginger in Japan. The warm winter climate of Miyazaki Prefecture allows intensive farming of root crops throughout the year. Since these crops are durable for long distance transportation, they are transported by ferry boats and marketed in big cities such as Osaka and Tokyo.

Intensive farming of root crops in Miyazaki takes place primarily on Andisols which occupy 89 percent of the total acreage of upland soils in Miyazaki Prefecture. The main Andisols are High-humic Andosols (43 percent), Humic Andosols (28 percent), Thick Humic Andosols (21 percent), and Thick High-humic Andosols (8 percent) (Division of Agricultural Administration and Fisheries, Miyazaki Pref., 1990). These soils are classified as Hapludands and Melanudands according to the Keys to Soil Taxonomy (Soil Survey Staff, 1992). They show excellent physical properties suitable for the intensive farming of root crops as follows:

(1) Friable consistence suitable for deep cultivation, ridging, and easy digging for harvesting,

(2) Stable granular soil structure suitable for repeated cultivation and mechanical mixing,

(3) Abundance of plant available water (Andisols are generally dry under intensive farming, except for the rainy season in early summer in Miyazaki), and

(4) Good drainage required to prevent hazards from wetness due to heavy continuous rainfall.

An example of a soil profile having excellent suitability for growing root crops (Kawano and Nonaka, 1975) has the following characteristics:

(1) The humus horizon is extremely thick and is rich in organic matter,

(2) The upper part of the humus horizon (0–20 cm) has a sandy loam texture and granular structure favoring easy tillage,

(3) The lower part of the humus horizon (20–95 cm) has a clay loam texture and can retain a large amount of plant available water, and

(4) A deep permeable horizon (below a depth of 95 cm) derived from pumiceous material that contributes to rapid drainage of excess soil water from continuous heavy rains.

8.5.3. Andisols for wetland rice

Andisols (Kurobokudo) used for rice farming in Japan are divided into Kurobokudo, Wet Kurobokudo and Gleyed Kurobokudo. These paddy soils have several limitations or hazards for wetland rice production and most of them are classified as capability class II or III. Management and fertilization of these soils for wetland rice were reviewed by Motomura (1988).

Kurobokudo with good drainage have several severe limitations such as high water permeability and inherent low fertility, especially phosphorus. Therefore, rice plants often show poor early growth under standard fertilization, resulting in delayed growth (Honya, 1961). This reduced growth makes the rice susceptible to severe attack by rice blast disease and cold damage prior to ripening. The most important management practices required for improving rice cultivation on Kurobokudo include control of water permeability, heavy applications of phosphate fertilizers, and growth of strong rice seedlings which can grow vigorously after transplanting (Honya, 1961).

Wet Kurobokudo are characterized by the presence of iron mottles formed by redoxy processes caused by irrigation and have some important limitations such as formation of toxic substances and low content of available nutrients, especially phosphorus. Therefore, control of irrigation water and heavy application of phosphate fertilizers are commonly employed to improve rice cultivation on these soils.

Gleyed Kurobokudo, formed under continuously reduced conditions, have some limitations for rice production such as accumulation of toxic substances and a low content of available nutrients. Since these soils are rarely drained and are subjected to strong reducing conditions throughout the growing season, toxic substances such as organic acids, hydrogen sulfide, and ferrous iron accumulate and may result in serious damage to rice roots. The slow mineralization of soil organic nitrogen causes the delayed growth of rice plants, resulting in outbreaks of rice blast disease and susceptibility to cold damage prior to ripening. Thus, the establishment of a drainage system is very important for rice cultivation on Gleyed Kurobokudo.

Water management is one of the most important practices for rice farming. Rice growth is successfully controlled by Japanese farmers through water management. Thus, high yielding rice fields require good drainage systems. The optimum water leakage was empirically determined to be 20–30 mm per day (Isozaki, 1961). In poorly drained soils with leakage rates less than the optimum, toxic substances accumulate. In contrast, since very large amounts of irrigation water are necessary for rice production in excessively drained soils, nutrient elements are easily leached.

Alternative use of Kurobokudo and Wet Kurobokudo for upland crops is a recent change in farming practices. For instance, soybean production has been especially successful in northeastern Japan. The existence of abundant plant available water and easy control of soil-borne diseases, nematodes, and weeds are considered to contribute to the high yields of soybeans, 4 to 6 ton per ha (Shoji, 1988).

REFERENCES

Adachi, T., 1985. A guide for improvement of soil physical and chemical properties. In: Y. Onikura (Editor), Soil and Water Conservation and Quality of Agricultural Materials. Hakuyusha, Tokyo, pp. 31–60 (in Japanese).

Agriculture, Forestry and Fisheries Council of Japan, 1981. Guide for management of upland fields to maintain soil productivity and to prevent soil injury caused by continuous cropping (in Japanese).

Akiyama, Y., 1979. Physical properties of soils on hilly lands and uplands. In: S. Terasawa (Editor), Soil Physical Properties and Plant Growth. Yokendo, Tokyo, pp. 71–80 (in Japanese).

Andisol TU database, 1992. Database on Andisols from Japan, Alaska, and Northwestern U.S.A. prepared by Soil Science Laboratory, Tohoku University (see Appendix 2).

Ando, H. and Shoji, S., 1986. Kinetic study on soil nitrogen mineralization and nitrogen immobilization of paddy soils. Jap. J. Soil Sci. Plant Nutr., 57: 1–7 (in Japanese).

Aomori Pref. Agricultural Experiment Station, 1972. Soil capability classification map of paddy and upland soils in northern Pacific Ocean side of Aomori Prefecture (in Japanese).

Ayres, A.A. and Hagihara, H.H., 1952. Available phosphorus in Hawaiian soil profiles. Hawaiian Planters' Record, 44: 81–99.

Bertsch, P.M. and Thomas, G.W., 1985. Potassium status of temperate region soils. In: R.D. Munson (Editor), Potassium in Agriculture. ASA, CSSA, SSSA, Madison, WI, pp. 131–162.

Chiba, A., Shirahata, H., Shinke, H., Chiba, Y., Furusawa, N., Uchida, S., Nakano, N., Sekizawa, N., Sato, K., Kurosawa, J., Takahashi, K. and Natsui, W., 1984. Studies on trace element deficiencies in Iwate Prefecture. Bull. Iwate Agricultural Experiment Station, 24: 75–184 (in Japanese, with English abstract).

Council of Agriculture Improvement in Miura Peninsula, 1988. Manual of soil improvement of field vegetables (in Japanese).

Dalal, R.C., 1977. Soil organic phosphorus. Adv. Agron., 29: 83–117.

Division of Agricultural Administration and Fisheries, Miyazaki Pref., 1990. General guidance of upland farming (in Japanese).

Fujita, T., Takahashi, C., Yoshida, S. and Shimizu, H., 1983. Coated granular fertilizer capable of controlling the effects of temperature on dissolution-out rate. United States Patent, 4, 369, 055.

Gandeza, A.T., Shoji, S. and Yamada, I., 1991. Simulation of crop response to polyolefin-coated urea, I. Field dissolution. Soil Sci. Soc. Am. J., 55: 1462–1467.

Hewitt, A.E., 1989. New Zealand soil classification (version 2.0). Tech. Record DN2, DSIR, New Zealand.

Honya, K., 1961. Studies on the improvement of rice plant cultivation in volcanic ash paddy fields in Tohoku district. Res. Bull. Tohoku Nat. Agricultural Experiment Station, 21: 1–43 (in Japanese, with English abstract).

Hue, N.V., Hirunburana, N. and Fox, R.L., 1988. Boron status of Hawaiian soils as measured by B sorption and plant uptake. Commun. Soil Sci. Plant Anal., 19: 517–528.

Isozaki, T., 1961. On the proper percolation rate. J. Agr. Engineering Soc. Jap., 24: 6–10 (in Japanese).

Ito, T., Shoji, S. and Saigusa, M., 1991. Classification of volcanic ash soils from Konsen district, Hokkaido, according to the last Keys to Soil Taxonomy (1990). Jap. J. Soil Sci. Plant Nutr., 62: 237–247 (in Japanese, with English abstract).

Kamata, H., 1977. Ca-Mg-K balance in volcanic ash soils to maximize vegetable production. In: Proc. International Seminar on Soil Environments and Fertilizer Management in Intensive Agriculture, Tokyo, Japan, pp. 496–499.

Kamata, H., 1978. Studies on soil classification and land utilization in Kanagawa Prefecture — On land classification for paddy and upland field. Bull. Agr. Res. Inst. of Kanagawa Pref., 119: 1–108 (in Japanese, with English abstract).

Kamata, H., Fujiwara, S., Suzuki, J. and Ogishi, R., 1986. Soil map of Yokosuka-Misaki, Kanagawa Prefecture (in Japanese).

Kamewada, K., 1987. Appropriate level and upper limit of available phosphorus. In: Research on the Appropriate Level and Upper Limit of Soil Nutrients, pp. 5–30 (in Japanese).

Kamprath, E.J. and Watson, M.E., 1980. Conventional soil and tissue tests for assessing the phosphorus status of soils. In: F.E. Khasawneh, et al. (Editors), The Role of Phosphorus in Agriculture. ASA, CSSA, and SSSA, Madison, WI, pp. 433–470.

Kawano, M. and Nonaka, S., 1975. Physical and chemical characterization and soil productivity of the upland soils in Miyazaki Prefecture, I. Soil survey reports and soil productive capability classification. Res. Bull. Miyazaki Pref. Agricultural Experiment Station, 9: 1–14 (in Japanese, with English abstract).

Kawasaki, H., Hayashi, K. and Katazaki, Y., 1991. Characteristics of the Handa pyroclastics flow abundant in available phosphorus. Kyushu Agri. Res., 53: 57 (in Japanese).

Kikuchi, K., 1981. Interpretative classification of the soils in Tokachi District — Its mapping and application to practical uses of soil improvement. Rep. Hokkaido Agricultural Experiment Station, 34: 1–118 (in Japanese, with English abstract).

Kimble, J.M. (Editor), 1986. Tour guide. First International Soil Correlation Meeting (ISCOM), Idaho, Washington, and Oregon, U.S.A., 20–31 July, 1986, pp. 38–77.

Kitamura, T., 1988. Relation of soil types and soil properties to growth and quality of Burly tobacco. Research Report (Cultural Practices Section), Japan Tobacco Inc., 3: 1–92 (in Japanese, with English abstract).

Kitamura, T., Shoji, S., Ogata, Y., Takeda, Y. and Akiya, T., 1986. Inorganic N supply to tobacco plant from the subsurface horizons of Andisols. Jap. J. Soil Sci. Plant Nutr., 57: 414–417 (in Japanese).

Kobayashi, M., Tsunoda, S., Tadaki, M. and Matsumura, S., 1964. Effects of basic and micronutrient elements on upland crops, II. Effects of the micronutrient elements (Mn, B and Cu) on the growth of vegetable crops, Bull. Gunma Agricultural Experiment Station, 5: 115–140 (in Japanese).

Kobayashi, S., 1979. Studies on occurrence of copper and zinc and their plant availability in Andisols. Ph.D. thesis, Faculty of Agriculture, Tohoku University (in Japanese).

Kobayashi, S. and Shoji, S., 1976. Distribution of copper and zinc in volcanic ashes. Soil Sci. Plant Nutr., 22: 401–408.

Kobayashi, Y., 1981. Studies of cobalt-deficient grasslands for ruminants in soils of volcanic ash origin, I. Distribution and characteristics of soil cobalt. Jap. J. Soil Sci. Plant Nutr., 52: 392–400 (in Japanese).

Kobayashi, Y., 1982a. Studies of cobalt-deficient grasslands for ruminants in soils of volcanic ash origin, III. Availability and nature of soil cobalt. Jap. J. Soil Sci. Plant Nutr., 53: 77–86 (in Japanese).

Kobayashi, Y., 1982b. Cobalt substitution from soils for ruminant nutrition. Jap. J. Soil Sci. Plant Nutr., 53: 176–182 (in Japanese).

Kosaka, J. and Abe, K., 1958. Organic phosphorus in the upland soils. Jap. J. Soil Sci. Plant Nutr., 28: 383–386 (in Japanese).

Kosaki, T., 1989. Soil fertility of arable land in Zone de Kabare in the Lake Kivu area. In: S. Hirose and M. Mubandu (Editors), Agriculture and Soils in Zaire. College of Agr. and Vet. Med., Nihon Univ., Tokyo, pp. 129–138 (in Japanese).

Kubota, T., 1976. Surface chemical properties of volcanic ash soil — especially on phenomenon and mechanism of irreversible aggregation of the soil by drying. Bull. Natl. Inst. Agric. Sci., 28B: 1–74 (in Japanese, with English abstract).

Kurashima, K., 1980. The distribution of K, Ca and Mg contents in soils and their composition in herbage on the mountainous sloping pasture, II. K, Ca and Mg contents and their balance in herbage. J. Jap. Grassl. Sci., 25: 354–361 (in Japanese, with English abstract).

Kurashima, K., Shoji, S. and Yamada, I., 1981. Mobilities and related factors of chemical elements in the topsoils of Andosols in Tohoku, Japan, I. Mobility sequence of major chemical elements. Soil Sci., 132: 300–307.

Kurosawa, J., 1970. Classification and improvement of volcanic ash soils in Iwate Prefecture. Bull. Iwate Agricultural Experiment Station, 14: 1–124 (in Japanese, with English abstract).

Kurosawa, J., Uchida, S., Nakano, N., Takahashi, T., Takahashi, W., Chiba, A., Sekizawa, N., Kikuchi, T., Yonezawa, R., Takahashi, R., Arakawa, Y., Shirahata, K. and Oikawa, Y., 1965. Studies on the copper deficient soils. Bull. Iwate Agricultural Experiment Station, 8: 1–32 (in Japanese, with English abstract).

Leamy, M.L., 1982. Andisols and agriculture in New Zealand. Draft of speech delivered in Japan.

Leamy, M.L., 1984. Andisols of the world. In: Congreso Internacional de Suelos Volcanicos. Comunicaciones, Universidad de La Laguna Secretariado de Publicaciones, Serie Informes 13, pp. 368–387.

Leamy, M.L., Clayden, B., Parfitt, R.L., Kinloch., D.I. and Childs, C.W., 1988. Final proposal of the International Committee on the Classification on Andisols (ICOMAND). New Zealand Soil Bureau, DSIR, Lower Hutt.

Lee, H.J., 1974. Trace elements in animal production. In: D.J.D. Nicholas and A.R. Egan (Editors), Trace Elements in Soil-Plant-Animal Systems. Academic Press Inc., New York, San Francisco, London, pp. 39–54.

Lucus, R.E. and Knezik, B.D., 1972. Climate and soil conditions promoting micronutrient deficiencies in plants. In: J.J. Mortvedt, P.M. Giordano, and W.L. Lindsay (Editors), Micronutrients in Agriculture. ASA, CSSA, and SSSA, Madison, WI, pp. 265–288.

Machida, H., Arai, F. and Moriwaki, H., 1981. Tephra carried over the Japan Sea. Kagaku (Science), 51:

562–569 (in Japanese).

Marschner, H., and Richter, C., 1974. Calcium transport in roots of maize and bean plants. Plant Soil, 40: 193–210.

Masui, J., Shoji, S. and Minami, K., 1972. Copper and zinc in volcanic ash soils. Soil Sci. Plant Nutr., 18: 31–38.

Meisinger, J.J., 1984. Evaluating plant-available nitrogen in soil-crop systems. In: R.D. Hauck (Editor), Nitrogen in Crop Production. ASA, CSSA, and SSSA, Madison, WI, pp. 391–416.

Meurisse, R.T., 1988. Forest productivity and the management of U.S. Andisols. In: D.I. Kinloch, S. Shoji, F.H. Beinroth and H. Eswaran (Editors), Proc. 9th Int. Soil Classification Workshop, Japan, 20 July to 1 August, 1987. Publ. by Jap. Committee for 9th Int. Soil Classification Workshop, for the Soil Management Support Services, Washington D.C., U.S.A., pp. 245–257.

Michaelson, G.J. and Ping, C.L., 1986. Extraction of phosphorus from the major agricultural soils of Alaska. Commun. Soil Sci. Plant Anal., 17: 275–297.

Michaelson, G.J. and Ping, C.L., 1987. Effects of P, K, and liming on soil pH, Al, Mn, K, and forage barley dry matter yield and quality for a newly-cleared Cryorthod. Plant and Soil, 104: 155–161.

Minami, K., 1970. Status of copper and zinc in Andisols. Ph.D. thesis, Faculty of Agriculture, Tohoku University, Sendai, Japan (in Japanese).

Minist. Agr. Forest. and Fish., Japan, 1979. General Report on Basic Surveys for Soil Conservation (in Japanese).

Minist. Agr. Forest. and Fish., Japan, 1983. General Report on Basic Surveys for Soil Conservation (in Japanese).

Miyake, M., 1983. Soils in Indonesia. Nettai Nouken Shuhou (Bull. Tropical Agr. Res.), 46: 17–34 (in Japanese).

Miyake, M., Ismunadi, M., Zulkarnaini, I. and Roechan, S., 1984. Phosphate response of rice in Indonesian paddy fields. Tech. Bull. Trop. Agr. Res. Center, Japan, 17: 1–78.

Mizota, C., 1987. Chemical and mineralogical characterization of soils derived volcanic ashes. In: S. Hirose (Editor), Agriculture and Soils in Kenya. College of Agr. and Vet. Med., Nihon Univ., Tokyo, pp. 110–123 (in Japanese).

Mohr, E.C.J., 1938. The relation between soil and population density in the Netherlands Indies. Compies Rendus du Congres International de Geographic, Amsterdam. Tome Deuxiem, Section IIIc, pp. 478–493.

Motomura, S., 1988. Management of Ando soils for rice. In: D.I. Kinloch, S. Shoji, F.H. Beinroth and H. Eswaran (Editors), Proc. 9th Int. Soil Classification Workshop, Japan, 20 July to 1 August, 1987. Publ. by Jap. Committee for 9th Int. Soil Classification Workshop, for the Soil Management Support Services, Washington D.C., U.S.A., pp. 276–292.

Naka Extension Office, Mito City. 1990. Edible burdock (extension material) (in Japanese).

Nanzyo, M., 1987. Formation of noncrystalline aluminum phosphate through phosphate sorption on allophanic Andosols. Comm. Soil Sci. Plant Anal., 18: 735–742.

Nishimune, A., 1984. Evaluation of the soil nitrogen supply in upland field crops in Tokachi, Hokkaido. Res. Bull. Hokkaido Nat. Agric. Exp. Sta., 140: 33–91 (in Japanese, with English abstract).

Nishiya, M., 1987. Soil factors determining Al toxicity of crops and amelioration of subsoil acidity. Master thesis, Faculty of Agriculture, Tohoku University, Sendai, Japan (in Japanese).

Nomoto, K., 1958. Volcanic ash soils. In: Research Division, Improvement Bureau, Ministry of Agriculture and Forestry (Editor), General in Soils and Fertilizers. Yokendo, Tokyo, pp. 368–373 (in Japanese).

Oda, D., Miwa, E. and Iwamoto, A., 1987. Compact database for soil analysis data in Japan. Jap. J. Soil Sci. Plant Nutr., 58: 112–132 (in Japanese).

Okubo, T., 1981. Perspective on agrotechnology of prevention of soil injury caused by continuous cropping. In: Farming Japan. Assoc. Intern. Coop. Agr. and Forest., Tokyo, Japan, pp. 52–58.

Otowa, M., 1986. Morphology and classification. In: K. Wada (Editor), Ando Soils in Japan. Kyushu

University Press, Fukuoka, Japan, pp. 3–20.

Otsuka, H., Briones, A.A., Daquiado, N.P. and Evangelio, F.A., 1988. Characteristics and genesis of volcanic ash soils in The Philippines. Tech. Bull. Trop. Agr. Res. Center, 24: 1–122.

Saigusa, M., 1991. Plant growth on acid soils with special reference to phytotoxic Al and subsoil acidity (review). Jap. J. Soil Sci. Plant Nutr., 62: 451–459 (in Japanese).

Saigusa, M., Saito, K. and Shoji, S., 1976. Occurrence of Cu and Zn within the profiles of soils formed from different parent ashes and corn response to these elements. Tohoku J. Agr. Res., 27: 12–19.

Saigusa, M., Shoji, S. and Takahashi, T., 1980. Plant root growth in acid Andosols from northeastern Japan, II. Exchange acidity Y_1 as a realistic measure of aluminum toxicity potential. Soil Sci., 130: 242–250.

Saigusa, M., Shoji, S. and Sakai, H., 1983. The effects of subsoil acidity of Andosols on the growth and nitrogen uptake of barley and wheat. Jap. J. Soil Sci. Plant Nutr., 54: 460–466 (in Japanese).

Saigusa. M., Shoji, S. and Nakaminami, H., 1987. Measurement of water retention at 15 bar tension by pressure membrane method and available moisture of Andosols. Jap. J. Soil Sci. Plant Nutr., 58: 374–377 (in Japanese).

Saigusa, M., Shoji, S., Goto, J., Misumi, Y. and Sakuratani, T., 1989. Effect of subsoil acidity on the growth and water absorption of sorghum plant. Bull. Exp. Farm., Tohoku Univ., 5: 13–17 (in Japanese).

Saigusa, M., Nishiya, M., Matsuyama, N., Shoji, S. and Abe, T., 1990a. Liming and calcium nutrition in acid Andisols. Bull. Exp. Farm., Tohoku Univ., 6: 33–38 (in Japanese).

Saigusa, M., Shoji, S., Ishimori, H. and Ito, T., 1990b. Determination of 15 bar water of Andisols by centrifuging method. Jap. J. Soil Sci. Plant Nutr., 61: 635–637 (in Japanese).

Saigusa, M., Shoji, S. and Otowa, M., 1991. Clay mineralogy of two Andisols showing a hydrosequence and its relationships to their physical and chemical properties. Pedologist, 35: 21–33.

Saito, M., 1990. Nitrogen mineralization parameters and its availability indices of soils in Tohoku district, their relationship. Jap. J. Soil Sci. Plant Nutr., 61: 265–272 (in Japanese, with English abstract).

Sekizawa, N., Uchida, S., Chiba, A., Nakano, N., Takahashi, R. and Sato, K., 1973. Studies on the zinc deficient soil. Bull. Iwate. Agricultural Experiment Station, 17: 25–77 (in Japanese, with English abstract).

Shiga, E. and Nakajima, T., 1985. Available water content in an Andosol and growth, nutrient uptake and quality of dried leaves of tobacco. In: Proc. Res. Morioka Tobacco Exp. Sta. 14–15 March, 1985, Morioka, Japan, pp. 25.1–25.6 (in Japanese).

Shoji, S., 1979. Volcanic ash soils in Tohoku. Bull. of Tohoku Council of Soils and Fertilizers, 15: 1–29 (in Japanese).

Shoji, S., 1986. Mineralogical characteristics: 1. Primary minerals. In: K. Wada (Editor), Ando soils in Japan. Kyushu University Press, Fukuoka, Japan, pp. 21–40.

Shoji, S., 1988. Wet Andisols. In: 5th International Soil Management Workshop for Classification and Management of Rice-growing Soils. Food and Fertilizer Technology Center. Rep. China, pp. 11.1–11.13.

Shoji, S. and Gandeza, A.T., 1992. Controlled Release Fertilizers with Polyolefin Resin Coating. Konno Printing, Sendai, Japan.

Shoji, S. and Ono, T., 1978. Physical and chemical properties and clay mineralogy of Andosols from Kitakami, Japan. Soil Sci., 125: 297–312.

Shoji, S. and Otowa, M., 1988. Distribution and significance of Andisols in Japan. In: D.I. Kinloch, S. Shoji, F.H. Beinroth and H. Eswaran (Editors), Proc. 9th Int. Soil Classification Workshop, Japan, 20 July to 1 August, 1987. Publ. by Jap. Committee for 9th Int. Soil Classification Workshop, for the Soil Management Support Services, Washington D.C., U.S.A., pp. 13–24.

Shoji, S., Kobayashi, S., Yamada, I. and Masui, J., 1975. Chemical and mineralogical studies on volcanic ashes, I. Chemical composition of volcanic ashes and their classification. Soil Sci. Plant Nutr., 21: 311–318.

Shoji, S., Saigusa, M. and Ebihara, M., 1980a. Cobalt content in volcanic ashes. Jap. J. Soil Sci. Plant Nutr., 51: 335–336 (in Japanese).

Shoji, S., Saigusa, M. and Takahashi, T., 1980b. Plant root growth in acid Andosols from northeastern Japan, I. Soil properties and root growth of burdock, barley, and orchard grass. Soil Sci., 130: 124–131.

Shoji, S., Saigusa, M. and Goto, J., 1986. Acidity to subsoils of Andisols and nitrogen uptake and growth of sorghum. Jap. J. Soil Sci. Plant Nutr., 57: 264–271 (in Japanese).

Shoji, S., Gandeza, A.T. and Kimura, K., 1991. Simulation of crop response to polyolefin-coated urea, II. Nitrogen uptake by corn. Soil Sci. Soc. Am. J., 55: 1468–1473.

Shoji, S., Nanzyo, M., Shirato Y. and Ito, T., 1993. Chemical kinetics of weathering in young Andisols from Northeastern Japan using soil age normalized to 10°C. Soil Sci., 155: 53–60.

Sims, J.T. and Johnson, G.V., 1991. Micronutrient soil tests. In: R.L. Luxmoore (Editor), Micronutrients in Agriculture, 2nd ed. SSSA Book Series, No.4. Madison, WI, pp. 427–476.

Smith, G.D., 1978. A preliminary proposal for reclassification of Andepts and some andic subgroups (the Andisol proposal, 1978). New Zealand Soil Bureau Record, DSIR, Lower Hutt, 96.

Soil Survey Staff, 1992. Keys to Soil Taxonomy, 5th edition. AID, USDA-SMSS Technical Monograph, No. 19. Blacksburg, Virginia.

Sparks, D.L. and Huang, P.M., 1985. Physical chemistry of soil potassium. In: R.P. Munson (Editor), Potassium in Agriculture. ASA, CSSA, and SSSA, Madison, WI, pp. 201–276.

Stanford, G., 1982. Assessment of soil nitrogen availability. In: F.J. Stevenson (Editor), Nitrogen in Agricultural Soils (Agronomy No.22). ASA, CSSA, and SSSA, Madison, WI, pp. 651–658.

Stanford, G. and Smith, S.J., 1972. Nitrogen mineralization potentials of soils. Soil Sci. Soc. Am. Proc., 36: 465–472.

Stevenson, F.J., 1982. Organic forms of soil nitrogen. In: F.J. Stevenson (Editor), Nitrogen in Agricultural Soils (Agronomy No. 22). ASA, CSSA, and SSSA, Madison, WI, pp. 67–122.

Stevenson, F.J. and Ardakani, M.S., 1972. Organic matter reactions involving micronutrients in soils. In: J.J. Mortvedt, P.M. Giordano and W.L. Lindsay (Editors), Micronutrients in Agriculture. ASA, CSSA, and SSSA, Madison, WI, pp. 79–114.

Sugihara, S., Konno, T. and Ishii, K., 1986. Kinetics of mineralization of organic nitrogen in soil. Bull. Nat. Inst. Agro-Environ. Sci., 1: 127–166 (in Japanese, with English abstract).

Sutanto, R., DeConinck, F. and Doube, M., 1988. Mineralogy, Charge Properties and Classification of Soils on Volcanic Materials and Limestone in Central Java (Indonesia). State University Ghent, Belgium, 233 pp.

Takesako, H., 1987. Changes in CEC with phosphorus application. In: Research on the Appropriate Level and Upper Limit of Soil Nutrients, pp. 41–49 (in Japanese).

Taylor, R.M. and McKenzie, R.M., 1966. The association of trace elements with manganese minerals in Australian soils. Aust. J. Soil Res., 4: 29–39.

Thomas, G.W. and Hargrove, W.L., 1985. The chemistry of soil acidity. In: F. Adams (Editor), Soil Acidity and Liming, 2nd edition. (Agronomy, No.12) ASA, CSSA, and SSSA, Madison, WI, 12: 1–56.

Toxopeus, M.R.J., 1970. Sulphur deficiency in grassland on young soils from volcanic ash and its possible prediction by laboratory tests. In: Proc. XII International Grassland Congress, pp. 345–346.

Truog, E., 1930. The determination of the readily available phosphorus of soils. J. Am. Soc. Agron., 22: 874–882.

Tsuji, T., 1980. Distribution of soil and plant sulphur in grassland and responses of pasture plants to sulphur fertilizer, I. Some factors affecting sulphur contents of native grassland soils. Jap. J. Soil Sci. Plant Nutr., 51: 210–220 (in Japanese).

Tsutsumi, M., Ohira, K. and Fujiwara, A., 1967. Copper deficiency in humus rich volcanic ash soil, I. Effects of copper and other microelements, lime, and compost supply on growth of barley. Jap. J.

Soil Sci. Plant Nutr., 38: 459–465 (in Japanese).

Tsutsumi, M., Ohira, K. and Fujiwara, A., 1968. Copper deficiency in humus rich volcanic ash soil, II. Growth of plants on each soil horizon and effect of copper supply. Jap. J. Soil Sci. Plant Nutr., 39: 121–125 (in Japanese).

Wada, K., 1959. Reaction of phosphate with allophane and halloysite. Soil Sci., 87: 325–330.

Wada, K., 1986. Part II. Database, Kurobokudo Co-operative Research Group. In: K. Wada (Editor), Ando Soils in Japan. Kyushu University Press, Fukuoka, Japan, pp. 115–276.

Wakatsuki, T., 1990. Distribution and characteristics of volcanic ash soils. In: Soil Resources in Tropical Africa. Assoc. for Intern. Corp. of Agr. and Forestry, Tokyo, pp. 114–123 (in Japanese).

Walker, T.W. and Syers, J.K., 1976. The fate of phosphorus during pedogenesis. Geoderma, 15: 1–19.

Yamada, I., 1988. Tephra as parent material. In: D.I. Kinloch, S. Shoji, F.H. Beinroth and H. Eswaran (Editors), Proc. 9th Int. Soil Classification Workshop, Japan, 20 July to 1 August, 1987. Publ. by Jap. Committee for 9th Int. Soil Classification Workshop, for the Soil Management Support Services, Washington D.C., U.S.A., pp. 509–519.

Yamada, I. and Shoji, S., 1975. Relationships between particle size and mineral composition of volcanic ashes. Tohoku J. Agr. Res., 26: 7–10.

Yamada, I. and Shoji, S., 1983a. Properties of volcanic glasses and relationships between properties of tephra and volcanic zones. Jap. J. Soil Sci. Plant Nutr., 54: 311–318 (in Japanese).

Yamada, I. and Shoji, S., 1983b. Alteration of volcanic glass of recent Towada ash in different soil environments of northeastern Japan. Soil Sci., 135: 316–321.

Yamamoto, T. and Miyasato, S., 1976. Studies on the soil productivity of upland fields — heavy application of phosphate mixture in Iwate volcanic ash soil. Bull. Tohoku Nat. Agric. Exp. Sta., 42: 53–92 (in Japanese, with English abstract).

Yasuda, T. and Fujii, Y., 1983. Phosphorus nutrition of tomato plant. Jap. J. Soil Sci. Plant Nutr., 54: 406–410 (in Japanese).

Yasuda, T. and Saito, G., 1982. Accumulation and availability of phosphorus in Andosols in the Tokachi district, Res. Bull. Hokkaido Nat. Agric. Exp. Stn., 133: 7–15 (in Japanese).

Yoshida, Y., Obata, N and Shindo, T., 1966. The effect of boron application on the growth of alfalfa on Yatsugatake acid humic volcanic ash soils: Results of 5 year field experiment. Jap. J. Soil Sci. Plant Nutr., 37: 516–521 (in Japanese).

Yoshiike, A., 1983. The status of fertilized phosphate accumulation in agricultural land soil. Jap. J. Soil Sci. Plant Nutr., 54: 255–261 (in Japanese).

Appendix 1

DESCRIPTION AND SELECTED PROPERTIES OF TSUKUBA SOIL, FINDLEY LAKE SOIL, ABASHIRI SOIL, AND YUNODAI SOIL

1. TSUKUBA SOIL (HYDRIC HAPLUDAND) (PLATE 10)

Source: Tour guide of 9th International Soil Classification Workshop, "Properties, Classification, and Utilization of Andisols and Paddy Soils", 20 July to 1 August, 1987. pp. 273–281.

DESCRIPTION OF TSUKUBA SOIL (HYDRIC HAPLUDAND)

NSSL ID #	86P0096
Soil Survey #	S85-FN-490-020
Location	Location is NIAES, Yatabe T. Ibaraki Pref., Japan. The site is at Tsukuba Institute.
	Latitude = 36°01′00″N, Longitude = 140°07′00″E
Slope characteristics	1% plane
Elevation	21 m M.S.L.
Air temperature (°C)	Ann: 13, Sum: 24, Win: 3
Soil temperature (°C)	Ann: 14, Sum: 25, Win: 4
Precipitation	1300 mm, Udic moisture regime
Drainage	Moderately well drained
Permeability	Moderate
Land use	Pasture land and native pasture
Erosion or deposition	None
Diagnostic horizons	0–29 cm Umbric, 29–200 cm Cambic
Described by	T.D. Thorson and Dr. George Holmgren
Date	September, 1985

Vegetation is *Imperata cylindrica* and *Miscanthus sinensis*. Age of material is 0–67 cm, 7000 years; 67–89 cm, 16,000 years; 89–200 cm, 22,000 years.

Ap	0–6 cm:	very dark grayish brown (10 YR 3/2) loam; brown to dark brown (10 YR 4/3) dry; massive; soft, very friable, weakly smeary, slightly sticky, slightly plastic; many very fine roots throughout and common fine roots throughout; many very fine interstitial pores; clear smooth boundary.
A	6–18 cm:	very dark grayish brown (10 YR 3/2) loam; brown to dark brown (10 YR 4/3) dry; weak medium angular blocky structure; hard, firm, weakly smeary, slightly sticky, slightly plastic; common very fine and fine roots throughout and few medium roots throughout; few very fine interstitial pores; gradual wavy boundary.

AB   18–29 cm: 80 percent very dark grayish brown (10 YR 3/2) and 20 percent brown to
                          dark brown (7.5 YR 4/4) loam; 80 percent brown to dark brown (10 YR 4/3)
                          and 20 percent strong brown (7.5 YR 5/6) dry; weak medium subangular
                          blocky structure; slightly hard, friable, weakly smeary, slightly sticky, slightly
                          plastic; common very fine and fine roots throughout and few medium roots
                          throughout; common very fine interstitial and few very fine discontinuous
                          tubular pores; gradual wavy boundary.

Bw1   29–50 cm: 80 percent brown to dark brown (7.5 YR 4/4) and 20 percent dark brown
                          (10 YR 3/3) clay loam; 80 percent strong brown (7.5 YR 5/6) and 20 per-
                          cent pale brown (10 YR 6/3) dry; weak medium subangular blocky struc-
                          ture; hard, firm, weakly smeary, sticky, slightly plastic; common fine roots
                          throughout and few very fine roots throughout; few very fine and fine dis-
                          continuous tubular pores; clear smooth boundary.

Bw2   50–67 cm: brown to dark brown (7.5 YR 4/4) clay loam; strong brown (7.5 YR 5/6) dry;
                          weak coarse subangular blocky structure; very hard, firm, sticky, slightly plas-
                          tic; few very fine and fine roots throughout and few medium roots through-
                          out; few very fine and fine discontinuous tubular pores; clear wavy boundary.

2Ab1   67–89 cm: dark yellowish brown (10 YR 4/4) clay loam; light yellowish brown (10 YR
                          6/4) dry; few fine distinct yellowish red (5 YR 4/6) mottles; weak medium
                          prismatic structure; very hard, very firm, sticky, plastic; few very fine and fine
                          roots throughout; common very fine continuous tubular pores; few patchy
                          faint black (10 YR 2/1) manganese or iron-manganese coatings on faces of
                          peds; clear wavy boundary. Structure parts to moderate medium and coarse
                          subangular blocky. Mottles are in pores.

3Ab2   89–127 cm: dark brown (10 YR 3/3) clay loam; yellowish brown (10 YR 5/4) dry; few fine
                          distinct yellowish red (5 YR 4/6) mottles; weak medium prismatic structure;
                          very hard, very firm, sticky, plastic; few very fine and fine roots through-
                          out; common very fine and fine continuous tubular pores; few patchy faint
                          black (10 YR 2/1) manganese or iron-manganese coats on faces or peds;
                          clear wavy boundary. Structure parts to moderate coarse subangular blocky.
                          Mottles are throughout.

3Bwb1   127–154 cm: brown to dark brown (7.5 YR 4/4) clay loam; strong brown (7.5 YR 5/6) dry;
                          common fine faint dark reddish brown (5 YR 3/3) and common medium
                          faint dark reddish brown (5 YR 3/3) mottles; weak medium prismatic struc-
                          ture; hard, firm, sticky, plastic, few very fine roots throughout; many very
                          fine and fine continuous tubular pores; many discontinuous distinct black
                          (10 YR 2/1) manganese or iron-manganese coats on faces of peds and in
                          pores; common discontinuous faint dark brown (10 YR 3/3) organic coats on
                          faces of peds; gradual wavy boundary. Structure parts to moderate medium
                          and coarse subangular blocky. Mottles are throughout.

3Bwb2   154–200 cm: brown to dark brown (7.5 YR 4/4) clay loam; strong brown (7.5 YR 5/6) dry;
                          common fine faint dark reddish brown (5 YR 3/3) mottles; weak coarse pris-
                          matic structure; hard, firm, sticky, plastic; many very fine and fine continuous
                          tubular pores; many discontinuous distinct black (10 YR 2/1) manganese or
                          iron-manganese coats on faces of peds and in pores; many discontinuous
                          faint dark brown (10 YR 3/3) organic coats on faces of peds. Mottles are in
                          pores.

TABLE A1.1

Selected physical and chemical properties (U.S. Department of Agriculture, Soil Conservation Service, National Soil Survey Laboratory, Lincoln, Nebraska 68508-386)

Depth (cm)	Horizon	Org. C (%)	Bulk density (g cm^{-3})	NH$_4$OAc-extractable bases (cmol$_c$ kg^{-1})					Extr. Al (cmol$_c$ kg^{-1})
				Ca	Mg	Na	K	SUM	
0– 6	Ap	8.32		5.0	4.9	–	0.6	10.5	1.0
6– 18	A	6.42	0.64	1.3	7.5	–	0.3	9.1	1.4
18– 29	AB	4.17		3.2	6.3	–	0.2	9.7	
29– 50	Bw1	2.30	0.51	4.6	5.5	–	0.1	10.2	
50– 67	Bw2	1.75		4.2	0.7	–	0.1	5.0	
67– 89	2Ab1	1.87	0.51	2.8	1.0	–	0.1	3.9	
89–127	3Ab2	1.81		1.3	0.8	–	0.1	2.2	0.1
127–154	3Bwb1	1.22		1.5	1.8	tr.	tr.	3.3	0.1
154–204	3Bwb2	1.01		2.9	2.3	0.7	tr.	5.9	

Depth (cm)	Horizon	pH		P ret. (%)	Acid oxalate extr. (%)			Water retention (%)	
		CaCl$_2$ 0.01 M 1:2	H$_2$O 1:1		Al	Fe	Si	15 bar Air-dried	15 bar Field-moist
0– 6	Ap	4.8	5.0	100	5.41	1.73	1.80	29.7	38.4
6– 18	A	4.8	5.1	100	5.55	1.55	1.90	28.2	49.4
18– 29	AB	5.2	5.6	100	6.60	2.64	2.34	30.1	69.2
29– 50	Bw1	5.5	5.8	100	7.32	2.88	3.01	28.8	83.8
50– 67	Bw2	5.7	6.0	100	7.21	1.49	2.89	34.4	93.9
67– 89	2Ab1	5.7	5.9	100	8.52	1.96	3.70	35.6	98.1
89–127	3Ab2	5.6	5.5	100	8.56	2.52	3.62	39.0	98.6
127–154	3Bwb1	5.5	5.5	100	6.26	1.53	2.53	36.1	87.3
154–204	3Bwb2	5.4	5.7	100	5.80	2.91	2.80	29.8	79.4

2. FINDLEY LAKE SOIL (ANDIC HUMICRYOD) (PLATE 11)

Source: University of Washington, 1984, Findley Lake Field Trip (September 3, 1984), and R. Dahlgren, 1987. Aluminum Biogeochemistry in a Subalpine *Abies amabilis* Ecosystem. Ph.D. dissertation submitted to University of Washington.

DESCRIPTION OF FINDLEY LAKE SOIL (ANDIC HUMICRYOD)

Location	T21N R1OE; SE 1/4 of NW 1/4 of Sec. 7, Findley Lake, Washington, U.S.A. Latitude = 47°19′N, Longitude = 121°35′W
Elevation	1128 m M.S.L.
Air temperature (°C)	Ann: 5.4
Precipitation	2730 mm
Snowpack	2–7 m
Drainage	Moderately well drained
Vegetation	Old growth of *Abies amabilis* and *Tsuga mertensiana* approximately 180 years old
Parent material	Andesitic glacial drift overlain by Mt. St. Helens Wn (450 yr B.P.), and Yn (3500 yr. B.P.) and Mt. Mazama (7000–6700 yr B.P.) tephras

Oi, e	7.5–5.0 cm:	forest litter consisting of conifer needles, cones, partially decomposed twigs and mosses making up a mat of recognizable plant remains; 1 to 5 cm thick.
Oa	5.0–0.0 cm:	very dark brown (10 YR 2/2); well decomposed organic material; few recognizable plant parts; greasy feel; material is light and fluffy when dry; few coarse, common medium and many fine to very fine roots; abrupt, smooth boundary; 4 to 10 cm thick.
E	0–8.0 cm:	grayish brown (10 YR 5/2); sandy loam; weak to moderate coarse and medium granular structure; soft, very friable; nonsticky, nonplasitc; few coarse and medium, common fine and many very fine roots; abrupt wavy boundary; 8 to 13 cm thick. (Developed in Mt. St. Helens Yn and Wn tephra)
2Bhs	8–18 cm:	dark reddish brown (5 YR 3/2 and 5 YR 2.5/2); gravelly silt loam; the Bhs contains up to 30% Bs1 material randomly and distinctly mixed throughout; primary structure is weak, thick, platy breaking into strong coarse and medium subangular blocky; hard, firm; slightly sticky, slightly plastic; common medium and many fine roots; cobbles compose 25-35% by volume; abrupt broken boundary. (Developed in Mt. St. Helens Yn and Mt. Mazama tephra)
3Bs1	18–43 cm	dark brown (7.5 YR 4/4) and dark yellowish brown (10 YR 4/6); cobbly silt loam; contains up to 40% Bhs material randomly and distinctly mixed throughout; structure is strong medium to fine subangular blocky; smeary; slightly hard, friable; slightly sticky, slightly plastic; common medium and fine roots; cobbles comprise 25-35% by volume; clear broken boundary. (Developed in Mt. Mazama tephra and glacial drift)
4Bs2	43–71 cm	yellowish brown (10 YR 5/6) and dark yellowish brown (10 YR 4/6); very gravelly loam; medium thick platy structure breaking into medium fine subangular blocky; slightly hard, friable; slightly sticky, slightly plastic; smeary; few very fine and fine roots; dark reddish brown (5 YR 3/2) organic rich material is found around stones and cobbles in this horizon; stones and cobbles compose 40-60% by volume, clear smooth boundary; 22-32 cm thick. (Horizon formed in glacial drift)

4BC   71–104 cm: yellowish brown (10 YR 5/4); very stony loam; moderate to strong medium subangular blocky; hard, friable; slightly sticky, slightly plastic; smeary; few fine roots; 35% stones and 15% cobbles by volume; stones are weathered andesitic clasts with a dark reddish brown (5 YR 3/2) organic rich deposit surrounding them; some stones contain a weathering rind of dark yellowish brown (10 YR 4/4) material; clear wavy boundary; 15-43 cm thick. (Formed in glacial drift)

4C   104 + cm: pale brown (10 YR 6/3) and light yellowish brown (10 YR 6/4) dense glacial till that breaks into an extremely gravelly sandy loam; common fine dark reddish brown (5 YR 3/2) mottles; massive, very hard, very firm; nonsticky, nonplastic; 45% subangular gravel, 25% subangular cobbles by volume.

TABLE A2.1

Selected chemical properties

Depth (cm)	Horizon	pH		P ret. (%)	C_p (%)	C_t (%)	Al_p (%)	Al_d (%)	Al_o (%)
		H_2O	KCl						
0– 8	E	3.8	3.4	5	0.4	1.8	0.04	0.05	0.10
8– 18	2Bhs	4.3	3.8	95	7.6	13.4	2.39	2.42	2.51
18– 43	3Bs1	4.9	4.4	99	2.6	4.8	1.19	2.44	4.88
43– 71	4Bs2	4.9	4.5	98	1.1	1.7	0.53	1.04	2.21
71–104	4BC	4.9	4.4	92	1.5	1.9	0.64	0.98	1.91
104+	4C	4.9	4.4	87	1.3	1.9	0.77	0.93	1.52

Depth (cm)	Horizon	Fe_p (%)	Fe_o (%)	Fe_d (%)	Si_o (%)	$(Al_o - Al_d)/Si_o$
0– 8	E	0.02	0.06	0.14	0.05	–
8– 18	2Bhs	1.25	1.68	2.48	0.11	–
18– 43	3Bs1	0.28	1.83	2.90	1.22	2.1
43– 71	4Bs2	0.07	0.54	1.80	0.60	2.0
71–104	4BC	0.08	0.38	1.35	0.48	2.0
104+	4C	0.18	0.46	1.29	0.31	1.9

C_p = pyrophosphate-extractable carbon; C_t = total carbon; Al_p and Fe_p = pyrophosphate extractable; Al_d and Fe_d = citrate buffered dithionite extractable; Al_o, Fe_o and Si_o = oxalate extractable.

3. ABASHIRI SOIL (TYPIC HAPLUDAND) (PLATE 12)

Source: Shoji, S., Hakamada, T. and Tomioka, E., 1990. Properties and classification of selected volcanic ash soils from Abashiri, northern Japan — Transition of Andisols to Mollisols. Soil Sci. Plant Nutr., 36: 409–423.

DESCRIPTION OF ABASHIRI SOIL (TYPIC HAPLUDAND)

Location	Abashiri (Kakkumi, Tsubetsu), Hokkaido, Japan
Physiography	Level tableland
Elevation	101 m M.S.L.
Air temperature (°C)	Ann: 5.1–5.9, Sum: 16.1–17.5, Win: −8.0 to −6.6
Precipitation	753–839 mm
Land use	Vegetable cultivation
Parent material	Ash and pumice flow deposits

TABLE A3.1

Morphological properties

Depth (cm)	Horizon	Color Moist	Color Dry	Struc- ture	Consistency moist	Boundary	Parent material [a]
0– 19	Ap	10 YR 1.7/1	7.5 YR 4/1	2csbk	fr s	a	Shari and Kpfl* (Me-a, Km-5a)
19– 48	2ABw	7.5 YR 3/3	10 YR 5/3	3msbk	fi vs	g	Bihoro
48– 65	2Bw	5 YR 4/6	10 YR 7/4	2mabk	fi vs	g	Bihoro
65– 81	3C1	5 YR 4/8	10 YR 6/4	1cabk	fi s	c	Kpfl
81–125+	3C2	10 YR 7/4	10 YR 8/4	m	fr s	c	Kpfl

[a] Me-a = Meakan-a ash; Km-5a = Kamui-nupuri-5a ash; Shari = Shari loam (ash flow deposit); Bihoro = Bihoro loam (ash flow deposit); Kpfl = Kussharo pumice flow; Kpfl* = secondary deposit of Kpfl.

TABLE A3.2

Selected physical, chemical and mineralogical properties

Horizon	Bulk density (g cm^{-3})	Water content Field-most 33 kPa	Water content Field-most 1500 kPa	Org. C (%)	Exchangeable bases (cmol$_c$ kg^{-1}) Ca	Mg	K	Na	CEC (cmol$_c$ kg^{-1})	Base sat. (%)
Ap	0.99	51	22	5.2	28.1	4.4	5.4	0.4	39.1	98
2ABw	1.10	51	26	2.2	17.9	2.4	8.2	0.3	31.3	92
2Bw	1.23	36	27	0.7	9.2	1.5	7.6	0.2	20.7	89
3C1	1.01	45	32	0.5	7.8	1.5	7.8	0.2	20.1	86
3C2	0.77	56	22	0.4	7.0	1.4	8.4	0.3	20.0	86

TABLE A3.2 (continued)

Horizon	pH H$_2$O	P ret. (%)	Oxalate extr. (%) [a]			Pyro extr. (%) [b]		Al/Si [c] (%)	Allophane [d] of clay fraction	Major components
			Al$_o$	Fe$_o$	Si$_o$	Al$_p$	Fe$_p$			
Ap	7.1	37	0.50	0.82	0.13	0.15	0.25	2.8	0.9	Glass, Halloysite
2ABw	7.2	59	0.75	1.58	0.22	0.16	0.36	2.8	1.6	Halloysite
2Bw	7.2	54	0.56	1.50	0.22	0.10	0.15	2.2	1.6	Halloysite
3C1	7.1	55	0.63	1.59	0.30	0.08	0.10	1.9	2.1	Halloysite
3C2	7.0	45	0.45	1.08	0.22	0.04	0.05	1.9	1.6	Halloysite

[a] Acid oxalate extractable components.
[b] Sodium pyrophosphate extractable components.
[c] Atomic ratio of $(Al_o - Al_p)/Si_o$.
[d] Allophane or allophane and imogolite content estimated using $(Si_o \times 7.14)$.

4. YUNODAI SOIL (ACRUDOXIC MELANUDAND) (PLATE 7)

Source: Shoji, S., Takahashi, T., Saigusa, M. and Yamada, I., 1987. Morphological properties and classification of ash-derived soils in South Hakkoda. Aomori Prefecture, Japan. Jpn. J. Soil Sci. Plant Nutr., 58: 638–646. (in Japanese); Tour guide of the 9th International Soil Classification Workshop. "Properties, Classification, and Utilization of Andisols and Paddy Soils", 20 July to 1 August, 1987, pp. 145–154.

SITE INFORMATION OF YUNODAI SOIL (PLATE 7)

Location	Yunodai, Towada, Aomori
	Latitude = 40°35′20″N. Longitude = 141°00′30″E
Classification	Medial over pumiceous, mesic, Acrudoxic Melanudand
Physiography	Hillside or mountainside in mountains
Geomorphic position	Middle third of shoulder noseslope
Slope characteristics	5% convex, southeast facing
Elevation	410 m M.S.L.
Air temperature (°C)	Ann: 9, Sum: 19, Win: −3
Soil temperature (°C)	Ann: 10, Sum: 20, Win: −2, Mesic temperature regime
Precipitation	1500 mm, udic moisture regime
Water table	Not observed
Drainage	Well drained
Permeability	Moderate
Land use	Forest land, not grazed
Vegetation	*Miscanthus sinensis* and mixed tree species
Erosion or deposition	Slight
Diagnostic horizon	0 to 60 cm Melanic, 60 to 79 cm Cambic
Described by	S. Shoji et al.

TABLE A4.1

Morphological properties

Depth (cm)	Hori-zon	Soil color (moist)	Texture	Structure	Consistence Fria-bility	Plasti-city	Sticki-ness	Bound-ary	Parent material
5– 0	Oi								
0–18	A1	7.5 YR 1.85/1.5	SL	2fgr	vfr	sp	ss	gw	T-a (1000 yr B.P.)
18–30	A2/C	7.5 YR 2/1 (A2)	SL	2fgr	fr	sp	ss	as	T-a (1000 yr B.P.)
		2.5 Y 5/3 (C)							
30–39	2Ab1	7.5 YR 1.85/1.5	SCL	1fsbk	fr	sp	ss	cw	Chu (5000 yr B.P.)
39–60	2Ab2	7.5 YR 1.7/1	SCL	1f-msbk	fr	sp	ss	cw	Chu (5000 yr B.P.)
60–79	2Bwb	7.5 YR 3/4	SL	1m-csbk	fr	sp	ss	ci	Chu (5000 yr B.P.)
79+	3C	7.5 YR 5/6	S	sgr	1	po	so		Chu (5000 yr B.P.)

TABLE A4.2

Characterization data of the Yunodai soil

Depth (cm)	Horizon	Total[a] (%)			Bulk density $(g\ cm^{-1})$	Water retention at 1500 kPa (%)		Organic carbon (%)
		Clay <2 μm	Silt 2–20 μm	Sand 20 μm–2 mm		Air-dry	Field moist	
5– 0	Oi							
0–18	A1	19.6	28.2	52.1	0.48	29.4	32.5	10.7
18–30	A2/C	13.4	28.5	58.1	0.57	24.2 (A2)	30.3 (A2)	5.3
30–39	2Ab1	22.7	26.5	50.8	–	20.2	47.6	6.5
39–60	2Ab2	25.8	23.0	51.2	0.56	20.2	47.6	6.5
60–79	2Bwb	17.9	18.3	63.8	0.61	16.1	41.5	3.6
79+	3C	3.6	1.4	95.0	0.60	2.7	8.0	0.6

Depth (cm)	Horizon	Melanic index	Ratio of humic acid to fulvic acid	Type of humic acid	KCl-extr. Al $(cmol_c\ kg^{-1})$	CEC $(cmol_c\ kg^{-1})$	Base sat. (%)
5– 0	Oi						
0–18	A1	1.67–1.61	1.15	B	2.71	21.9	8
18–30	A2/C	–	0.97	A	1.06	12.8	6
30–39	2Ab1	–	1.30	A	0.75	16.1	7
39–60	2Ab2	1.65	1.20	A	0.55	16.4	7
60–79	2Bwb	1.73	0.43	A	tr	10.2	6
79+	3C		0.11	P	tr	1.6	7

Depth (cm)	Horizon	pH H_2O	Acid oxalate extr. (%)				Allophane (%)	P. ret. (%)
			Al_o	Fe_o	$Al_o + Fe_o/2$	Si_o		
5– 0	Oi							
0–18	A1	4.9	1.38	0.70	1.72	0.22	1.8	90
18–30	A2/C	5.3	1.44	0.63	1.78	0.40	3.2	89
30–39	2Ab1	5.5	2.58	0.97	3.07	0.86	6.9	96
39–60	2Ab2	5.6	3.34	1.12	3.90	1.22	9.8	97
60–79	2Bwb	5.7	3.29	0.94	3.76	1.49	11.9	96
79+	3C	6.2	1.04	0.20	1.14	0.53	4.2	57

[a] Dispersed by sonication and pH adjustment.

Appendix 2

ANDISOL TU DATABASE, 1992

In examining chemical and physical characteristics of volcanic ash soils, the authors prepared the Andisol Tohoku University (TU) database. The soil samples recorded are mostly from Tohoku and Hokkaido districts in Japan and Alaska and Oregon in the United States. These soils are mostly uncultivated soils with udic soil moisture regime and cryic to mesic soil temperature regime. The total number of pedons and soil horizons used are 70 and 440, respectively. All the analytical data were cited from the following papers:

Ito, T., Shoji, S. and Saigusa, M., 1991. Classification of volcanic ash soils Konsen, Hokkaido, according to the last keys to Soil Taxonomy (1990). Jap. J. Soil Sci. Plant Nutr., 62: 237–347 (in Japanese, with English abstract).

Ping, C.L., Shoji, S. and Ito, T., 1988. Properties and classification of three volcanic ash-derived pedons from Aleutian Islands and Alaska Peninsula, Alaska. Soil Sci. Soc. Am. J., 52: 455–462.

Ping, C.L.,Shoji, S., Ito, T., Takahashi, T. and Moore, J.P., 1989. Characteristics and classification of volcanic-ash derived soils in Alaska. Soil Sci., 148: 8–28.

Saigusa, M., Shoji, S. and Otowa, M., 1991. Clay mineralogy of two Andisols showing a hydrosequence and its relationships to their physical and chemical properties. Pedologist, 35: 21–33.

Saigusa, M. and Shoji, S., 1986. Surface weathering in Zao tephra dominated by mafic glass. Soil Sci. Plant Nutr., 32: 617–628.

Shoji, S., Ito, T., Saigusa, M. and Yamada, I., 1985. Properties of nonallophanic Andosols from Japan. Soil Sci., 140: 264–277.

Shoji, S., Suzuki, Y. and Saigusa, M., 1987. Clay mineralogical and chemical properties of nonallophanic Andepts (Andisols) from Oregon, U.S.A. Soil Sci. Soc. Am. J., 51: 986–990.

Shoji, S., Takahashi, T., Ito, T. and Ping, C.L., 1988. Properties and classification of selected volcanic ash soils from Kenai Peninsula, Alaska, U.S.A. Soil Sci., 145: 395–413.

Shoji, S., Hakamada, T. and Tomioka, E., 1990. Properties and classification of selected volcanic ash soils from Abashiri, northern Japan — transition of Andisols to Mollisols. Soil Sci. Plant Nutr., 36: 409–423.

Takahashi, T., Shoji, S. and Sato. A., 1989. Clayey Spodosols and Andisols showing a biosequential relation from Shimokita peninsula, northeastern Japan. Soil Sci., 148: 204–218.

Takahashi, T., 1990. Genesis, properties, and classification of volcanic ash-derived Spodosols and Andisols. Bull. Akita Pref. Coll. Agric., 16: 53–124, (in Japanese, with English abstract).

Appendix 3

SELECTED THERMODYNAMIC DATA AND UNCERTAINTIES USED IN MODELING MINERAL STABILITY RELATIONSHIPS IN FIG. 5.12

Species	Chemical composition	G°_f (kJ mole^{-1})	Uncertainty (kJmole^{-1})	Ref.[*]
Kaolinite	$Al_2Si_2O_5(OH)_4$	−3799.364	±4.017	1
Halloysite	$Al_2Si_2O_5(OH)_4$	−3780.713	±3.010	1
Gibbsite	$Al(OH)_3$	−1154.889	±1.213	1
Imogolite	$Al_2SiO_3(OH)_4$	−2926.7	±1.100	2
Houston black	$(Si_{7.40}Al_{0.60})$			
clay smectite	$(Al_{2.69}Mg_{0.54}Fe_{0.81})O_{20}(OH)_4$	−10183.26	±3.347	3
Quartz	SiO_2	−856.288	±1.100	1
Amorphous silica	SiO_2	−850.559	±2.134	1
H_2O (aq)		−237.141	±0.084	1
H_4SiO_4		−1308.000	±1.700	1
Al (aq)		−489.4	±1.400	1
Mg (aq)		−454.80	±1.670	1
Fe^{3+} (aq)		−4.60	±1.000	1

[*] References: (1) Robie et al., 1978; (2) Farmer et al., 1979; (3) Carson ct al., 1976.

REFERENCES

Carson, C.D., Kittrick, J.A., Dixon, J.B. and McKee, T.R., 1976. Stability of soil smectite from a Houston black clay. Clays Clay Miner., 24: 151–155.

Farmer, V.C., Smith, B.F.L. and Tait, J.M., 1979. The stability, free energy and heat of formation of imogolite. Clay Miner., 14: 103–107.

Robie, R.A., Hemingway, B.S. and Fisher, J.R., 1978. Thermodynamic properties of minerals and related substances at 298.15 K and 1 bar (10^5 Pascals) pressure and at higher temperatures. U.S. Geol. Surv. Bull. 1452., 456 pp.

Appendix 4

REACTIONS AND EQUILIBRIUM CONSTANTS (25°C) USED IN MODELING MINERAL STABILITY RELATIONSHIPS IN FIG. 5.12

Reaction	Log K
$Al_2SiO_3(OH)_4$ (imogolite) $+ 6H^+ = Al^{3+} + H_4SiO_4 + 3H_2O$	12.53 ± 0.6
$Al(OH)_3$ (gibbsite) $+ 3H^+ = Al^{3+} + 3H_2O$	8.05 ± 0.33
$Al_2Si_2O_5(OH)_4$ (kaolinite) $+ 6H^+ = 2Al^{3+} + 2H_4SiO_4 + H_2O$	5.71 ± 1.04
$Al_2Si_2O_5(OH)_4$ (halloysite) $+ 6H^+ = 2Al^{3+} + 2H_4SiO_4 + H_2O$	8.97 ± 0.94
$(Si_{7.40}Al_{0.60})(Al_{2.69}Mg_{0.54}Fe_{0.81})O_{20}(OH)_4$ (smectite) $+ 5.6H_2O + 14.40H^+ = 7.40H_4SiO_4 + 3.29Al^{3+} + 0.54Mg^{2+} + 0.81Fe^{3+}$	4.80 ± 2.43
SiO_2 (amorphous silica) $+ 2H_2O = H_4SiO_4$	-2.95 ± 0.48
SiO_2 (quartz) $+ 2H_2O = H_4SiO_4$	-3.95 ± 0.36

REACTIONS AND EQUILIBRIUM CONSTANTS (25°C) USED IN MODELING MINERAL STABILITY RELATIONSHIPS IN FIG. 3.12

Reaction		log K
$Al_2Si_2O_5(OH)_4 \text{ (kaolinite)} + 6H^+ = 2Al^{3+} + 2H_4SiO_4 + H_2O$		
$Al(OH)_3 \text{ (gibbsite)} + 3H^+ = Al^{3+} + 3H_2O$		
$Al_3Si_2O_{10}(OH)_2 \text{ (pyrophyllite)} + 6H^+ + 4H_2O = 3Al^{3+} + 4H_4SiO_4$		
$KAl_3Si_3O_{10}(OH)_2 \text{ (muscovite)} + 10H^+ = K^+ + 3Al^{3+} + 3H_4SiO_4$		
$Mg_5Al_2Si_3O_{10}(OH)_8 + 16H^+ = 5Mg^{2+} + 2Al^{3+} + 3H_4SiO_4 + 6H_2O$		
$SiO_2 \text{ (quartz)} + 2H_2O = H_4SiO_4$		
$SiO_2 \text{ (amorph.)} + 2H_2O = H_4SiO_4$		

REFERENCES INDEX

SUBJECT INDEX